Contents

T0264922

Preface to the Twelfth Edition

Introduction

IBM® SPSS Statistics software ("SPSS®") is a statistical package produced by IBM, Inc. Prior to 2009, SPSS was a separate company and produced statistical software under the SPSS and PASW names. IBM SPSS Statistics is designed to perform a wide range of statistical procedures. As with any other powerful program for the computer, there are certain conventions and techniques that must be mastered for efficient use of the software and to obtain consistently correct answers. By providing detailed, step-by-step guidance illustrated with examples, this book will help you attain such mastery.

In addition to showing you how to enter data and obtain results, this book explains how to select appropriate statistics and present the results in a form that is suitable for use in a research report in the social or behavioral sciences. For instance, the section on the independent *t* test shows how to state (i.e., phrase) the results of both a significant and an insignificant test.

Audience

This book is ideal as a supplement to traditional introductory- and intermediate-level statistics textbooks. It can also be used as a statistics refresher manual in a research methods course. Finally, students can use it as a desk reference guide in a variety of workplace settings after they graduate from college.

SPSS Statistics is an incredibly powerful program, and this text is not intended to be a comprehensive user's manual. Instead, the emphasis is on the procedures normally covered in introductory- and intermediate-level courses in statistics and research methods.

Organization

This book is divided into nine chapters plus several useful appendices. The first two chapters deal with the basic mechanics of using the SPSS program. Each of the remaining chapters focuses on a particular class of statistics.

Each chapter contains several short sections. For the most part, these sections are self-contained. However, students are expected to master the SPSS basics in Chapters 1 and 2 before attempting to learn the skills presented in the rest of the book. Except for the skills in the first two chapters, this book can be used in a nonlinear manner. Thus, an instructor can assign the first two chapters early in a course and then assign other sections in whatever order is appropriate.

Appendix A contains a discussion of effect size. Appendix B contains datasets that are needed for the practice exercises interspersed throughout this book. Appendix C provides the sample data files that are used throughout this book. Appendix D provides information on choosing the appropriate statistical test. The Glossary in Appendix E provides definitions of most of the statistical terms used in this book. Because it is assumed that this text is being used in conjunction with a main statistics textbook, the

How to Use SPSS®

How to Use SPSS® is designed with the novice computer user in mind and for people who have no previous experience using SPSS. Each chapter is divided into short sections that describe the statistic being used, important underlying assumptions, and how to interpret the results and express them in a research report.

The book begins with the basics, such as starting SPSS, defining variables, and entering and saving data. It covers all major statistical techniques typically taught in beginning statistics classes, such as descriptive statistics, graphing data, prediction and association, parametric inferential statistics, nonparametric inferential statistics, and statistics for test construction.

More than 275 screenshots (including sample output) throughout the book show students exactly what to expect as they follow along using SPSS. The book includes a glossary of statistical terms and practice exercises. A complete set of online resources including video tutorials and output files for students, and PowerPoint slides and test bank questions for instructors, make *How to Use SPSS®* the definitive, field-tested resource for learning SPSS.

New to this edition:

- Fully updated to the reflect SPSS version 29. Every screen shot has been recaptured.
- New video supplements for all practice exercises.
- References to significance levels have been updated to reflect the new SPSS output format.
- Effect size is now shown in output for many procedures and reference to some effect size has been moved from Appendix A to be more integrated into the chapters. Sample results sections now also include effect size where SPSS directly calculates effect size.
- A new section covering the EXPLORE command has been added to Chapter 3.

Brian C. Cronk is Professor of Psychology, SUNY Buffalo State University, USA (PhD in Psychology 1993, University of Wisconsin-Milwaukee).

How to Use SPSS®

A Step-By-Step Guide to Analysis and Interpretation

TWELFTH EDITION

Brian C. Cronk

Routledge
Taylor & Francis Group
NEW YORK AND LONDON

Designed cover image: matejmo / Getty Images

Twelfth edition published 2024
by Routledge
605 Third Avenue, New York, NY 10158

and by Routledge
4 Park Square, Milton Park, Abingdon, Oxon, OX14 4RN

Routledge is an imprint of the Taylor & Francis Group, an informa business

First edition published by Pyrczak Publishing 1999
Eleventh edition published by Routledge 2020

ISBN: 978-1-032-58519-2 (hbk)
ISBN: 978-1-032-58235-1 (pbk)
ISBN: 978-1-003-45046-7 (ebk)

DOI: 10.4324/9781003450467

Typeset in Palatino
by KnowledgeWorks Global Ltd.

Access the Instructor and Student Resources: www.routledge.com/cw/cronk

Glossary definitions are brief and are designed to serve only as reminders. Appendix F provides a text-based version of the decision tree on the inside front cover.

SPSS Versions

This text is designed to work with SPSS Version 29, although it should work well for any recent version of SPSS.

SPSS also offers several modules. The screenshots in this text were taken using a "full" version of the software (all modules active and installed). Most institutions have at least the Base, Regression, and Advanced modules installed. With these three modules, or if you are using the *SPSS Statistics GradPack* or *Premium Gradpacks*, you can conduct all of the analyses in this text. If you are using the *SPSS Statistics Base GradPack*, some procedures will not be available. When this is the case, the text will clearly indicate that another product is necessary.

If you have different add-on modules installed, the menu bars shown in the illustrations of this text may be slightly different. If your menu bars look different, or if you cannot locate a menu item for a command presented in this book, it may be that your institution supports different modules. Ask your instructor for additional guidance.

Availability of SPSS

Some institutions purchase site licenses from IBM to provide the software at no charge to their faculty and, sometimes, to their students. If your institution has not purchased a site license, you should still be able to purchase the student version of the software in your campus bookstore at a price discounted for the educational community. Either of these will be a fully functional version of the software intended for your personal use. However, the student version is limited to 50 variables and 1,500 cases. If you need software with more functionality, you can purchase the *SPSS Statistics GradPack* from a variety of sources. A quick Internet search for "SPSS graduate pack" will bring up some choices. Please note that you can generally lease SPSS for a period of a single semester if you would like. You can also purchase a month-to-month license of the IBM SPSS Statistics Cloud product.

Conventions

The following conventions have been used throughout this book:

- Items in **bold** are defined in the Glossary in Appendix E.
- Items in *italics* are either buttons or menus from the SPSS program, or they are statistical symbols.
- Items in ALL CAPITAL LETTERS are either acronyms or the names of variables in the SPSS data file.

Screenshots

Screenshots have been used extensively throughout the text as a visual representation of what is described. In some instances, there may be minor differences between screenshots shown in the text and those on the student's own screen. For instance, a screenshot captured while working with a dataset that has been previously saved to the hard drive will show the filename and dataset number, whereas a dataset that has not yet been saved will simply show "*Untitled" and "[DataSet0]." Because no two Windows computers are configured exactly alike, and because different modules and software versions can produce slightly modified screenshots, such minor differences are unavoidable.

Practice Exercises

Practice exercises are included for each skill presented. In addition, the skills acquired in this text can be used in practice exercises in other statistics texts or workbooks. In this respect, an excellent complement to this text is *Real Data: A Statistics Workbook Based on Empirical Data.*[1]

Acknowledgments

This book is dedicated to the students in my Introductory Psychological Statistics and Research Team courses when I was a faculty member at Missouri Western State University. While teaching those courses, I became aware of the need for an SPSS manual that did more than simply tell students how to start the program and enter data. I am deeply indebted to Wendy Schweigert at Bradley University, who first showed me the power, simplicity, and usefulness of statistics. I would also like to thank the hundreds of instructors who adopted previous editions of this text and provided me with constructive feedback. Of course, this text would not have been possible without the support of Kate and my children Jonathan and Katherine. A lot has changed since the first edition of this text. My son had just been born as I wrote the proposal for the first edition. By the time you buy this book he will have his Master's Degree.

Brian C. Cronk

Chapter 1
Getting Started

Section 1.1 Starting SPSS

Startup procedures for SPSS will differ slightly, depending on the configuration of the machine on which it is installed. If you have difficulty finding it, look for an *IBM SPSS Statistics* section of your *Start Menu*. This text uses screenshots from the Windows version of SPSS. The MacOS and Unix versions will have the same functionality but could appear differently than what is depicted herein.

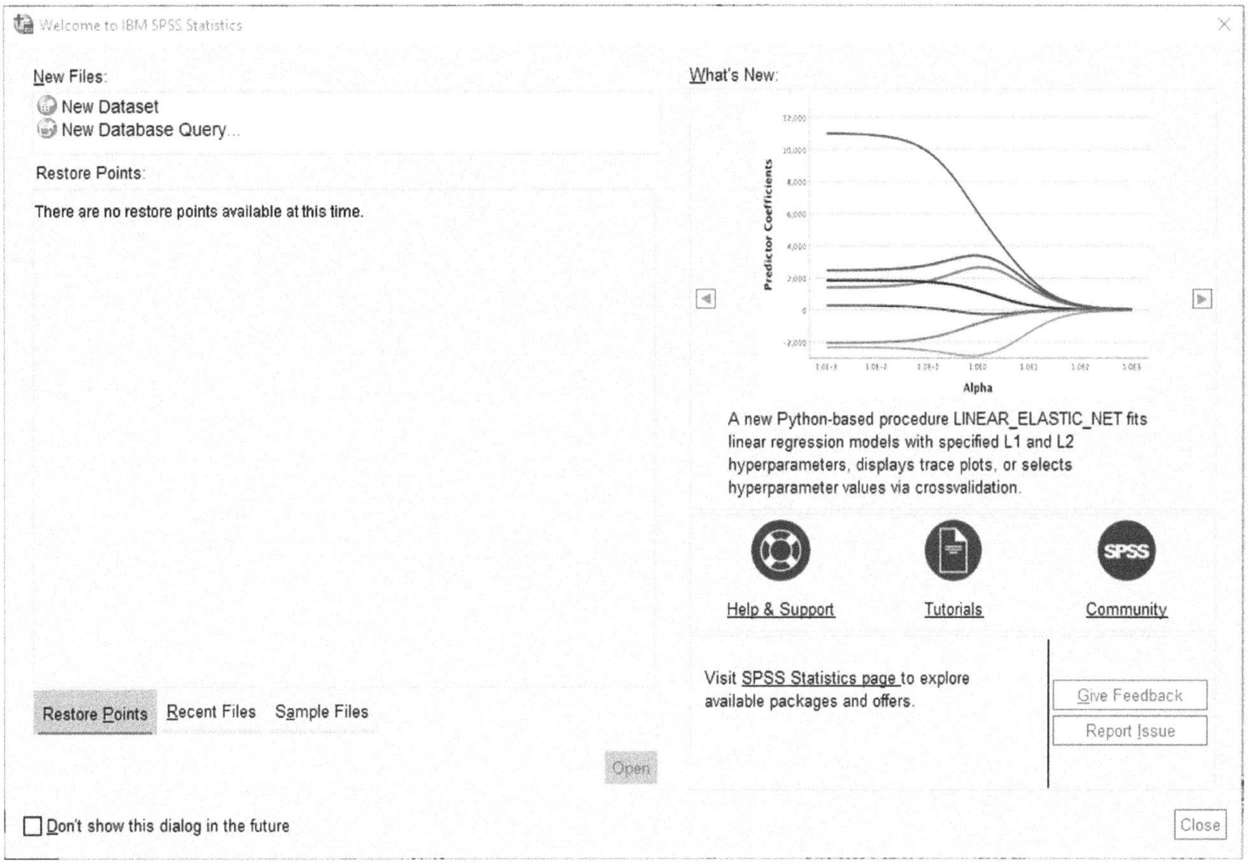

When SPSS is started, you may be presented with the **dialog box** above, depending on the options your system administrator selected for your version of the program. If you have the **dialog box**, click *New Dataset* and *Open*, which will present a blank **data window**.[1]

If you were not presented with the **dialog box** above, SPSS should open automatically with a blank **data window**.

The **data window** and the **output window** provide the basic interface for SPSS. A blank **data window** is shown on page 3.

DOI: 10.4324/9781003450467-1

Section 1.2 Entering Data

One of the keys to success with SPSS is knowing how it stores and uses your data. To illustrate the basics of data entry with SPSS, we will use Example 1.2.1.

Example 1.2.1 A survey was given to several students from four different classes (Tues/Thurs mornings, Tues/Thurs afternoons, Mon/Wed/Fri mornings, and Mon/Wed/Fri afternoons). The students were asked whether or not they were "morning people" and whether or not they worked. This survey also asked for their final grade in the class (100% being the highest grade possible). The response sheets from two students are presented below:

Response Sheet 1				
ID	4593			
Day of class	____	MWF	X____	TTh
Class time	____	Morning	X____	Afternoon
Are you a morning person?	____	Yes	X____	No
Final grade in class	85%			
Do you work outside school?	____	Full-time	____	Part-time
	X	No		
Response Sheet 2				
ID	1901			
Day of class	X____	MWF	____	TTh
Class time	X____	Morning	____	Afternoon
Are you a morning person?	X____	Yes	____	No
Final grade in class	83%			
Do you work outside school?	____	Full-time	X____	Part-time
	____	No		

Our goal is to enter the data from the two students into SPSS for use in future analyses. The first step is to determine the variables that need to be entered. Any information that can vary among participants is a variable that needs to be considered. Example 1.2.2 lists the variables we will use.

Example 1.2.2

ID
Day of class
Class time
Morning person
Final grade
Whether or not the student works outside school

In the SPSS **data window**, columns represent variables, and rows represent participants. Therefore, we will be creating a data file with six columns (variables) and two rows (students/participants).

Section 1.3 Defining Variables

Before we can enter any data, we must first enter some basic information about each variable into SPSS. For instance, variables must first be given names that

- begin with a letter, and
- do not contain a space.

Thus, the variable name "Q7" is acceptable, while the variable name "7Q" is not. Similarly, the variable name "PRE_TEST" is acceptable, but the variable name "PRE TEST" is not. Capitalization does not matter, but variable names are capitalized in this text to make it clear when we are referring to a variable name, even if the variable name is not necessarily capitalized in screenshots.

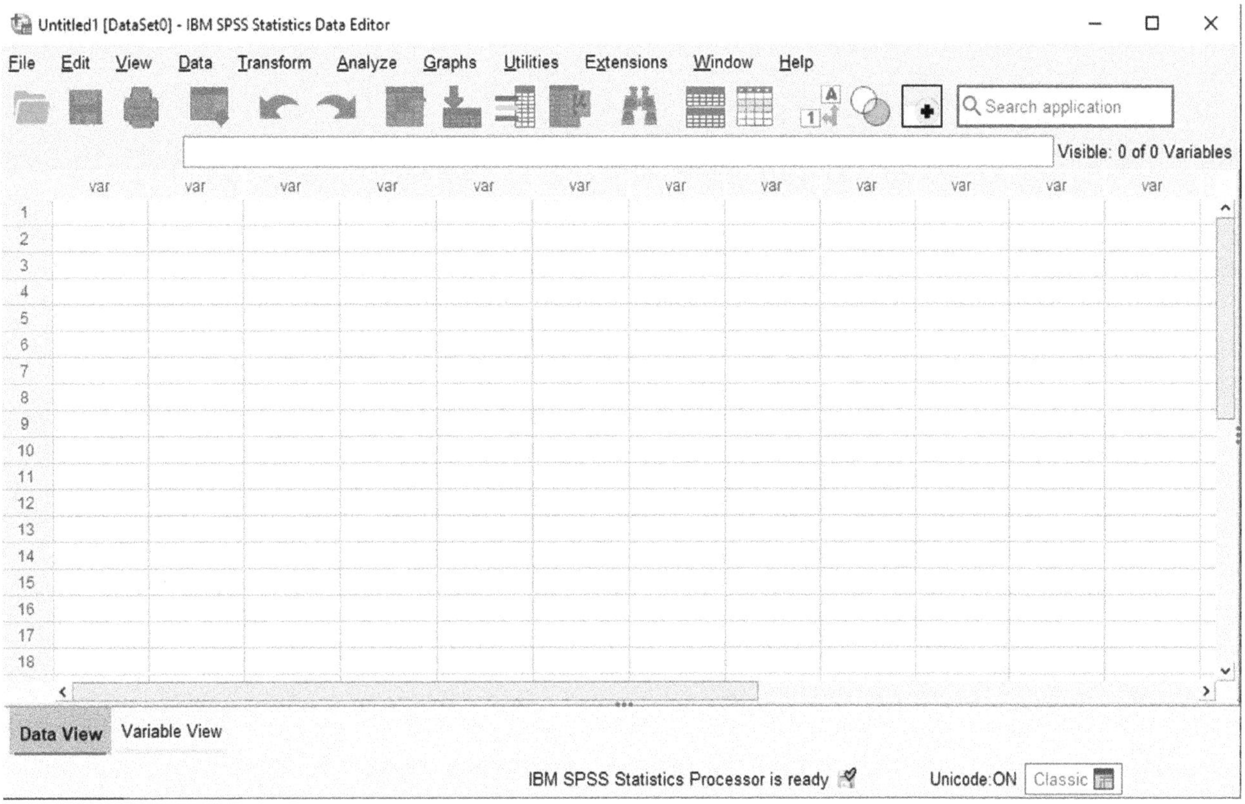

To define a variable, click on the *Variable View* tab at the bottom of the main screen. This will show you the *Variable View* window. To return to the *Data View* window, click on the *Data View* tab.

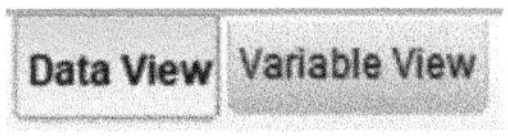

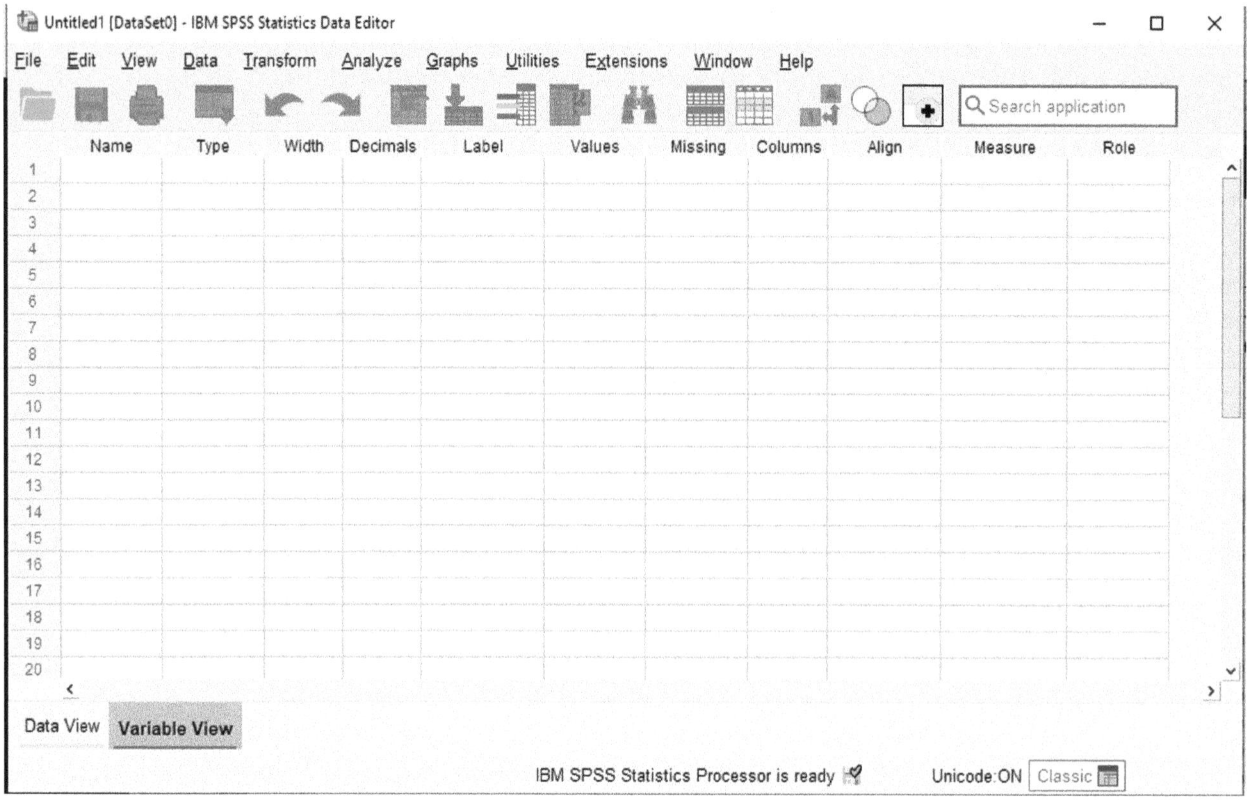

From the *Variable View* screen, SPSS allows you to create and edit all of the variables in your data file. Each column represents some property of a variable, and each row represents a variable. All variables must be given a name. To do that, click on the first empty cell in the *Name* column and type a valid SPSS variable name. The program will then fill in default values for most of the other properties.

One useful function of SPSS is the ability to define variable and value labels. Variable labels allow you to associate a description with each variable.

Value labels allow you to associate a description with each value of a variable. For instance, for most procedures, SPSS requires numerical values. Thus, for data such as the day of the class (i.e., Mon/Wed/Fri and Tues/Thurs), we need to first code the values as numbers. We can assign the number 1 to Mon/Wed/Fri and the number 2 to Tues/Thurs. To help us keep track of the numbers we have assigned to the values, we use value labels.

To assign value labels, click in the cell you want to assign values to in the *Values* column (in this case, for Variable 2). This will bring up a small gray button (shown below). Click on that button to bring up the Value Labels **dialog box**.

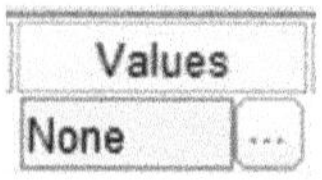

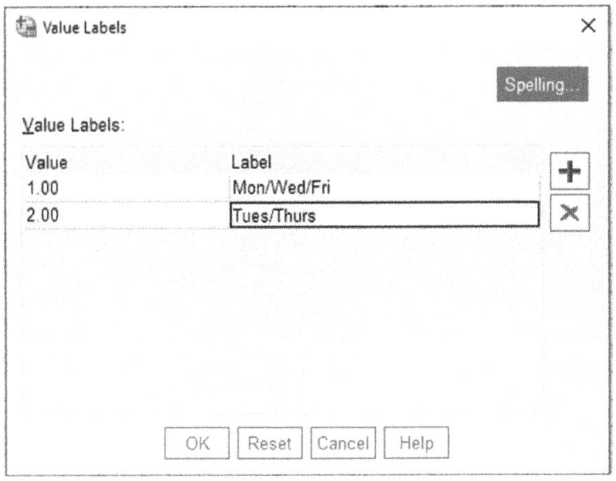

When you enter a value label, you must click the green plus sign after each entry. This will create a new row for you to enter a value label into. When all labels have been added, click *OK* to return to the *Variable View* window.

In addition to naming and labeling the variable, you have the option of defining the variable type. To do so, simply click on the *Type*, *Width*, or *Decimals* columns in the *Variable View* window. The default value is a numeric field that is eight digits wide with two decimal places displayed. If your data are more than eight digits to the left of the decimal place, they will be displayed in scientific notation (e.g., the number 2,000,000,000 will be displayed as 2.00E + 09).[2] SPSS maintains accuracy beyond two decimal places, but all output will be rounded to two decimal places unless otherwise indicated in the *Decimals* column.

There are several other options available in this screen, which are beyond the scope of this text. In our example, we will be using numeric variables with all the default values.

Practice Exercise

Create a data file for the six variables and two sample students presented in Example 1.2.1. Name your variables: ID, DAY, TIME, MORNING, GRADE, and WORK. You should code DAY as 1 = Mon/Wed/Fri, 2 = Tues/Thurs. Code TIME as 1 = morning, 2 = afternoon. Code MORNING as 0 = No, 1 = Yes. Code WORK as 0 = No, 1 = Part-time, 2 = Full-time. Be sure you enter value labels for the different variables. Note that because value labels are not appropriate for ID and GRADE (because the values themselves serve as labels), these are not coded. When complete, your *Variable View* window should look like the screenshot below.

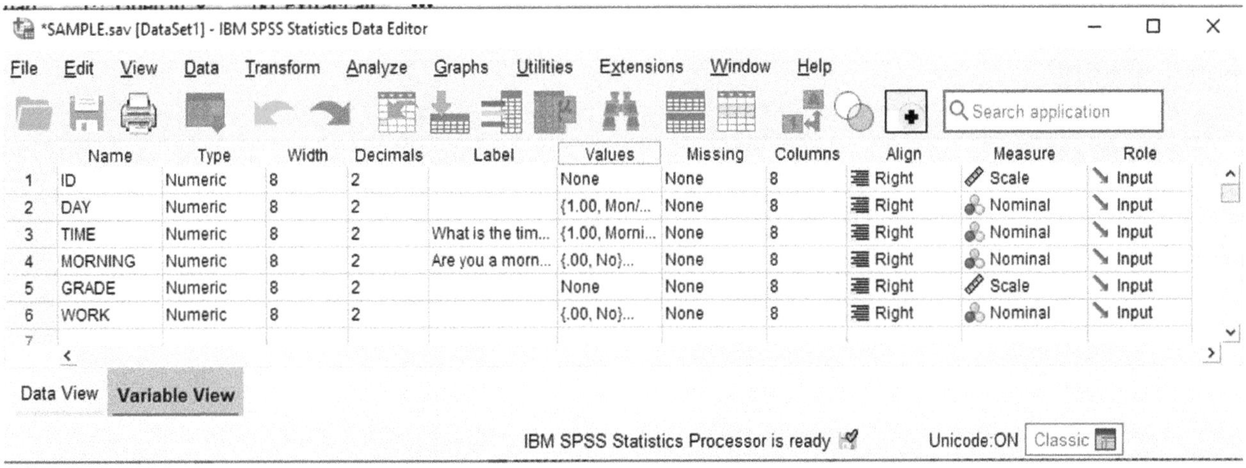

Click on the *Data View* tab to open the data-entry screen. Enter data horizontally, beginning with the first student's ID number. Enter the code for each variable in the appropriate column. To enter the GRADE variable value, enter the student's class grade.

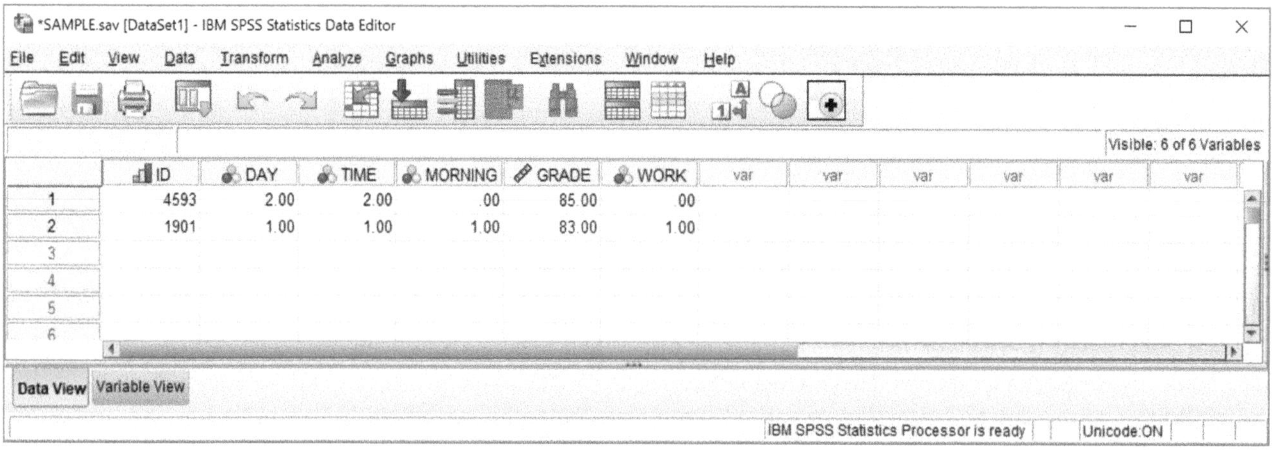

The previous **data window** can be changed to look like the screenshot on the next page by clicking on the *Value Labels* icon (see below). In this case, the cells display value labels rather than the corresponding codes. If data are entered in this mode, it is not necessary to enter codes, as clicking the button that appears in each cell as the cell is selected will present a drop-down list of the predefined labels. You may use whichever method you prefer.

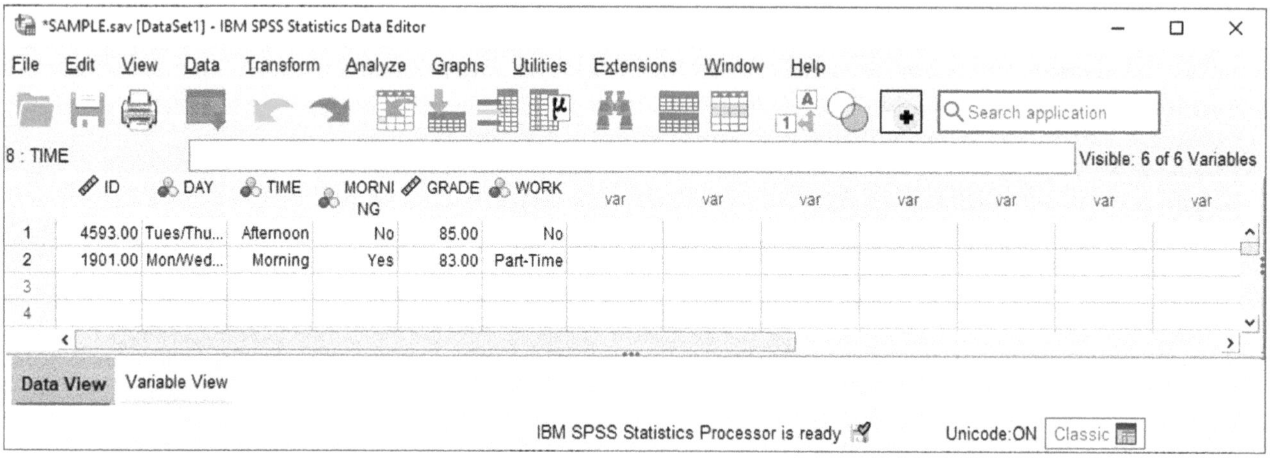

Instead of clicking the *Value Labels* icon, you may toggle between views by clicking *Value Labels* under the *View* menu.

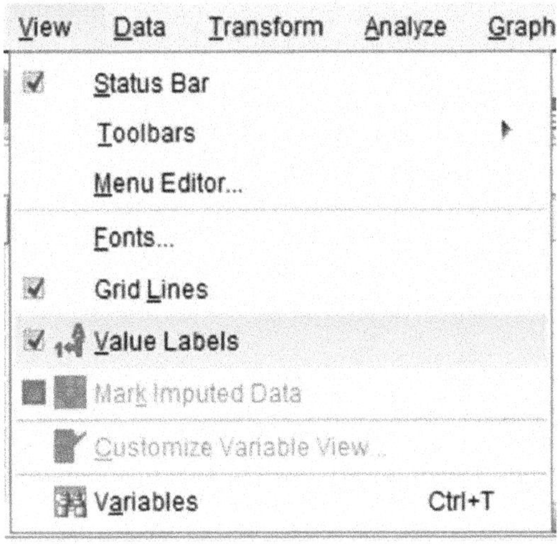

Section 1.4 Loading and Saving Data Files

Once you have entered your data, you will need to save it with a unique name so that you can retrieve it when necessary for later use.

Loading and saving SPSS data files works in the same way as most Windows-based software. Under the *File* menu, there are *Open*, *Save*, and *Save As* commands. SPSS data files have a ".sav" extension, which is added by default to the end of the filename (that is, do not type ".sav" after the filename; SPSS will add it automatically). This tells Windows that the file is an SPSS data file. Other SPSS extensions include ".spv" for saved output files and ".sps" for saved syntax files.

Save Your Data

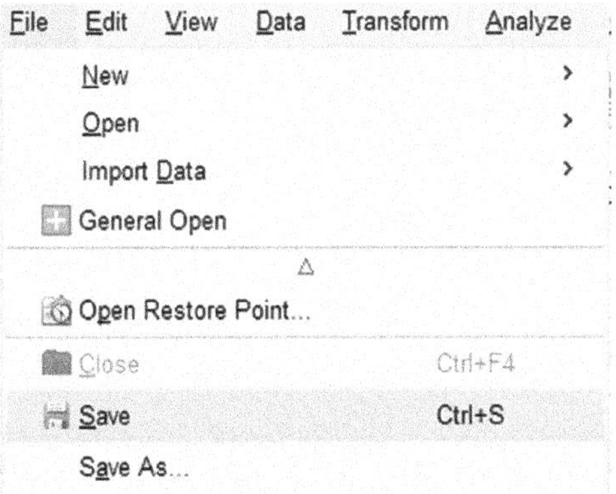

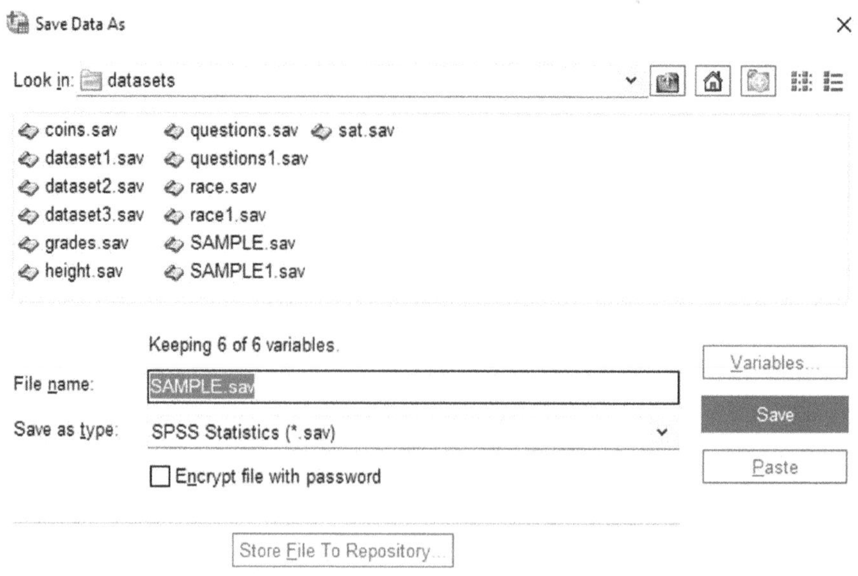

When you save your data file (by clicking *File*, then clicking *Save* or *Save As* to specify a unique name), pay special attention to where you save it. You will probably want to save your data on a removable USB drive so that you can take the file with you.

Load Your Data

When you load your data (by clicking *File*, then clicking *Open*, then *Data*, or by clicking the open file folder icon), you get a similar window. This window lists all files with the ".sav" extension. If you have trouble locating your saved file, make sure you are looking in the right directory.

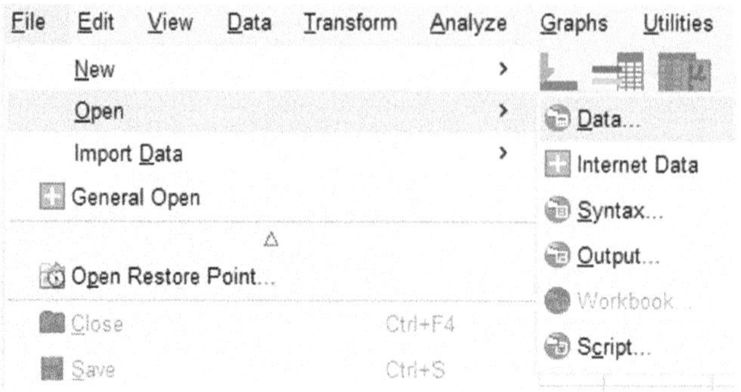

Practice Exercise

To be sure that you have mastered saving and opening data files, name your sample data file "SAMPLE" and save it to a removable storage medium. Once it is saved, SPSS will display the name of the file at the top of the **data window**.

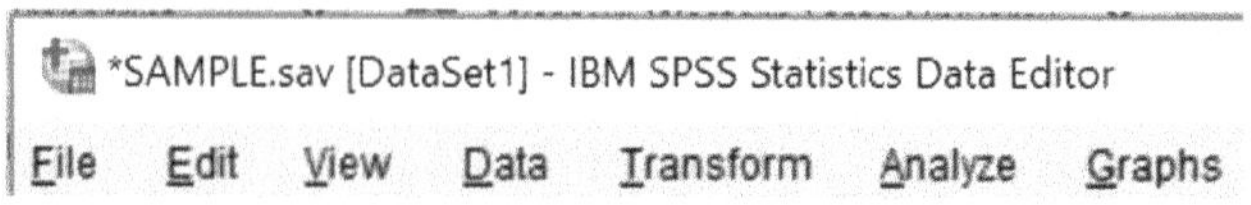

It is wise to save your work frequently, in case of computer crashes. Note that filenames may be uppercase or lowercase. In *this* text, uppercase is used for clarity. In naming files, though, screenshots may show lowercase only.

After you have saved your data, exit SPSS (by clicking *File*, then *Exit*). Restart SPSS and load your data by selecting the "SAMPLE.sav" file you just created.

Section 1.5 Running Your First Analysis

Any time you open a **data window**, you can run any of the analyses available. To get started, we will calculate the students' average grade. (With only two students, you can easily check your answer by hand, but imagine a data file with 10,000 student records.)

The majority of the available statistical tests are under the *Analyze* menu. This menu displays all the options available for your version of the SPSS program (the menus in this book were created with SPSS Statistics Version 29). Other versions may have slightly different sets of options.

To calculate a **mean** (average), we are asking the computer to summarize our dataset. Therefore, we run the command by clicking *Analyze*, then *Descriptive Statistics*, then *Descriptives*.

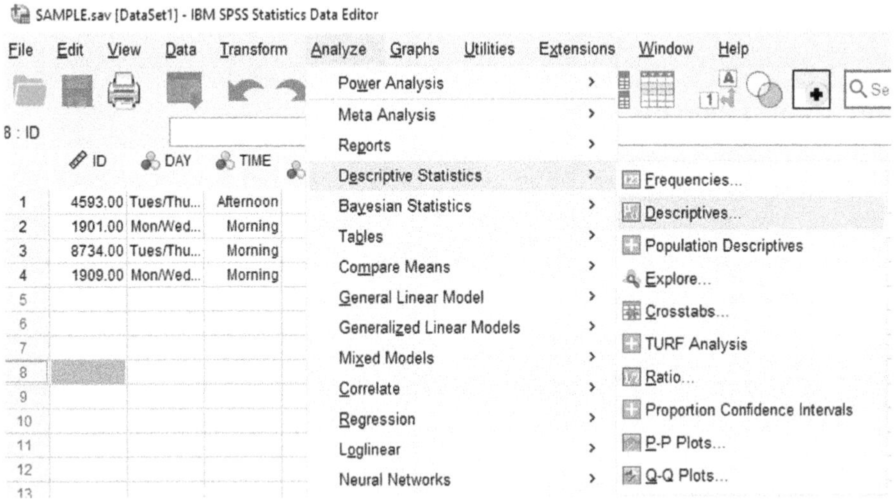

This brings up the Descriptives **dialog box**. Note that the left side of the box contains a list of all the variables in our data file. On the right is an area labeled *Variable(s)*, where we can specify the variables we would like to use in this particular analysis.

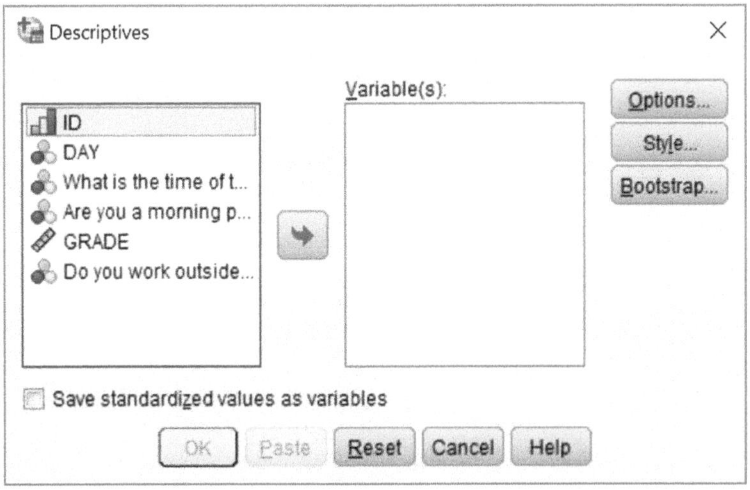

We want to compute the **mean** for the variable called GRADE. Thus, we need to select the variable name in the left window (by clicking on it). To transfer it to the right window, click on the right arrow between the two windows. The arrow always points to the window opposite the highlighted item and can be used to transfer selected variables in either direction. Note that double-clicking on the variable name will also transfer the variable to the opposite window. Standard Windows conventions of "Shift" clicking or "Ctrl" clicking to select multiple variables can be used as well. Note: Some configurations of SPSS show the variable names, and others show the variable labels (if any). This can be changed under *Edit → Options → General.*

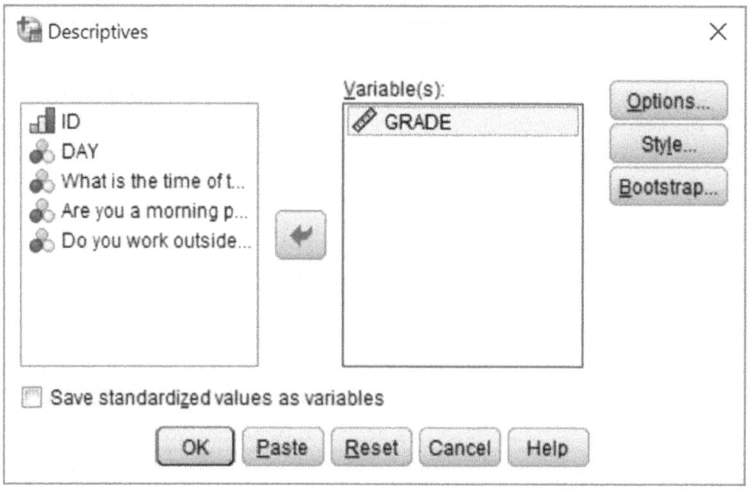

When we click on the *OK* button, the analysis will be conducted, and we will be ready to examine our output.

Section 1.6 Examining and Printing Output Files

After an analysis is performed, the output is placed in the **output window**, and the **output window** becomes the active window. If this is the first analysis you have conducted since starting SPSS, then a new **output window** will be created. If you have run previous analyses and saved them, your output is added to the end of your previous output.

To switch back and forth between the **data window** and the **output window**, select the desired window from the *Window* menu bar. Alternately, you can select the window using the taskbar at the bottom of the screen.

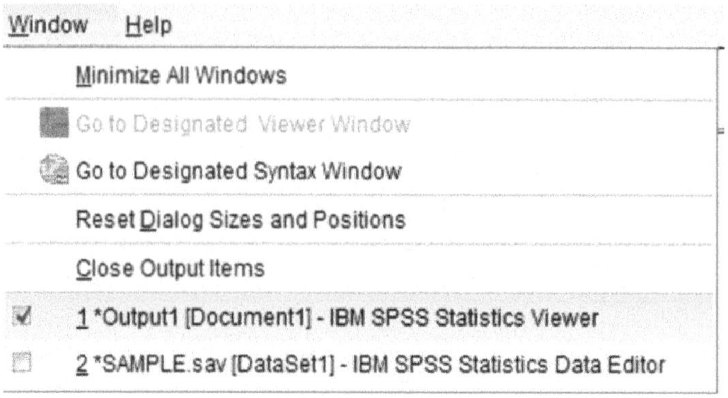

The **output window** is split into two sections. The left section is an outline of the output (SPSS refers to this as the *outline view*). The right section is the output itself.

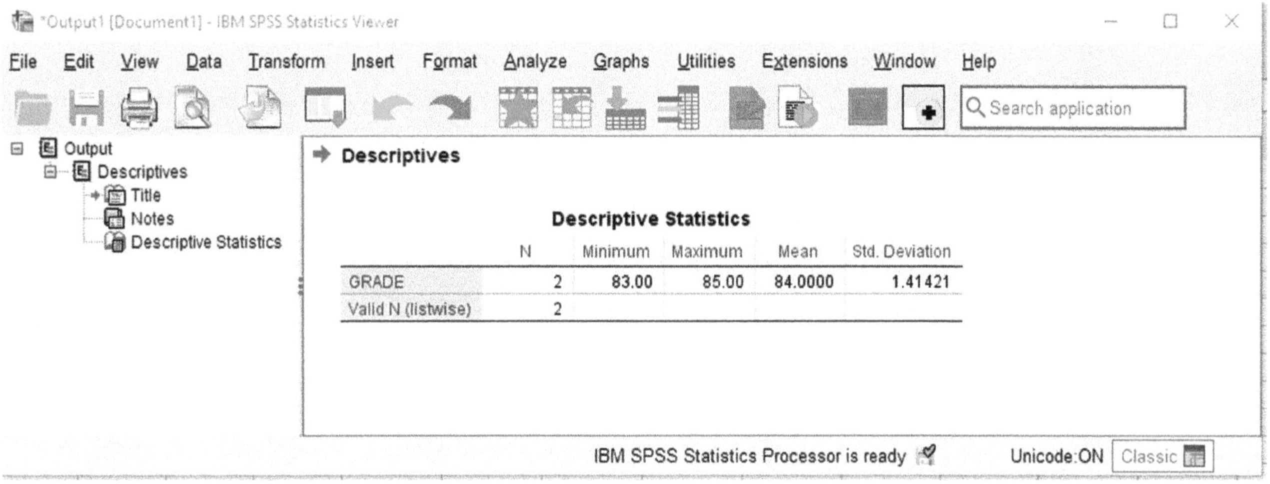

The section on the left of the **output window** provides an outline of the entire **output window**. All of the analyses are listed in the order in which they were conducted. Note that this outline can be used to quickly locate a section of the output. Simply click on the section you would like to see, and the right window will jump to the appropriate place.

Clicking on a statistical procedure also selects all of the output for that command. By pressing the *Delete* key, that output can be deleted from the **output window**. This is a quick way to ensure that the **output window** contains only the desired output. Output can also be selected and pasted into a word processor or spreadsheet by clicking *Edit*, then *Copy* to copy the output. You can then switch to your word processor and click *Edit*, then *Paste.*

To print your output, simply click *File*, then *Print*, or click on the printer icon on the toolbar. You will have the option of printing all of your output or just the currently selected section. Be careful when printing! Each time you run a command, the output is added to the end of your previous output. Thus, you could be printing a very large output file containing information you may not want or need.

One way to ensure that your **output window** contains only the results of the current command is to create a new **output window** just before running the command. To do this, click *File*, then *New*, then *Output.* All your subsequent commands will go into your new **output window**.

You can also save your output files as SPSS format files (.spv extension). Note that SPSS saves whatever window you have open. If you are on a **data window**, you will save your data. If you are on an **output window**, it will save your output.

Practice Exercise

Load the sample data file you created earlier (SAMPLE.sav). Run the *Descriptives* command for the variable GRADE, and print the output. Next, select the **data window** and print it.

Section 1.7 Modifying Data Files

Once you have created a data file, it is really quite simple to add additional cases (rows/participants) or additional variables (columns). Consider Example 1.7.1.

Example 1.7.1 Two more students provide you with surveys. Their information is as follows:

Response Sheet 3				
ID	8734			
Day of class	____	MWF	X	TTh
Class time	X	Morning	____	Afternoon
Are you a morning person?	____	Yes	X	No
Final grade in class	80%			
Do you work outside school?	____	Full-time	____	Part-time
	X	No		
Response Sheet 4				
ID	1909			
Day of class	X	MWF	____	TTh
Class time	X	Morning	____	Afternoon
Are you a morning person?	X	Yes	____	No
Final grade in class	73%			
Do you work outside school?	____	Full-time	X	Part-time
	____	No		

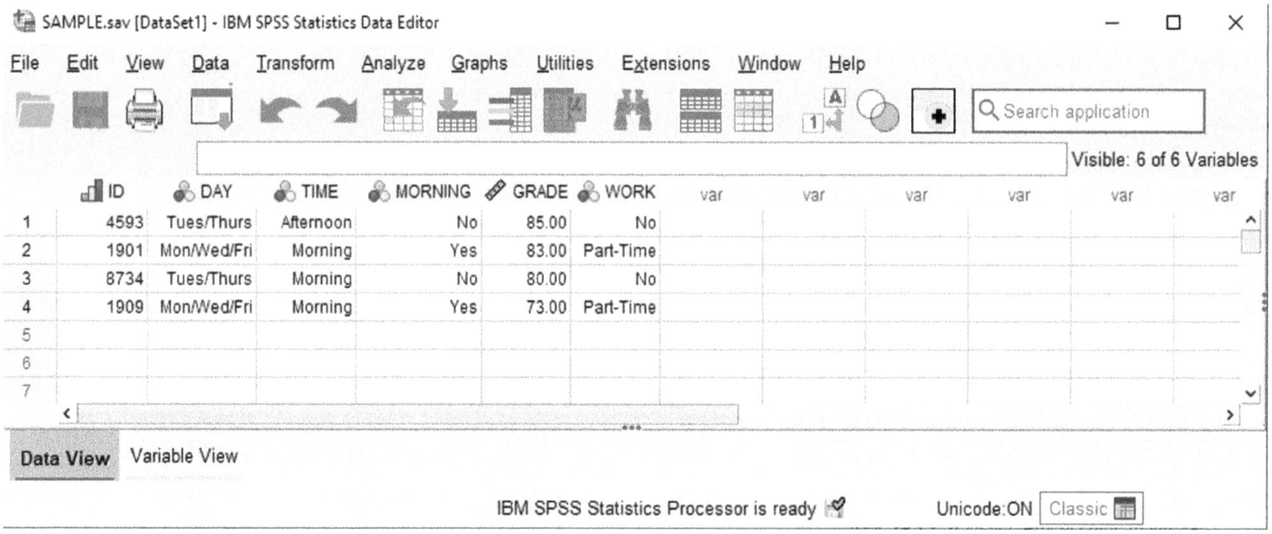

To add these data, simply place two additional rows in the *Data View* window (after loading your sample data). Note that as new participants are added, the row numbers become bold. When done, the screen should look like the screenshot above.

New variables can also be added. For example, if the first two participants were given special training on time management, and the two new participants were not, the data file can be changed to reflect this additional information. The new variable could be called TRAINING (whether or not the participant received training), and it would be coded so that 0 = No and 1 = Yes. Thus, the first two participants would be assigned a "1" and the last two participants a "0." To do this, switch to the *Variable View* window, then add the TRAINING variable to the bottom of the list. Then switch back to the *Data View* window to update the data.

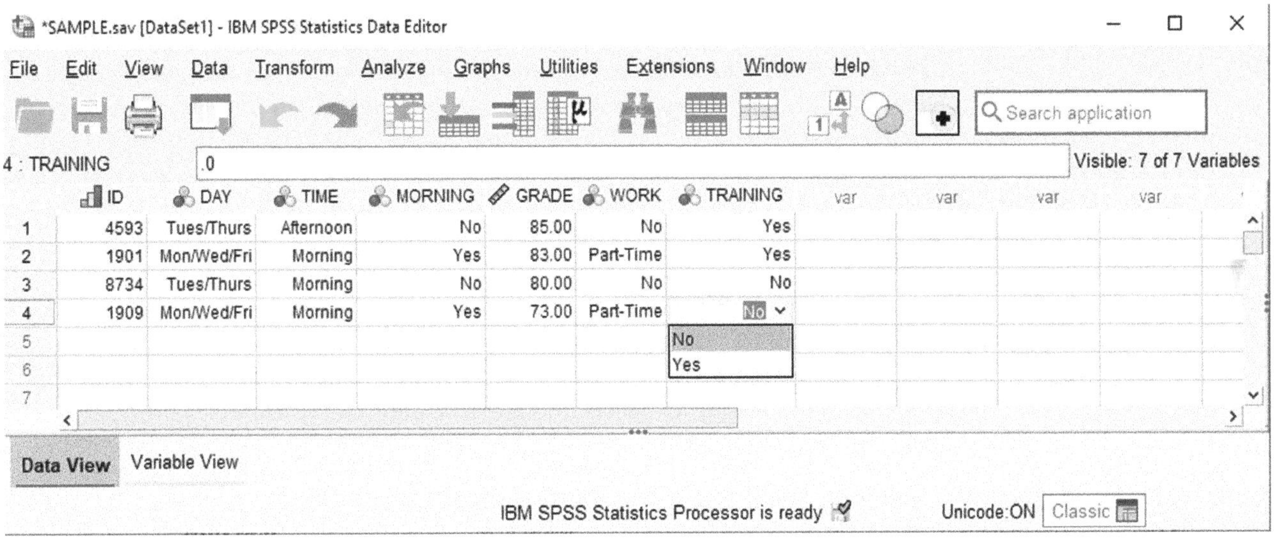

Adding data and variables are logical extensions of the procedures we used to originally create the data file. Save this new data file. We will be using it again later in this book.

Practice Exercise

Follow the previous example (where TRAINING is the new variable). Make the modifications to your SAMPLE.sav data file and save it.

Chapter 2
Entering and Modifying Data

In Chapter 1, we learned how to create and save a simple data file, perform a basic analysis, and examine the output. In this section, we will go into more detail about variables and data.

Section 2.1 Variables and Data Representation

In SPSS, variables are represented as columns in the data file. Participants are represented as rows. Thus, if we collect four pieces of information from 100 participants, we will have a data file with 4 columns and 100 rows.

Measurement Scales

There are four types of measurement scales: **Nominal**, **ordinal**, **interval**, and **ratio**. While the measurement scale will determine which statistical technique is appropriate for a given set of data, SPSS generally does not discriminate. Thus, we start this section with this warning: If you ask it to, SPSS may conduct an analysis that is not appropriate for your data. For a more complete description of these four measurement scales, consult your statistics text or the Glossary in Appendix E.

SPSS allows you to indicate which types of data you have when you define your variable. You do this using the *Measure* column. You can indicate *Scale, Ordinal,* or *Nominal* (SPSS does not distinguish between **interval** and **ratio** scales).

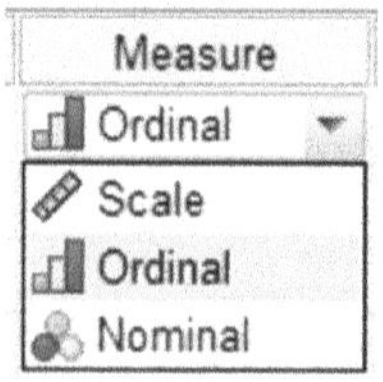

Look at the SAMPLE.sav data file we created in Chapter 1. We calculated a **mean** for the variable GRADE. GRADE was measured on a **ratio scale**, and the **mean** is an acceptable summary statistic (assuming that the distribution is **normal**).

We could have had SPSS calculate a **mean** for the variable TIME instead of GRADE. If we did, we would get the output presented here.

DOI: 10.4324/9781003450467-2

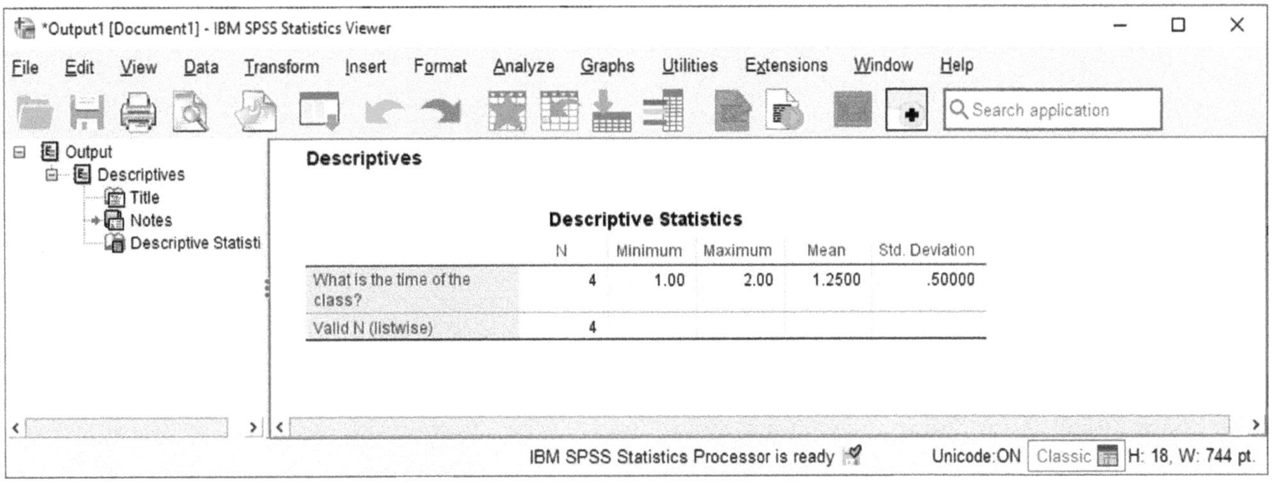

Descriptive Statistics

	N	Minimum	Maximum	Mean	Std. Deviation
What is the time of the class?	4	1.00	2.00	1.2500	.50000
Valid N (listwise)	4				

The output indicates that the average TIME was 1.25. Remember that TIME was coded as an ordinal variable (1 = morning class, 2 = afternoon class). Though the **mean** is not an appropriate statistic for an **ordinal scale**, SPSS calculated it anyway. The importance of considering the type of data cannot be overemphasized. Just because SPSS will compute a statistic for you does not mean that you should use it. Later in the text, when specific statistical procedures are discussed, the conditions under which they are appropriate will be addressed. Please note that there are some procedures (e.g., graphs and nonparametric tests) where SPSS limits what you can do based on the measurement scale. However, more often than not, it is up to the user to make that decision.

Missing Data

Often, participants do not provide complete data. For example, for some students, you may have a pretest score but not a posttest score. Perhaps one student left one question blank on a survey, or perhaps she did not state her age. Missing data can weaken any analysis. Often, a single missing answer can eliminate a subject from all analyses.

	q1	q2	total
1	2.00	2.00	4.00
2	3.00	1.00	4.00
3	4.00	3.00	7.00
4	2.00	.	.
5	1.00	2.00	3.00

If you have missing data in your dataset, leave that cell blank. In the example shown above, the fourth subject did not complete Question 2 (q2). Note that the total score (which is calculated from both questions) is also blank because of the missing data for Question 2. SPSS represents missing data in the **data window** with a period (although you should not enter a period—just leave it blank). It is NOT good practice to create a filler value (e.g., "999" or "0") to represent blank scores, because SPSS will see it as a value with meaning, whereas it will treat truly blank values as missing.

Section 2.2 Selection and Transformation of Data

We often have more data in a data file than we want to include in a specific analysis. For instance, our sample data file contains data from four participants, two of whom received special training and two of whom did not. If we wanted to conduct an analysis using only the two participants who did not receive the training, we would need to specify the appropriate subset.

Selecting a Subset

We can use the *Select Cases* command to specify a subset of our data. The *Select Cases* command is located under the *Data* menu. When you select this command, the **dialog box** below will appear. (Note the icons next to the variable names that indicate that all variables were defined as being measured on a **nominal scale** except grade, which was defined as scale.)

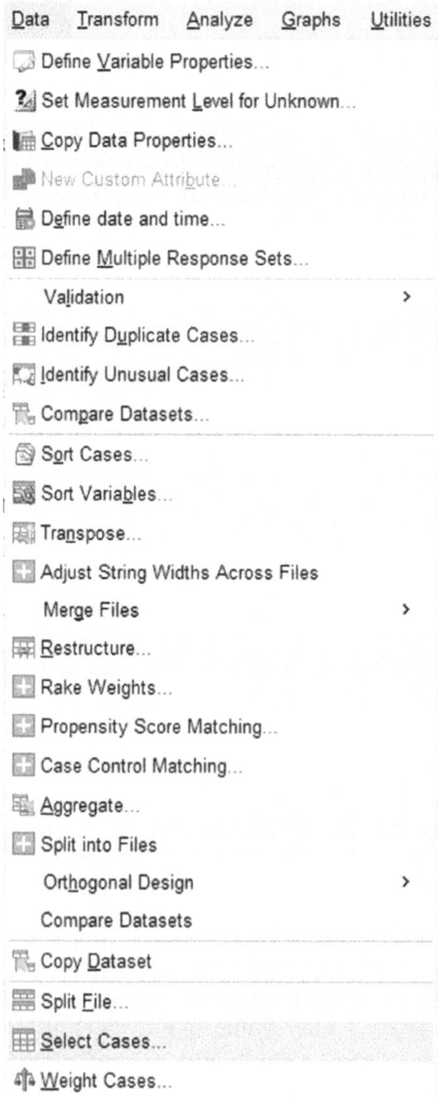

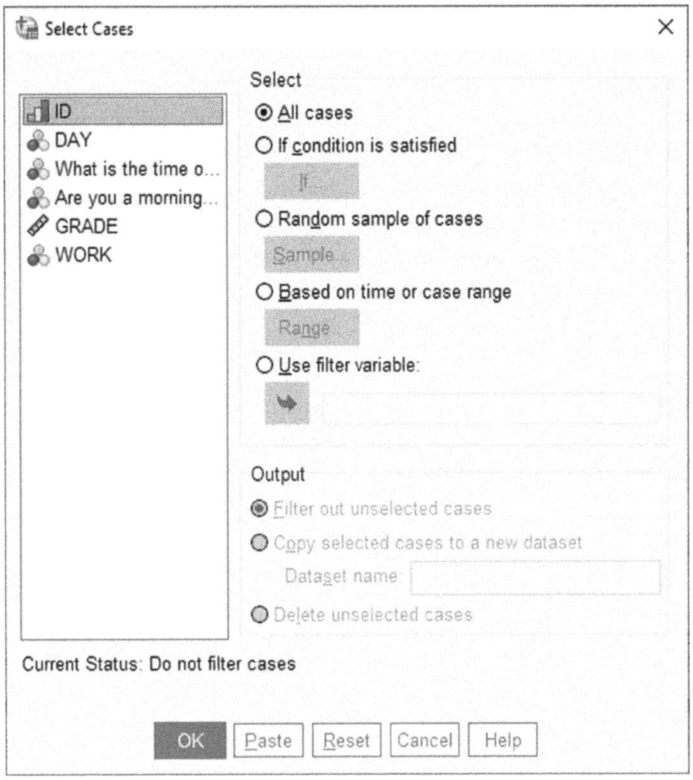

You can specify which cases (participants) you want to select by using the selection criteria, which appear on the right side of the Select Cases **dialog box**. By default, *All cases* will be selected. The most common way to select a subset is to click *If condition is satisfied*, then click on the button labeled *If*. This will bring up a new **dialog box** that allows you to indicate which cases you would like to use.

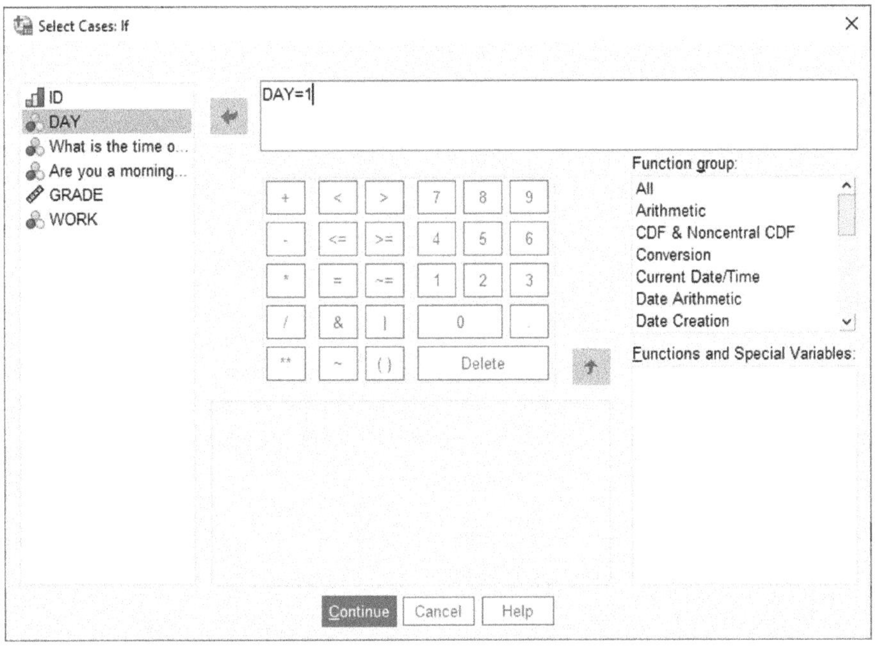

You can enter the logic used to select the subset in the upper section. If the logical statement is true for a given case, then that case will be selected. If the logical statement is false, that case will not be selected. For instance, you can select all cases that were coded as Mon/Wed/Fri by entering the formula DAY = 1 in the upper-left part of the window. If DAY is 1, then the statement will be true, and SPSS will select the case. If DAY is anything other than 1, the statement will be false, and the case will not be selected. Once you have entered the logical statement, click *Continue* to return to the Select Cases **dialog box**. Then, click *OK* to return to the **data window**.

After you have selected the cases, the **data window** will slightly change. The cases that were not selected will be marked with a diagonal line through the case number. For instance, for our sample data, the first and third cases are not selected. Only the second and fourth cases are selected for this subset.

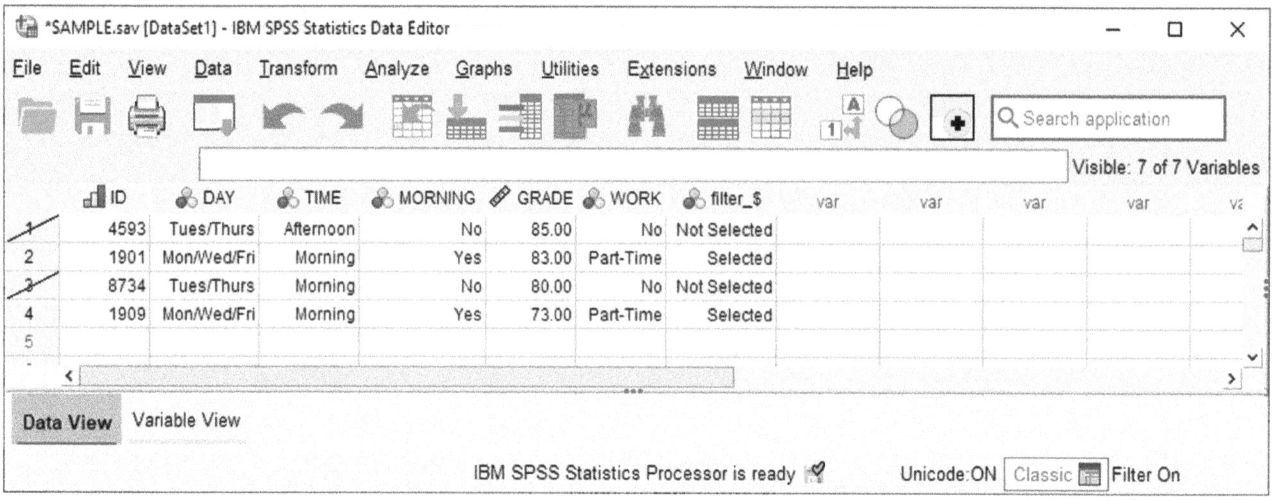

An additional variable will also be created in your data file. The new variable is called FILTER_$ and indicates whether a case was selected or not.

If we calculate a **mean** GRADE using the subset we just selected, we will receive the output here. Note that we now have a **mean** of 78.00 with a sample size (*N*) of 2 instead of 4.

Descriptive Statistics

	N	Minimum	Maximum	Mean	Std. Deviation
GRADE	2	73.00	83.00	78.0000	7.07107
Valid N (listwise)	2				

Be careful when you select subsets. *The subset remains in effect until you run the command again and select all cases.* You can tell if you have a subset selected because the bottom of the **data window** will indicate that a filter is on. In addition, when you examine your output, *N* will be less than the total number of records in your dataset if a subset is selected. The diagonal lines through some cases will also be evident when a subset is selected. Be careful not to save your data file with a subset selected, as this can cause considerable confusion later.

Computing a New Variable

SPSS can also be used to compute a new variable or manipulate your existing variables. To illustrate this, we will create a new data file. This file will contain data for four participants and three variables (Q1, Q2, and Q3). The variables represent the number of points each participant received on three different questions. Now enter the data shown on the screen below. When done, save this data file as "QUESTIONS.sav." We will be using it again in later chapters.

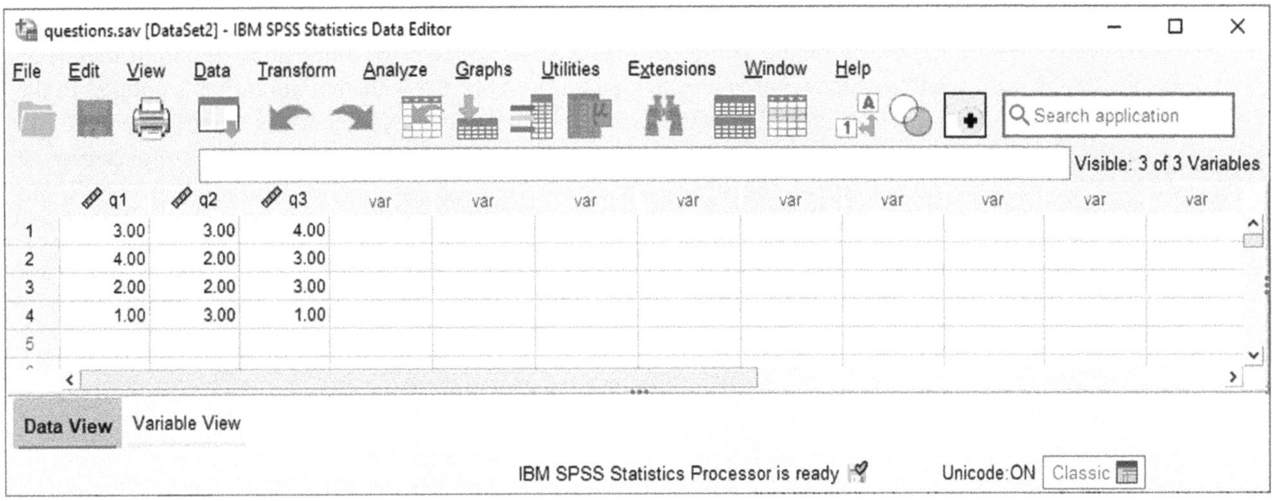

Now you will calculate the total score for each subject. We could do this manually, but if the data file were large, or if there were a lot of questions, this would take a long time. It is more efficient (and more accurate) to have SPSS compute the totals for you. To do this, click *Transform* and then click *Compute Variable*.

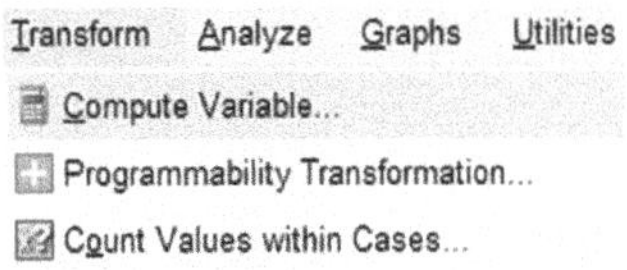

After clicking the *Compute Variable* command, we get the **dialog box** shown below.

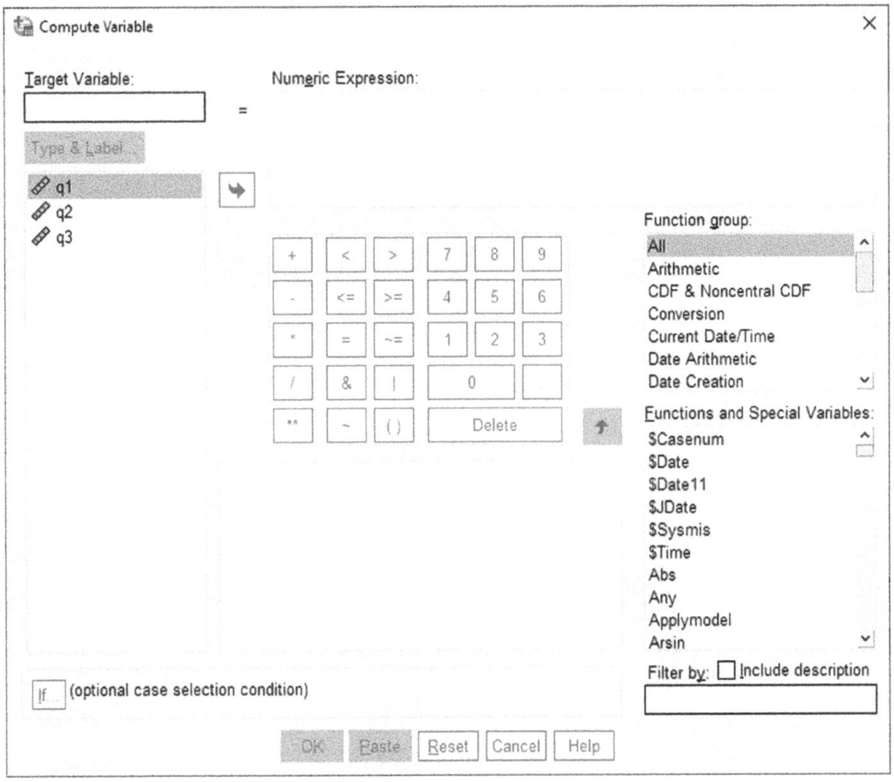

The blank field marked *Target Variable* is where we enter the name of the new variable we want to create. In this example, we are creating a variable called TOTAL, so type the word *total*.

Note that there is an equals sign between the *Target Variable* blank and the *Numeric Expression* blank. These two blank areas are the two sides of an equation that SPSS will calculate. For instance, total = Q1 + Q2 + Q3 is the equation that is entered in the sample presented here (screenshot shown above). Note that it is possible to create any equation here simply by using the number and operational keypad at the bottom of the **dialog box**. When we click *OK*, SPSS will create a new variable called TOTAL and make it equal to the sum of the three questions.

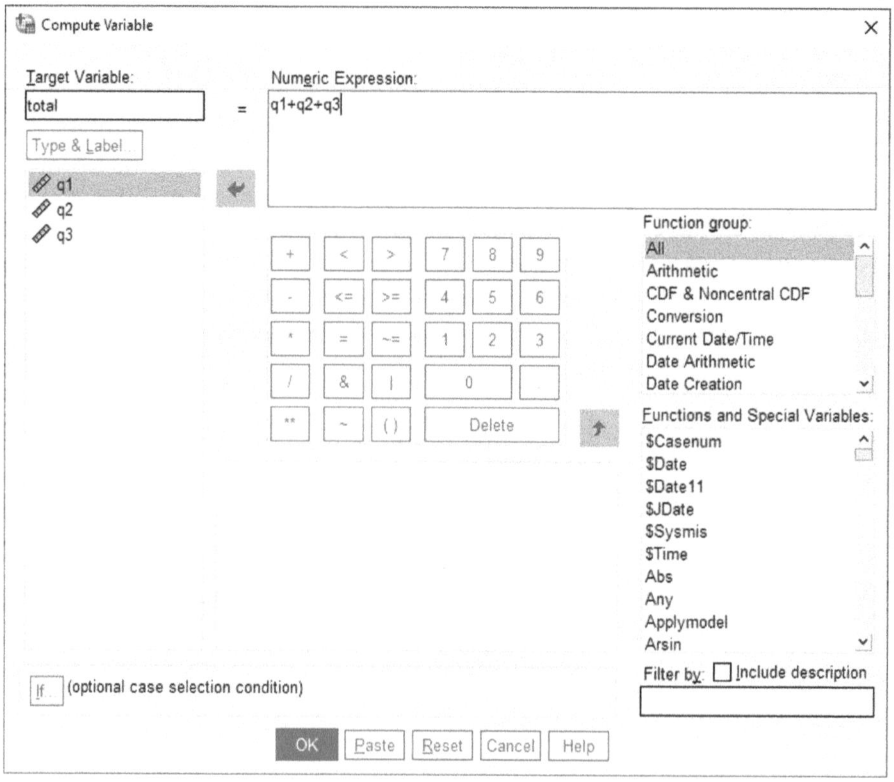

Save your data file again so that the new variable will be available for future sessions.

Recoding a Variable—Different Variable

SPSS can create a new variable based upon data from another variable. Say we want to split our participants on the basis of their total score. We want to create a variable called GROUP, which is coded 1 if the total score is low (less than or equal to 8) or 2 if the total score is high (9 or larger). To do this, we click *Transform*, then *Recode into Different Variables*.

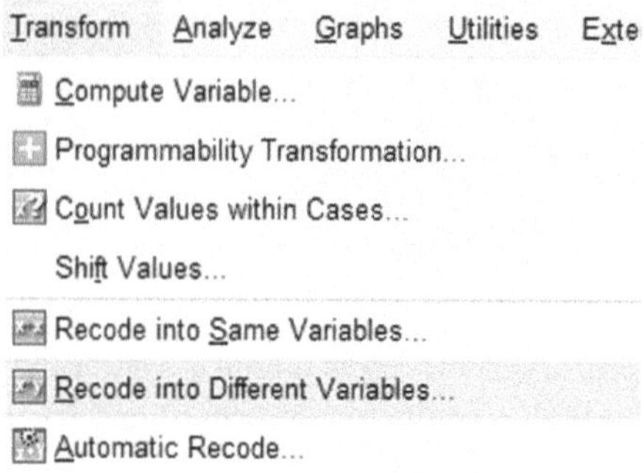

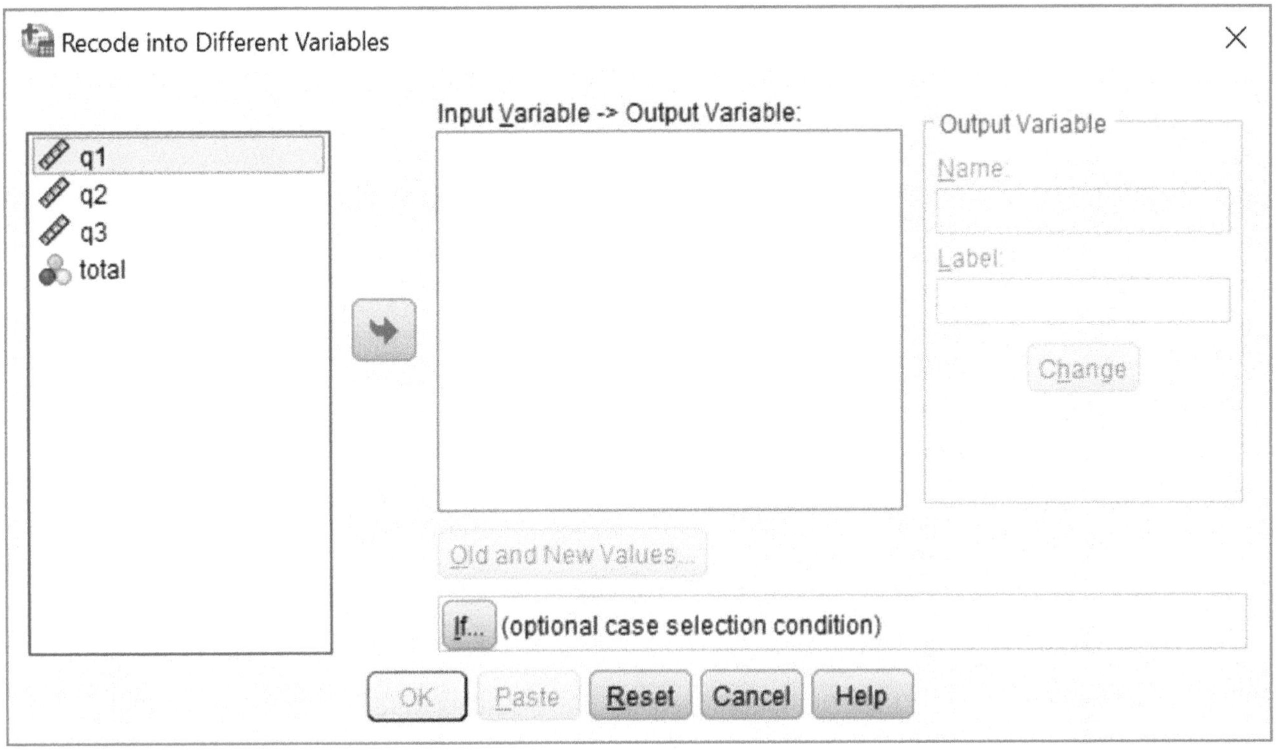

This will bring up the Recode into Different Variables **dialog box** shown above. Transfer the variable TOTAL to the middle blank. Type *group* in the *Name* field under *Output Variable*. Click *Change*, and the middle blank will show that TOTAL is becoming GROUP, as shown below.

Click *Old and New Values*. This will bring up the Recode **dialog box** below.

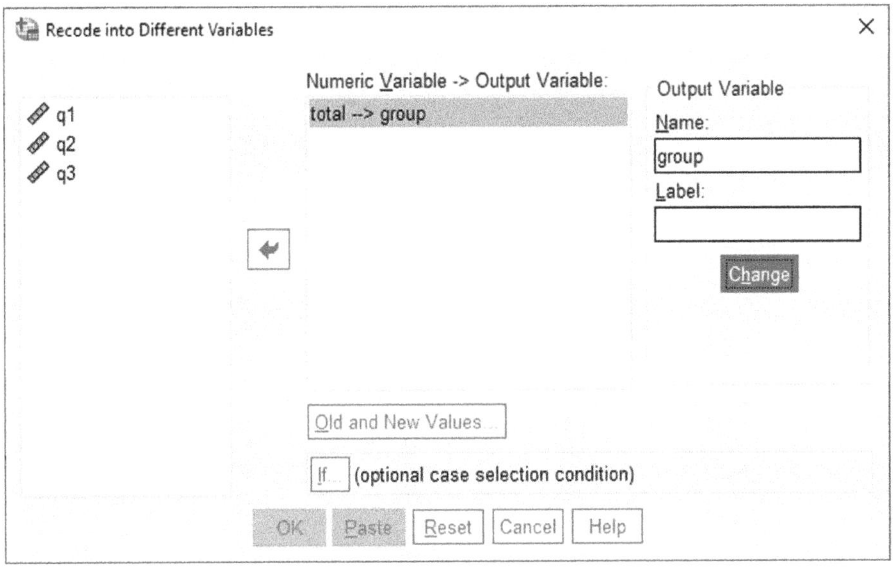

In the example shown here, we have entered a 9 in the *Range, value through HIGHEST* field, and a 2 in the *Value* field under *New Value*. When we click *Add*, the blank on the right displays the recoding formula. We next entered an 8 on the left in the *Range, LOWEST through value* blank, and a 1 in the *Value* field under *New Value*. Click *Add*, then *Continue*.

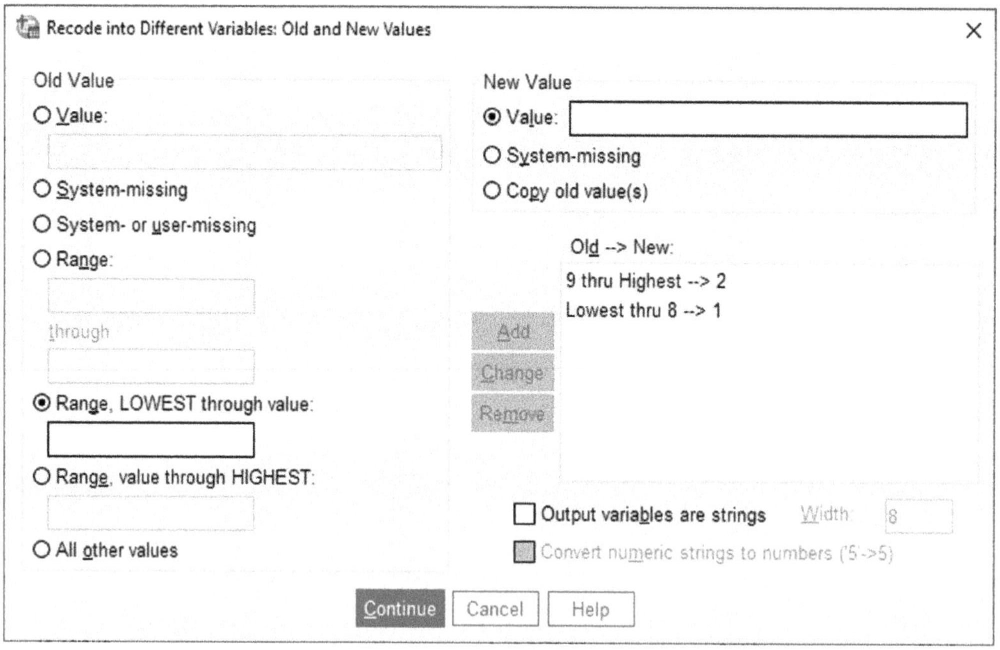

Click *OK*. You will be redirected to the **data window** shown below. A new variable (GROUP) will have been added and coded as 1 or 2, based on TOTAL.

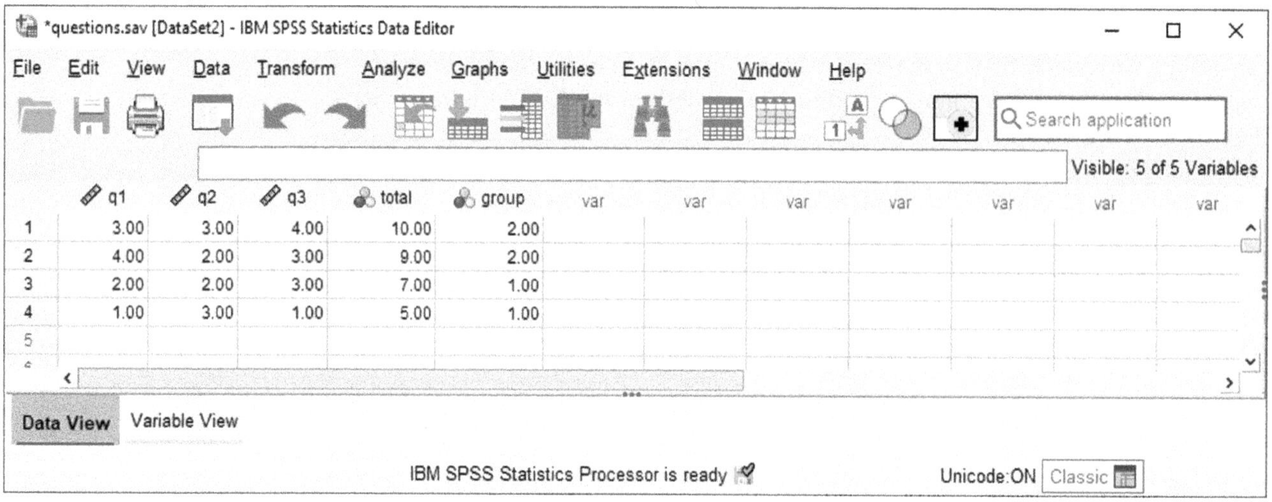

Chapter 3
Descriptive Statistics

In Chapter 2, we discussed many of the options available in SPSS for dealing with data. Now we will discuss ways to summarize our data. The procedures used to describe and summarize data are called **descriptive statistics**.

Section 3.1 Frequency Distributions and Percentile Ranks for a Single Variable

Description

The *Frequencies* command produces frequency distributions for the specified variables. The output includes the number of occurrences, percentages, valid percentages, and cumulative percentages. The valid percentages and the cumulative percentages comprise only the data that are not designated as missing.

The *Frequencies* command is useful for describing samples where the **mean** is not useful (e.g., **nominal** or **ordinal** scales). It is also useful as a method of getting the feel for your data. It provides more information than just a **mean** and **standard deviation** and can be useful in determining **skew** and identifying **outliers**. A special feature of the command is its ability to determine **percentile ranks**.

Assumptions

Cumulative percentages and **percentiles** are valid only for data that are measured on at least an **ordinal scale**. Because the output contains one line for each value of a variable, this command works best on variables with a relatively small number of values.

Drawing Conclusions

The *Frequencies* command produces output that indicates both the number of cases in the sample of a particular value and the percentage of cases with that value. Thus, conclusions drawn should relate only to describing the numbers or percentages of cases in the sample. If the data are at least ordinal in nature, conclusions regarding the cumulative percentage and/or **percentiles** can be drawn.

SPSS Data Format

The SPSS data file for obtaining frequency distributions requires only one variable, and that variable can be of any type.

DOI: 10.4324/9781003450467-3

Creating a Frequency Distribution

This example uses the CAR_SALES.sav data file that comes with SPSS. If you are using SPSS Version 29, you can access it when the program first starts by selecting *Sample Files* and then locating the file.

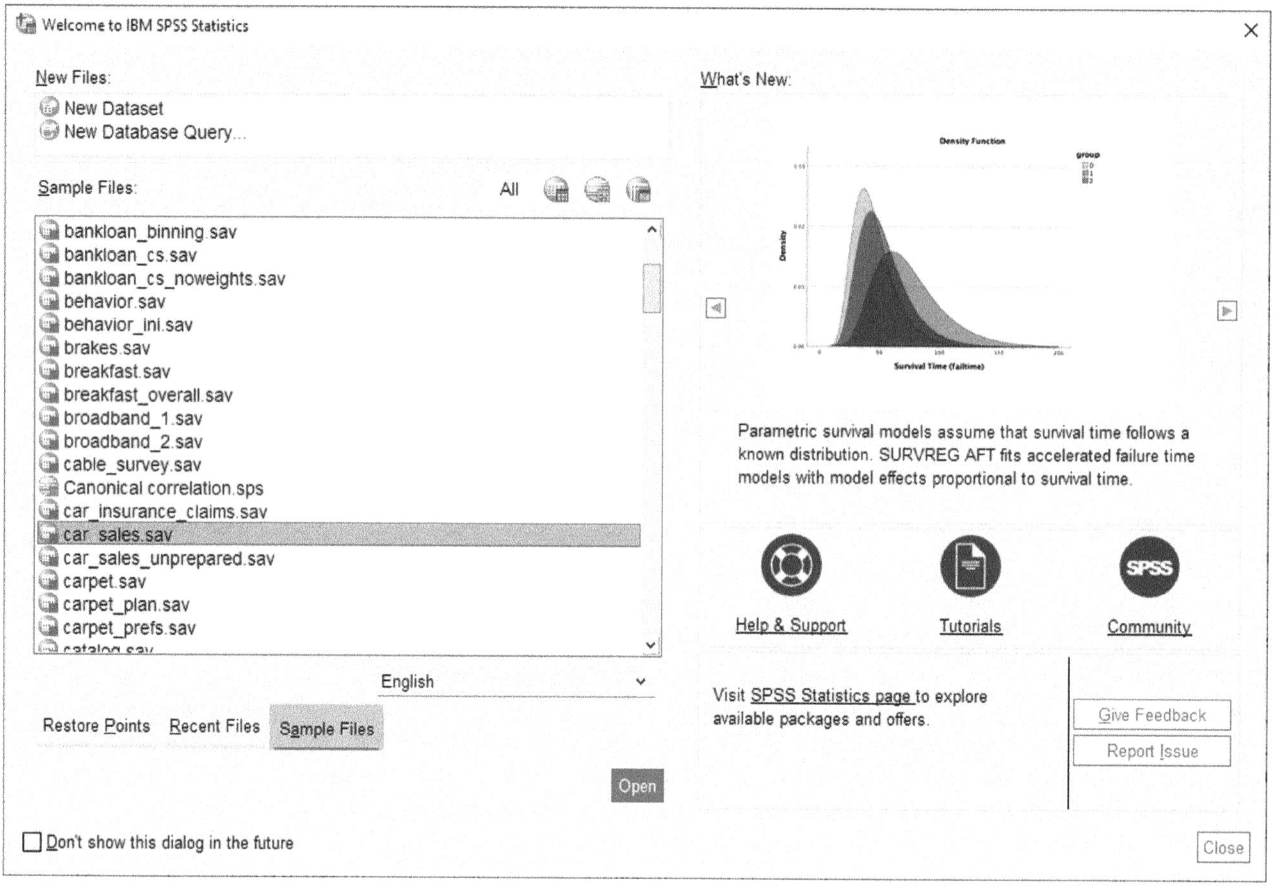

To run the *Frequencies* command, click *Analyze*, then *Descriptive Statistics*, then *Frequencies*. This will bring up the main **dialog box** below. Transfer the variable for which you would like a frequency distribution into the *Variable(s)* blank shown to the right (MANUFACTURER in this case). Be sure that the *Display frequency tables* option is checked.

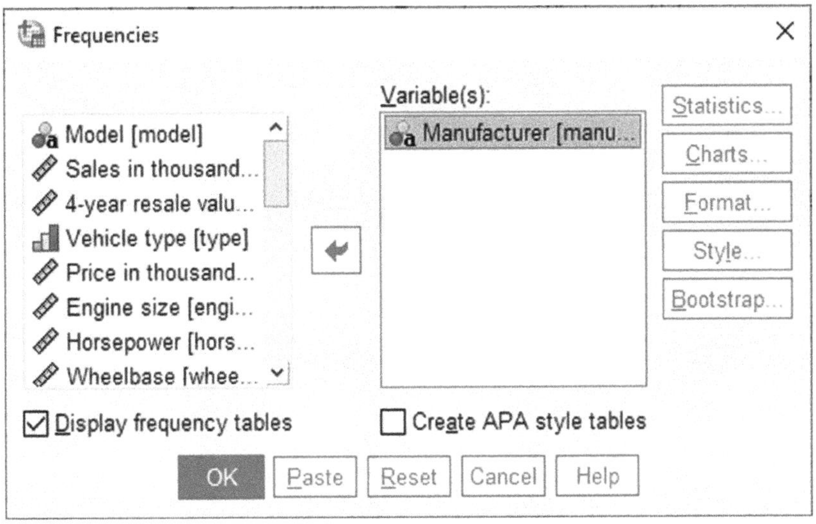

Note that the **dialog boxes** in SPSS show both the type of variable (the icon immediately to the left of the variable name) and the variable labels if they are entered. Thus, the variable MANUFACT shows up in the **dialog box** as *Manufacturer.*

Click *OK* to run the test.

Output for a Frequency Distribution

The output consists of two sections. The first section indicates the number of records with **valid data** for each variable selected. Records with a blank score are listed as missing. In this example, the data file contained 157 records.

The second section of the output contains a cumulative frequency distribution for each variable selected. At the top of the section, the variable label is given. The output itself consists of five columns. The first column lists the values of the variable sorted in alphabetical order. There is a row for each value of your variable, and additional rows are added at the bottom for the Total and Missing data. The second column gives the frequency of each value, including missing values. The third column gives the percentage of all records (including records with missing data) for each value. The fourth column, labeled *Valid Percent,* gives the percentage of records (without including records with missing data) for each value. If there were any missing values, these values would be larger than the values in column three because the total number of records would have been reduced by the number of records with missing values. The final column gives cumulative percentages. Cumulative percentages indicate the percentage of records with a score equal to or smaller than the current value. Thus, the last value is always 100 percent. These values are equivalent to **percentile ranks** for the values listed. Note: SPSS will provide cumulative percentages even for nominal variables and other variables where it does not make sense. For example, the variable MANUFACTURER is nominal.

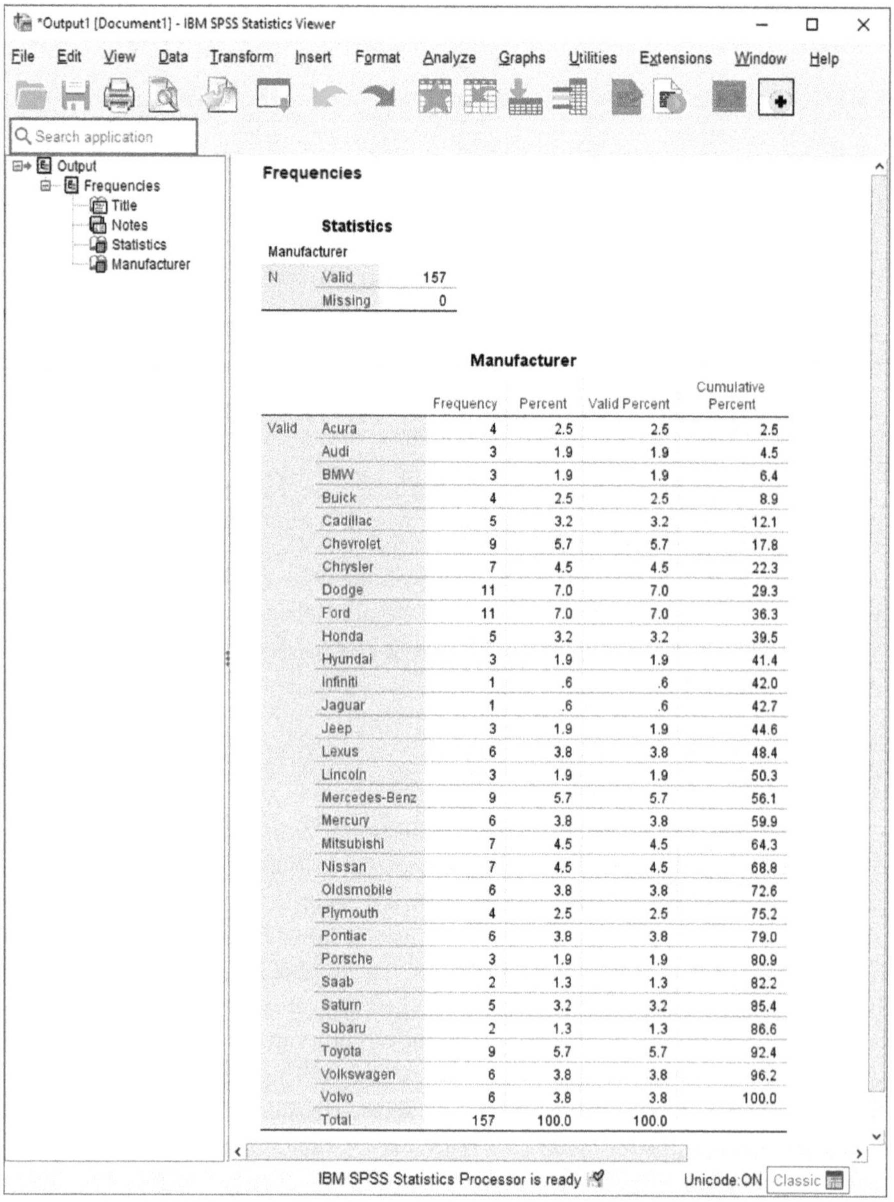

Frequencies

Statistics

Manufacturer

N	Valid	157
	Missing	0

Manufacturer

		Frequency	Percent	Valid Percent	Cumulative Percent
Valid	Acura	4	2.5	2.5	2.5
	Audi	3	1.9	1.9	4.5
	BMW	3	1.9	1.9	6.4
	Buick	4	2.5	2.5	8.9
	Cadillac	5	3.2	3.2	12.1
	Chevrolet	9	5.7	5.7	17.8
	Chrysler	7	4.5	4.5	22.3
	Dodge	11	7.0	7.0	29.3
	Ford	11	7.0	7.0	36.3
	Honda	5	3.2	3.2	39.5
	Hyundai	3	1.9	1.9	41.4
	Infiniti	1	.6	.6	42.0
	Jaguar	1	.6	.6	42.7
	Jeep	3	1.9	1.9	44.6
	Lexus	6	3.8	3.8	48.4
	Lincoln	3	1.9	1.9	50.3
	Mercedes-Benz	9	5.7	5.7	56.1
	Mercury	6	3.8	3.8	59.9
	Mitsubishi	7	4.5	4.5	64.3
	Nissan	7	4.5	4.5	68.8
	Oldsmobile	6	3.8	3.8	72.6
	Plymouth	4	2.5	2.5	75.2
	Pontiac	6	3.8	3.8	79.0
	Porsche	3	1.9	1.9	80.9
	Saab	2	1.3	1.3	82.2
	Saturn	5	3.2	3.2	85.4
	Subaru	2	1.3	1.3	86.6
	Toyota	9	5.7	5.7	92.4
	Volkswagen	6	3.8	3.8	96.2
	Volvo	6	3.8	3.8	100.0
	Total	157	100.0	100.0	

Determining Percentile Ranks

The *Frequencies* command can be used to provide a number of **descriptive statistics**, as well as a variety of percentile values (including **quartiles**, cut points, and scores corresponding to a specific **percentile rank**). Let us look at statistics for the variable PRICE.

To obtain either the descriptive or percentile functions of the *Frequencies* command, click the *Statistics* button at the top right of the main **dialog box**.

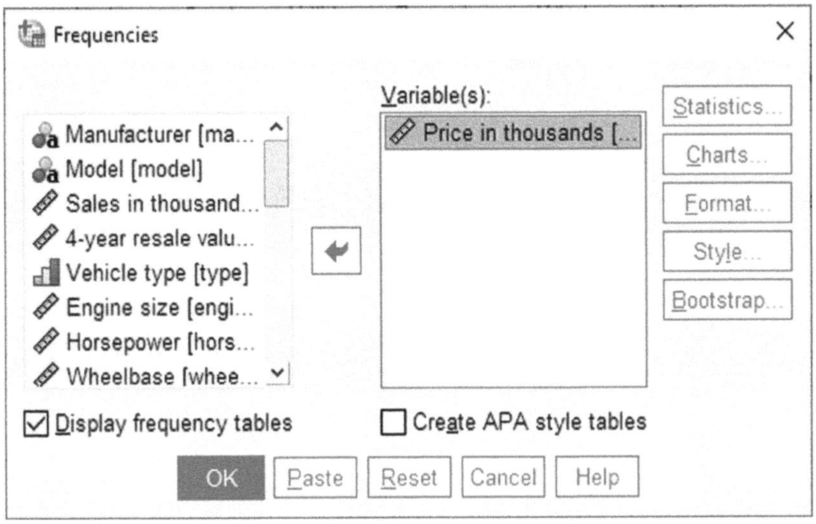

This brings up the Frequencies: Statistics **dialog box** below. Note that the *Central Tendency* and *Dispersion* sections of this box are useful for calculating values, such as the **median** or **mode**, that cannot be calculated with the *Descriptives* command (see Section 3.3). Check any additional desired statistic by clicking on the blank next to it. (In our example here, let us determine the **quartiles**, 80th **percentile**, **mean**, and **median**.)

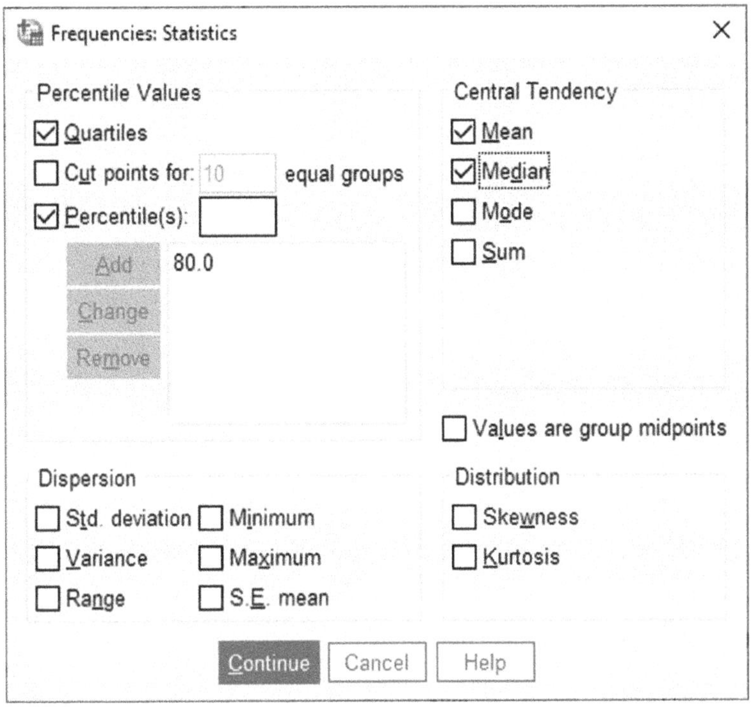

For **percentiles**, enter the desired **percentile rank** in the blank to the right of the *Percentile(s)* label. Then, click *Add* to add it to the list of **percentiles** requested. Once you have selected all your required statistics, click *Continue* to return to the main **dialog box**. Click *OK*.

Output for Percentile Ranks

Statistics

Price in thousands

N	Valid	155
	Missing	2
Mean		27.39075
Median		22.79900
Percentiles	25	17.89000
	50	22.79900
	75	31.96500
	80	36.21020

The Statistics **dialog box** adds on to the first section of output from the *Frequencies* command. The output contains a row for each piece of information you requested. In the previous example, we checked *Quartiles* and asked for the 80th **percentile**. Thus, the output contains rows for the 25th, 50th, 75th, and 80th **percentiles**.

Practice Exercise

Using Practice Dataset 1 in Appendix B, create a frequency distribution table for the mathematics skills scores. Determine the mathematics skills score at which the 60th percentile lies.

Section 3.2 Frequency Distributions and Percentile Ranks for Multiple Variables

Description

The *Crosstabs* command produces frequency distributions for multiple variables. The output includes the number of occurrences of each combination of **levels** of each variable. It is possible to have the command give percentages for any or all variables.

The *Crosstabs* command is useful for describing samples where the **mean** is not useful (e.g., **nominal** or **ordinal scales**). It is also useful as a method for getting a feel for your data.

Assumptions

Because the output contains a row or column for each value of a variable, this command works best on variables with a relatively small number of values.

SPSS Data Format

The SPSS data file for the *Crosstabs* command requires two or more variables. Those variables can be of any type.

Running the Command

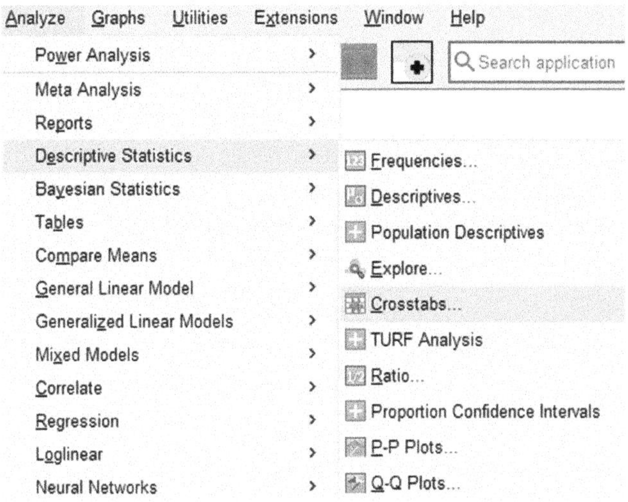

This example uses the SAMPLE.sav data file, which you created in Chapter 1. To run the procedure, click *Analyze*, then *Descriptive Statistics*, then *Crosstabs*. This will bring up the main Crosstabs **dialog box**, which is shown below.

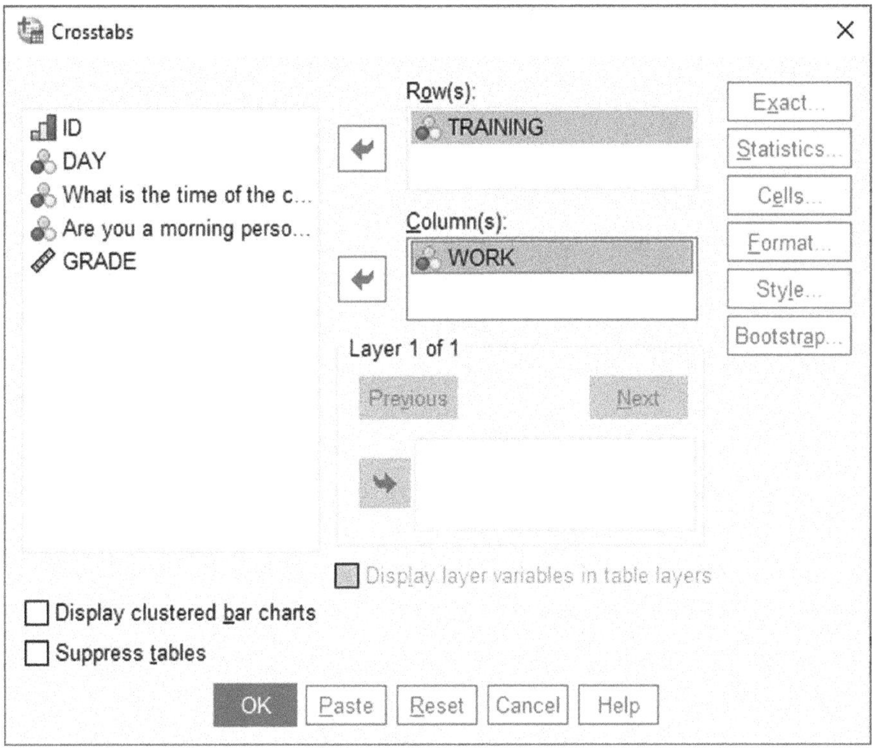

The **dialog box** initially lists all variables on the left and contains two blank boxes labeled *Row(s)* and *Column(s)*. Enter one variable (TRAINING) in the *Row(s)* box. Enter the second (WORK) in the *Column(s)* box. To analyze more than two variables, you would enter the third, fourth, and so on, in the unlabeled area (just under the *Layer* indicator).

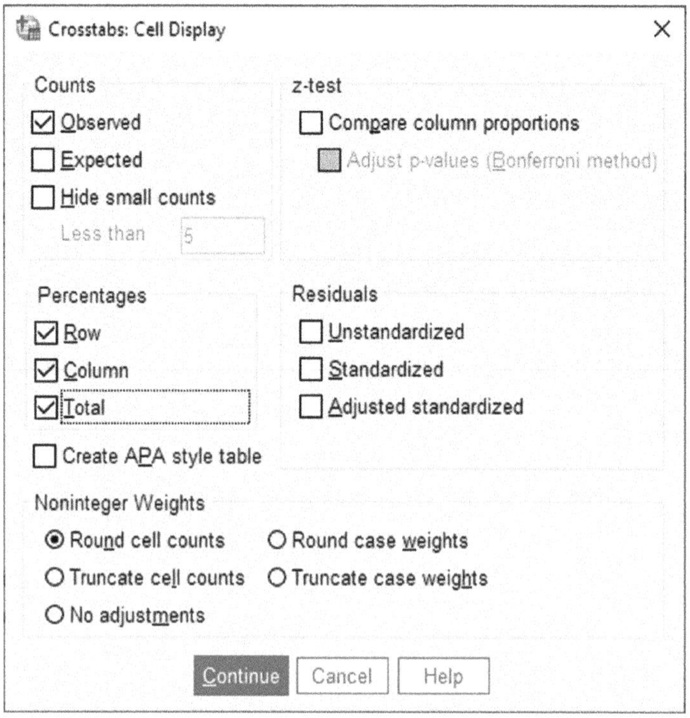

The *Cells* button allows you to specify percentages and other information to be generated for each combination of values. Click *Cells*, and you will get the box shown above, with *Observed* and *Round cell counts* selected by default.

For the example presented here, check *Row*, *Column*, and *Total* percentages. Then click *Continue*. This will return you to the Crosstabs **dialog box**. Click *OK* to run the analysis.

Interpreting Crosstabs Output

The output consists of a contingency table. Each level of WORK is given a column. Each level of TRAINING is given a row. In addition, a row is added for total, and a column is added for total.

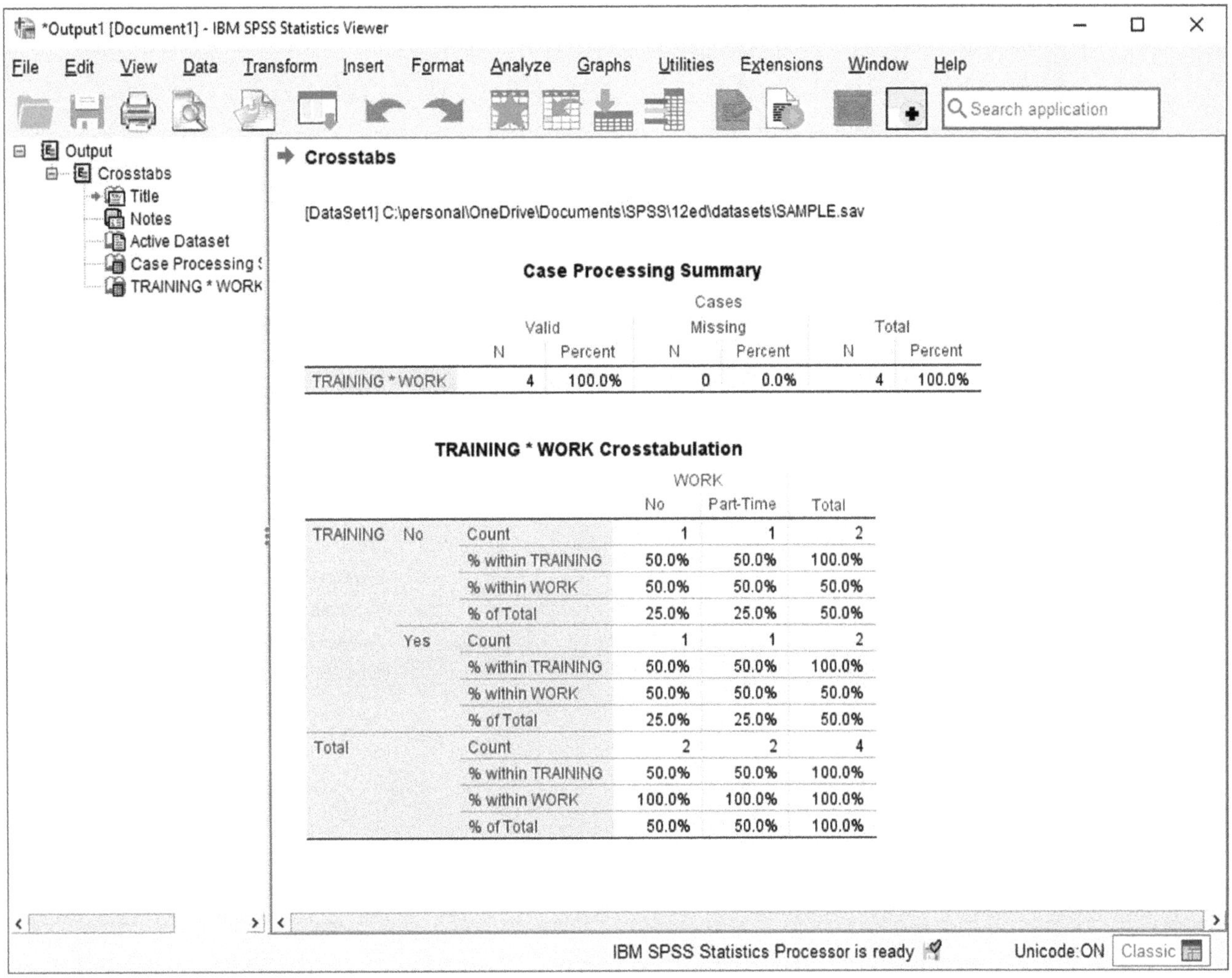

Case Processing Summary

	Cases					
	Valid		Missing		Total	
	N	Percent	N	Percent	N	Percent
TRAINING * WORK	4	100.0%	0	0.0%	4	100.0%

TRAINING * WORK Crosstabulation

			WORK		
			No	Part-Time	Total
TRAINING	No	Count	1	1	2
		% within TRAINING	50.0%	50.0%	100.0%
		% within WORK	50.0%	50.0%	50.0%
		% of Total	25.0%	25.0%	50.0%
	Yes	Count	1	1	2
		% within TRAINING	50.0%	50.0%	100.0%
		% within WORK	50.0%	50.0%	50.0%
		% of Total	25.0%	25.0%	50.0%
Total		Count	2	2	4
		% within TRAINING	50.0%	50.0%	100.0%
		% within WORK	100.0%	100.0%	100.0%
		% of Total	50.0%	50.0%	100.0%

Each cell contains the number of participants (e.g., one participant received no training and does not work; two participants received no training, regardless of employment status).

The percentages for each cell are also shown. Row percentages add up to 100 percent horizontally. Column percentages add up to 100 percent vertically. For instance, of all the individuals who had no training, 50 percent did not work and 50 percent worked part-time (using the "% within TRAINING" row). Of the individuals who did not work, 50 percent had no training and 50 percent had training (using the "% within WORK" row).

Practice Exercise

Using Practice Dataset 1 in Appendix B, create a contingency table using the *Crosstabs* command. Determine the number of participants in each combination of the variables SEX and MARITAL. What percentage of participants are married? What percentage of participants are male and married?

Section 3.3 Measures of Central Tendency and Measures of Dispersion for a Single Group

Description

Measures of central tendency are values that represent a typical member of the sample or population. The three primary types of measures are the **mean**, **median**, and **mode**. Measures of dispersion tell you the variability of your scores. The primary types are the **range** and the **standard deviation**. Together, a measure of central tendency and a measure of dispersion provide a great deal of information about the entire dataset.

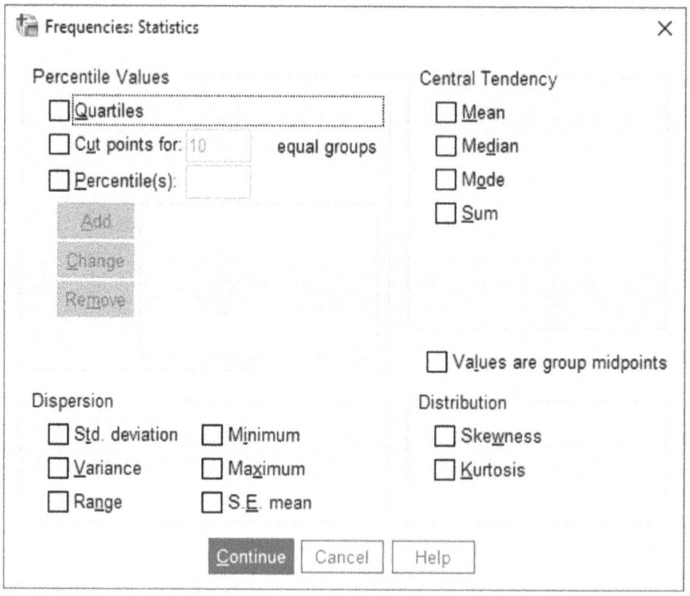

We will discuss these measures of central tendency and measures of dispersion in the context of the *Descriptives* command. Note that many of these statistics can also be calculated with several other commands (e.g., the *Frequencies* or *Compare Means* commands are required to compute the **mode** or **median**—the *Statistics* option for the *Frequencies* command is shown above).

Assumptions

Each measure of central tendency and measure of dispersion has different assumptions associated with it. The **mean** is the most powerful measure of central tendency, and it has the most assumptions. For instance, to calculate a **mean**, the data must be measured on an **interval** or **ratio scale**. In addition, the distribution should be normally distributed or, at least, not highly skewed. The **median** requires at least **ordinal** data. Because the **median** indicates only the middle score (when scores are arranged in order), there are no assumptions about the shape of the distribution. The **mode** is the weakest measure of central tendency. There are no assumptions for the **mode**.

The **standard deviation** is the most powerful measure of dispersion, but it, too, has several requirements. It is a mathematical transformation of the **variance** (the **standard deviation** is the square root of the **variance**). Thus, if one is appropriate, the other is also appropriate. The **standard deviation** requires data measured on an **interval** or **ratio scale**. In addition, the distribution should be normal. The **range** is the weakest measure of dispersion. To calculate a **range**, the variable must be at least **ordinal**. For **nominal scale** data, the entire frequency distribution should be presented as a measure of dispersion.

Drawing Conclusions

A measure of central tendency should be accompanied by a measure of dispersion. Thus, when reporting a **mean**, you should also report a **standard deviation**. When presenting a **median**, you should also state the **range** or interquartile **range**. The interquartile **range** can be determined using the *Frequencies* command.

SPSS Data Format

Only one variable is required.

Running the Command

The *Descriptives* command is the command you will most likely use for obtaining measures of central tendency and measures of dispersion. This example uses the SAMPLE.sav data file we used in the previous chapters.

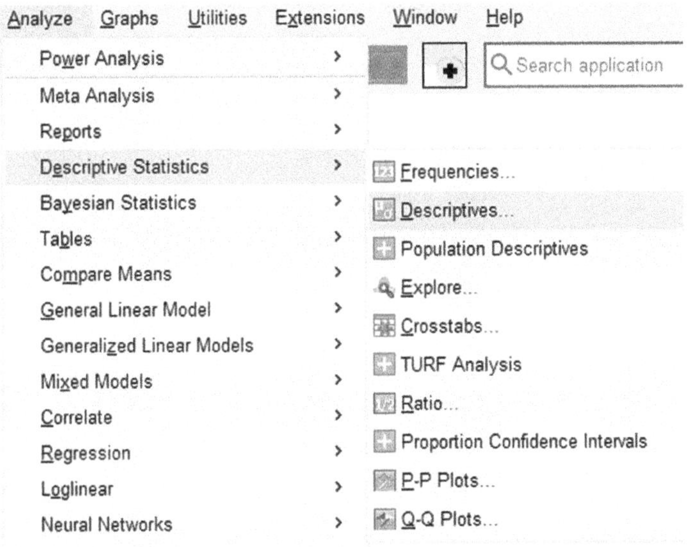

To run the command, click *Analyze*, then *Descriptive Statistics*, then *Descriptives*. This will bring up the main **dialog box** for the *Descriptives* command. Any variables you would like information about can be placed in the right blank by double-clicking them or by selecting them and then clicking on the arrow. Move the variable GRADE to the right for this example.

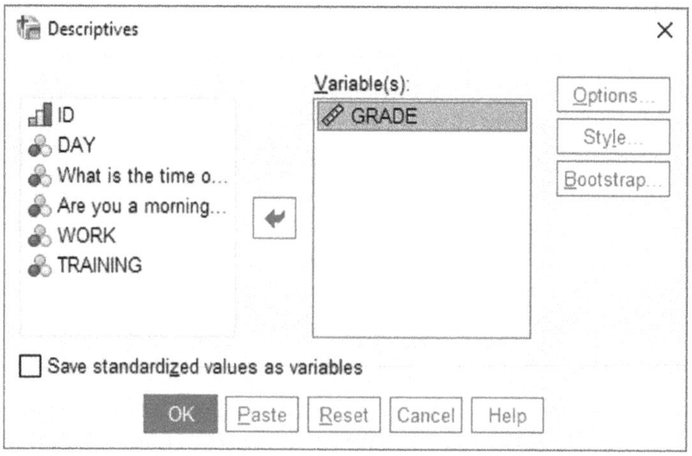

By default, you will receive the *N* (number of cases/participants), the minimum value, the maximum value, the **mean**, and the **standard deviation**. Note that some of these may not be appropriate for the types of data you have selected.

If you would like to change the default statistics that are given, click *Options* in the main **dialog box**. You will be given the Options **dialog box** shown below. Click *Continue* or *Cancel* to close the *Options* box. Then click *OK*.

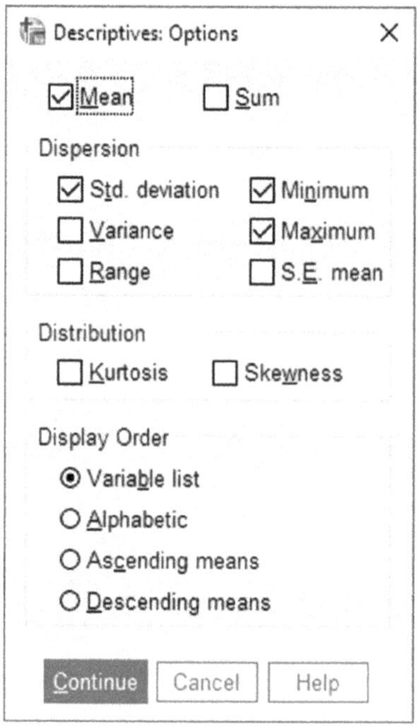

Reading the Output

The output for the *Descriptives* command is quite straightforward. Each type of output requested is presented in a column, and each variable is given in a row. The output presented below is for the sample data file. It shows that we have one variable (GRADE) and that we obtained the *N*, minimum, maximum, **mean**, and **standard deviation** for this variable.

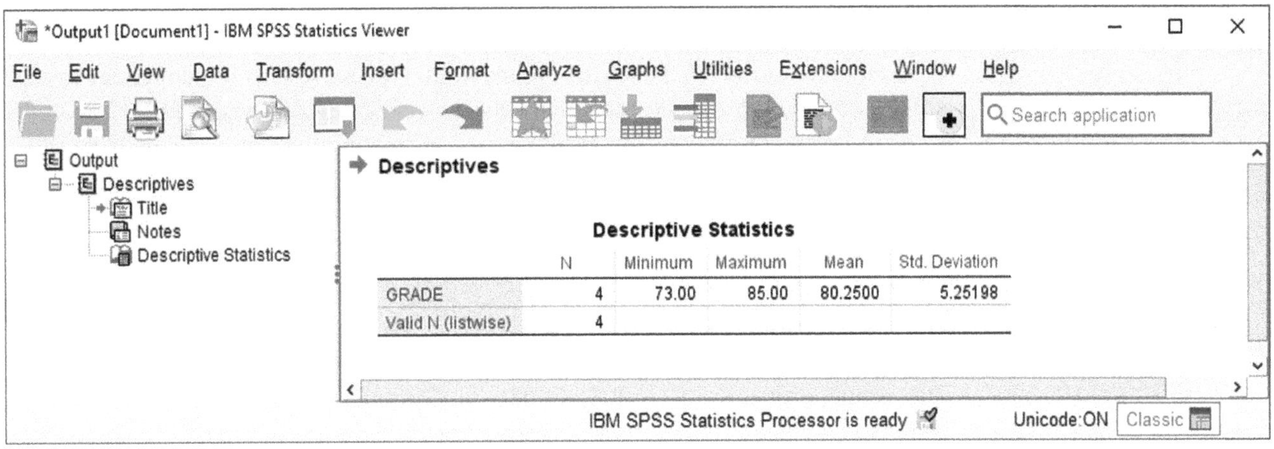

Descriptive Statistics

	N	Minimum	Maximum	Mean	Std. Deviation
GRADE	4	73.00	85.00	80.2500	5.25198
Valid N (listwise)	4				

Practice Exercise

Using Practice Dataset 1 in Appendix B, obtain the **descriptive statistics** for the age of the participants. What is the **mean**? The **median**? The **mode**? What is the **standard deviation**? Minimum? Maximum? The **range**? (Refer to Section 3.1 for assistance with **median** and **mode**.)

Section 3.4 Measures of Central Tendency and Measures of Dispersion for Multiple Groups

Description

The measures of central tendency discussed earlier are often needed not only for the entire dataset but also for several subsets. One way to obtain these values for subsets is to use the data-selection techniques discussed in Chapter 2 and apply the *Descriptives* command to each subset. An easier way to perform this task is to use the *Means* command. The *Means* command is designed to provide **descriptive statistics** for subsets of your data.

Assumptions

The assumptions discussed in the section titled "Measures of Central Tendency and Measures of Dispersion for a Single Group" (Section 3.3) also apply to multiple groups.

Drawing Conclusions

A measure of central tendency should be accompanied by a measure of dispersion. Thus, when giving a **mean**, you should also report a **standard deviation**. When presenting a **median**, you should also state the **range** or interquartile **range**.

SPSS Data Format

Two variables in the SPSS data file are required. One represents the **dependent variable** and will be the variable for which you receive the **descriptive statistics**. The other is the **independent variable** and will be used in creating the subsets. Note that while SPSS calls this variable an **independent variable**, it may not meet the strict criteria that define a true **independent variable** (e.g., treatment manipulation). Thus, some SPSS procedures refer to it as the **grouping variable**.

Running the Command

This example uses the SAMPLE.sav data file you created in Chapter 1. The *Means* command is run by clicking *Analyze*, then *Compare Means*, then *Means*.

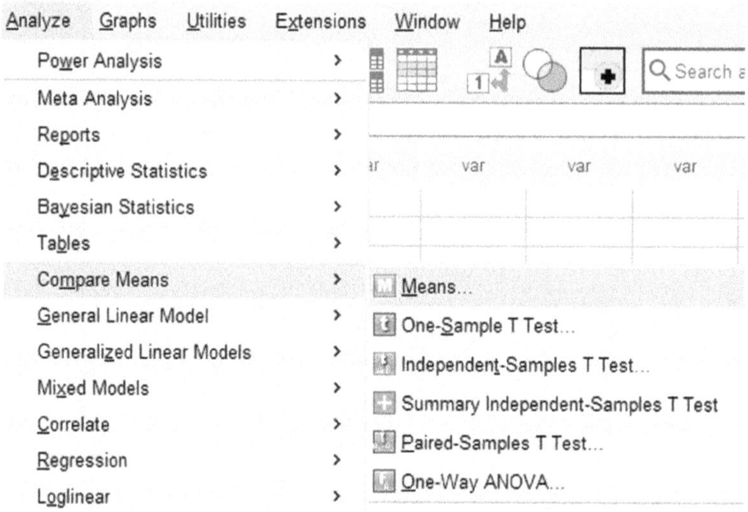

This will bring up the main **dialog box** for the *Means* command. Place the selected variable in the blank field labeled *Dependent List*. Place the **grouping variable** in the box labeled *Independent List*. In this example, through use of the SAMPLE.sav data file, measures of central tendency and measures of dispersion for the variable GRADE will be given for each level of the variable MORNING.

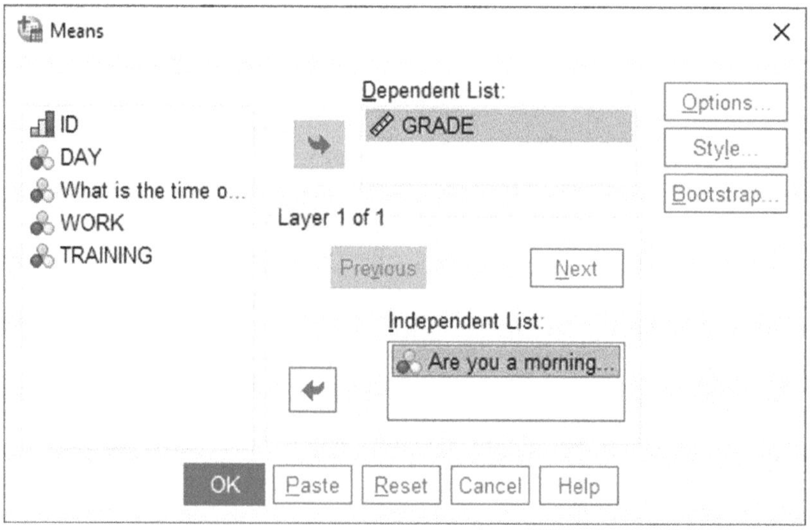

By default, the **mean**, number of cases, and **standard deviation** are given. If you would like additional measures, click *Options* and you will be presented with the **dialog box** shown here. You can opt to include any number of measures.

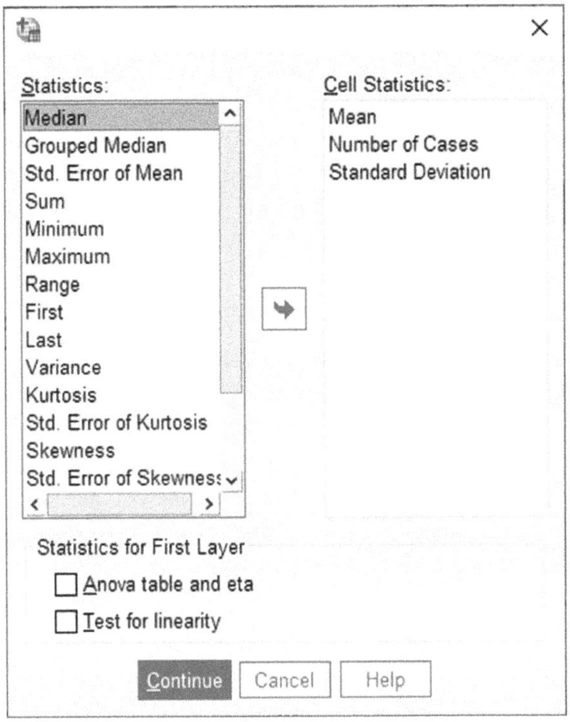

Reading the Output

The output for the *Means* command is split into two sections. The first section, called a **case processing summary**, gives information about the data used. In our sample data file, there are four students (cases), all of whom were included in the analysis.

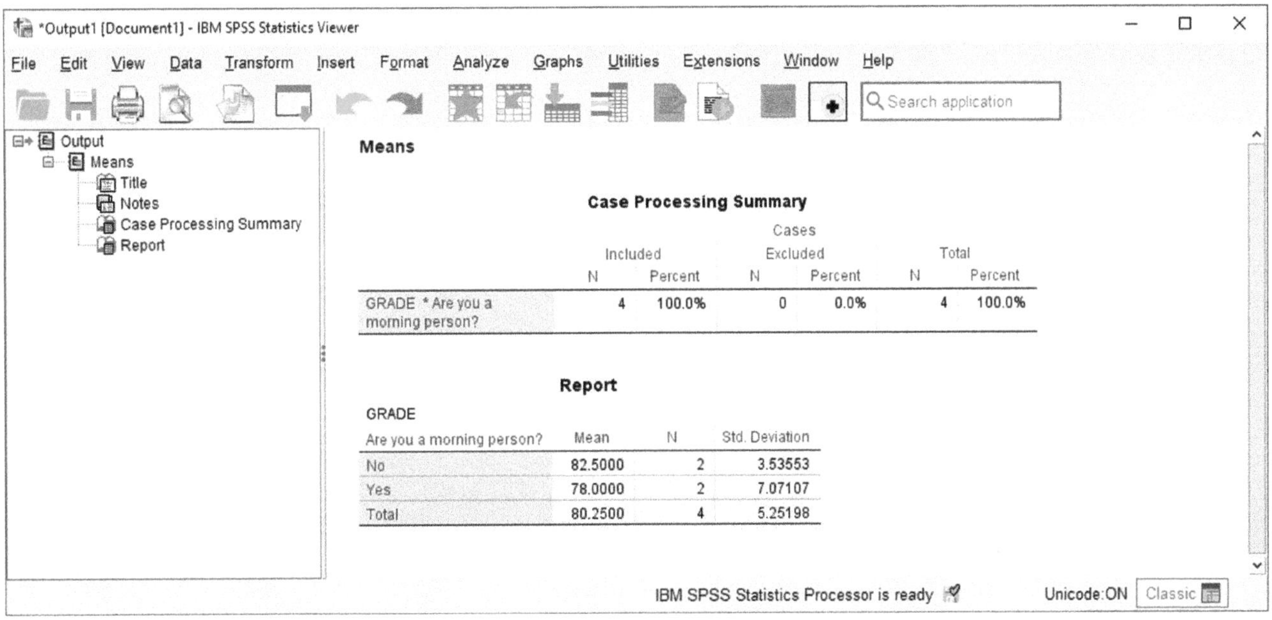

This report lists the name of the **dependent variable** (GRADE) at the top. Every **level** of the **independent variable** (MORNING) is shown in a row in the table. In this example, the **levels** are 0 and 1, labeled No and Yes. Note that if a variable is labeled, the labels will be used instead of the raw values.

The summary statistics given in the report correspond to the data, where the **level** of the **independent variable** is equal to the row heading (e.g., No, Yes). Thus, two participants were included in each row. Morning people had a **mean** GRADE of 78.0 and non-morning people had a **mean** GRADE of 82.5.

An additional row has been added, named Total. That row contains the combined data, and the values are the same as they would be if we had run the *Descriptives* command for the variable GRADE.

Extension to More Than One Independent Variable

If you have more than one **independent variable**, SPSS can break down the output even further. Rather than adding more variables to the *Independent List* section of the **dialog box**, you need to add them in a different layer. If you click *Next*, you will be presented with *Layer 2 of 2*, and you can select a second **independent variable** (e.g., TRAINING). Note that SPSS indicates with which layer you are working. Now, when you run the command (by clicking *OK*), you will be given summary statistics for the variable GRADE by each level of MORNING and TRAINING.

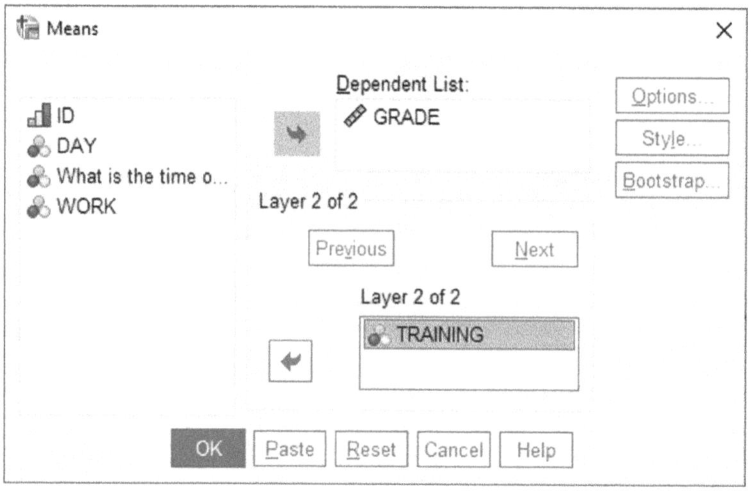

Your output will look like the output shown below. You now have two main sections (No and Yes), along with the Total. Now, however, each main section is broken down into subsections (No, Yes, and Total).

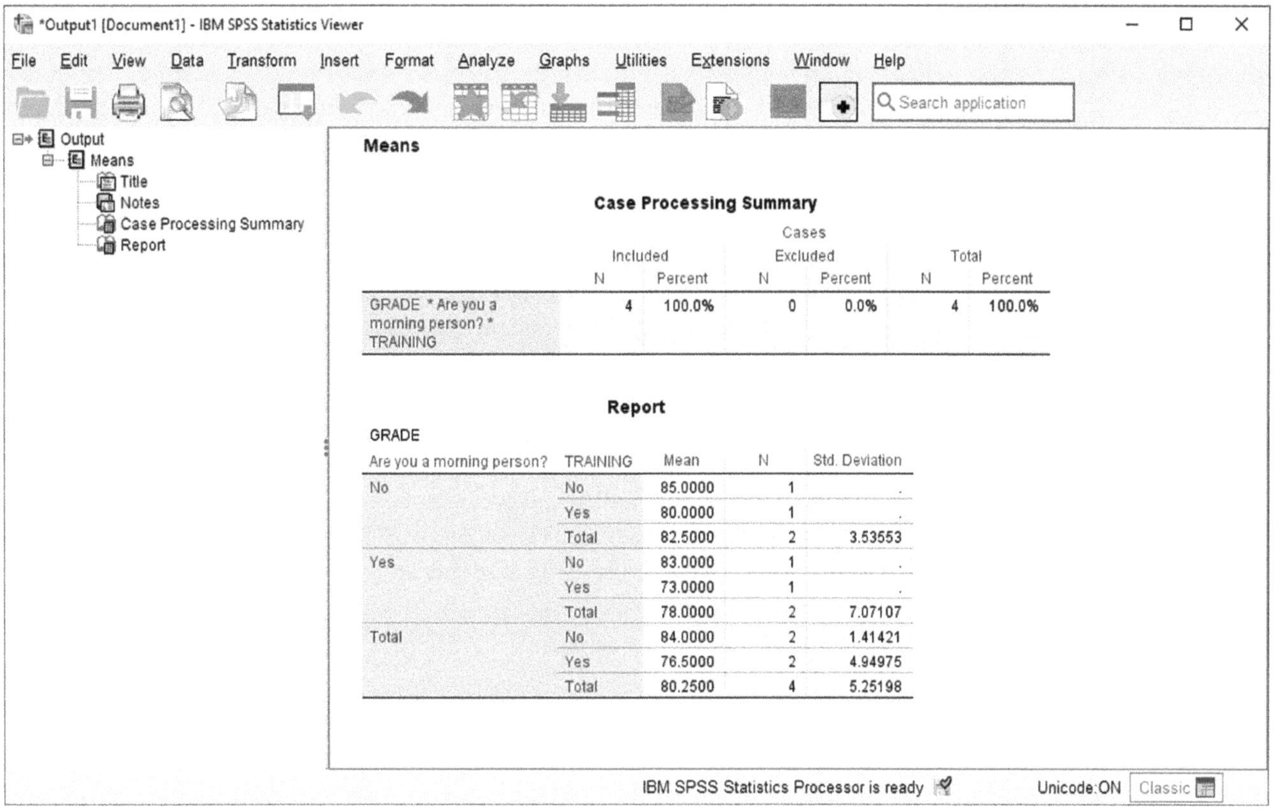

Means

Case Processing Summary

	Cases					
	Included		Excluded		Total	
	N	Percent	N	Percent	N	Percent
GRADE * Are you a morning person? * TRAINING	4	100.0%	0	0.0%	4	100.0%

Report

GRADE

Are you a morning person?	TRAINING	Mean	N	Std. Deviation
No	No	85.0000	1	.
	Yes	80.0000	1	.
	Total	82.5000	2	3.53553
Yes	No	83.0000	1	.
	Yes	73.0000	1	.
	Total	78.0000	2	7.07107
Total	No	84.0000	2	1.41421
	Yes	76.5000	2	4.94975
	Total	80.2500	4	5.25198

The variable you used in Level 1 (MORNING) is the first one listed, and it defines the main sections. The variable you had in Level 2 (TRAINING) is listed second. Thus, the first row represents those participants who were not morning people and who received training. The second row represents participants who were not morning people and did not receive training. The third row represents the total for all participants who were not morning people.

Note that **standard deviations** are not given for all of the rows. This is because there is only one participant per cell in this example. One problem with using many subsets is that it increases the number of participants that are required to obtain meaningful results. See a research design text or your instructor for more details.

Practice Exercise

Using Practice Dataset 1 in Appendix B, compute the **mean** and **standard deviation** of ages for each value of marital status. What is the average age of the married participants? The single participants? The divorced participants?

Section 3.5 Standard Scores

Description

Standard scores allow the comparison of different scales by transforming the scores into a common scale. The most common standard score is the *z*-score. A *z*-score is based on a **standard normal distribution** (i.e., a **mean** of 0 and a **standard deviation** of 1). A *z*-score, therefore, represents the number of **standard deviations** above or below the **mean** (e.g., a *z*-score of –1.5 represents a score 1½ **standard deviations** below the **mean)** a given score is.

Assumptions

The **standard normal distribution** is the basis for *z*-scores as normally used. Therefore, the distributions that are converted to *z*-scores should be normally distributed, and the scales should be either **interval** or **ratio**.

Drawing Conclusions

Conclusions based on *z*-scores consist of the number of **standard deviations** above or below the **mean** a score is. For instance, a student scores 85 on a mathematics exam in a class that has a **mean** of 70 and **standard deviation** of 5. The student's test score is 15 points above the class **mean** (85 − 70 = 15). The student's *z*-score is 3 because she scored three **standard deviations** above the **mean** (15 ÷ 5 = 3). If the same student scores 90 on a reading exam, with a class **mean** of 80 and a **standard deviation** of 10, the *z*-score will be 1.0 because she is one **standard deviation** above the **mean**. Thus, even though her raw score was higher on the reading test, she actually did better in relation to other students on the mathematics test because her *z*-score was higher on that test.

SPSS Data Format

Calculating *z*-scores requires only a single variable in SPSS. That variable must be numerical.

Running the Command

This example uses the sample data file (SAMPLE.sav) created in Chapters 1and 2. Computing *z*-scores is a component of the *Descriptives* command. To access it, click *Analyze*, then *Descriptive Statistics*, then *Descriptives*.

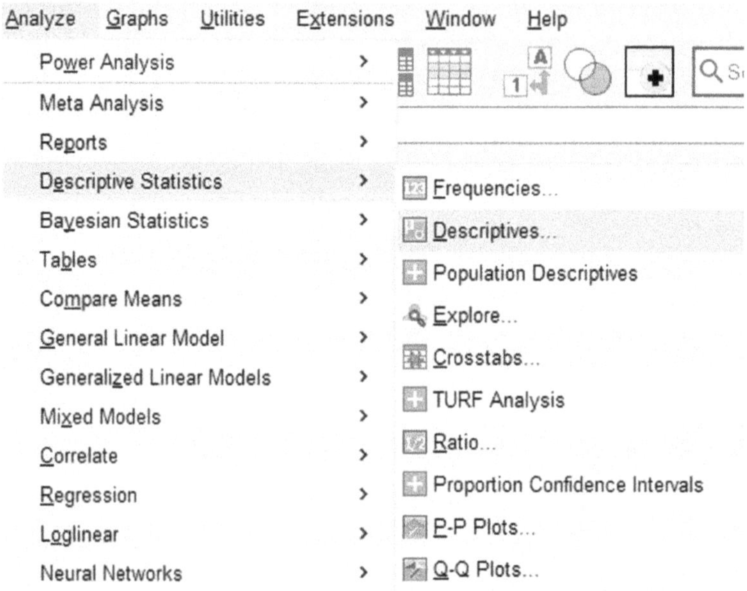

This will bring up the standard **dialog box** for the *Descriptives* command. Note the checkbox in the bottom-left corner labeled *Save standardized values as variables.*

Check this box and move the variable GRADE into the right-hand blank. Then, click *OK* to complete the analysis. You will be presented with the standard output from the *Descriptives* command. Note that the *z*-scores are not listed. They were inserted into the **data window** as a new variable.

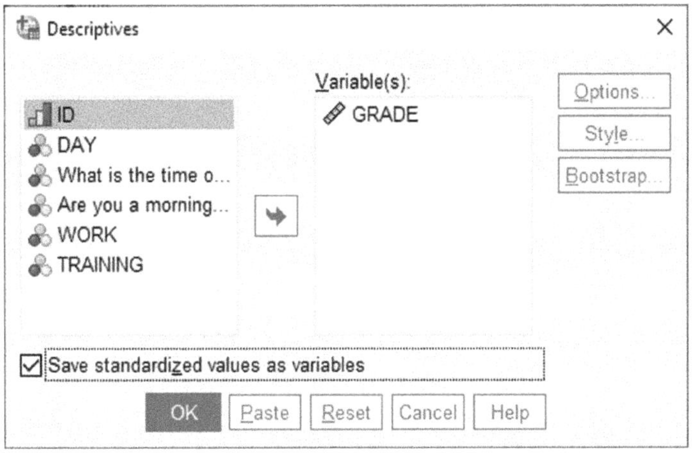

Switch to the *Data View* window and examine your data file. Note that a new variable, called ZGRADE, has been added. When you asked SPSS to save standardized values, it created a new variable with the same name as your old variable preceded by a Z. The *z*-score is computed for each case and placed in the new variable.

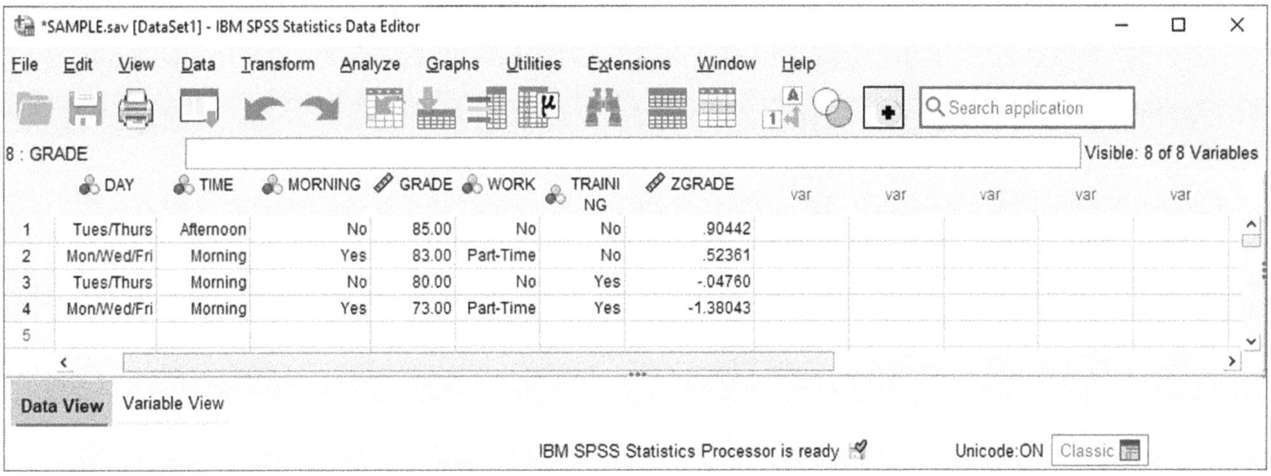

Reading the Output

After you conducted your analysis, the new variable was created. You can perform any number of subsequent analyses on the new variable.

Practice Exercise

Using Practice Dataset 2 in Appendix B, determine the *z*-score that corresponds to each employee's salary. Determine the **mean** *z*-scores for salaries of male employees and female employees. Determine the **mean** *z*-score for salaries of the total sample.

Section 3.6 Exploring Data

Description

The *Explore* command is a useful tool for getting a descriptive overview of a dataset. It can provide basic descriptive statistics and a visual representation for a variable, or a variable broken down by levels of another variable.

Running the Command

This example uses the sample data file (SAMPLE.sav) created in Chapters 1and 2. To run the *Explore* command, click *Analyze*, then *Descriptive Statistics*, then *Explore*.

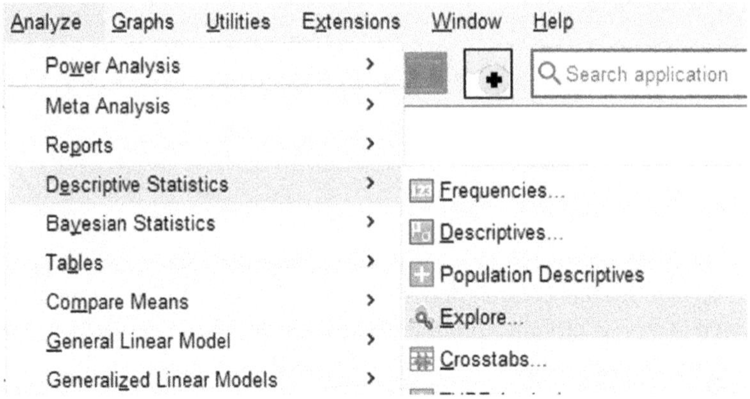

The standard **dialog box** will come up and you can choose any variables you would like information on as your *Dependent List*. In the example here we are going to ask for descriptive statistics on GRADE broken down by each level of WORK (if you want to break down data by levels of another variable place that other variable in the *Factor List* section).

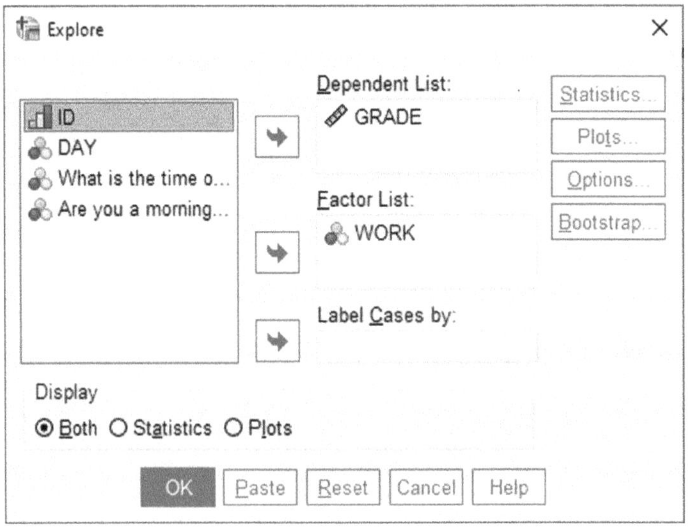

Reading the Output

The first part of the output will consist of a wide variety of descriptive statistics for your variable (in this case, broken down by each level of WORK).

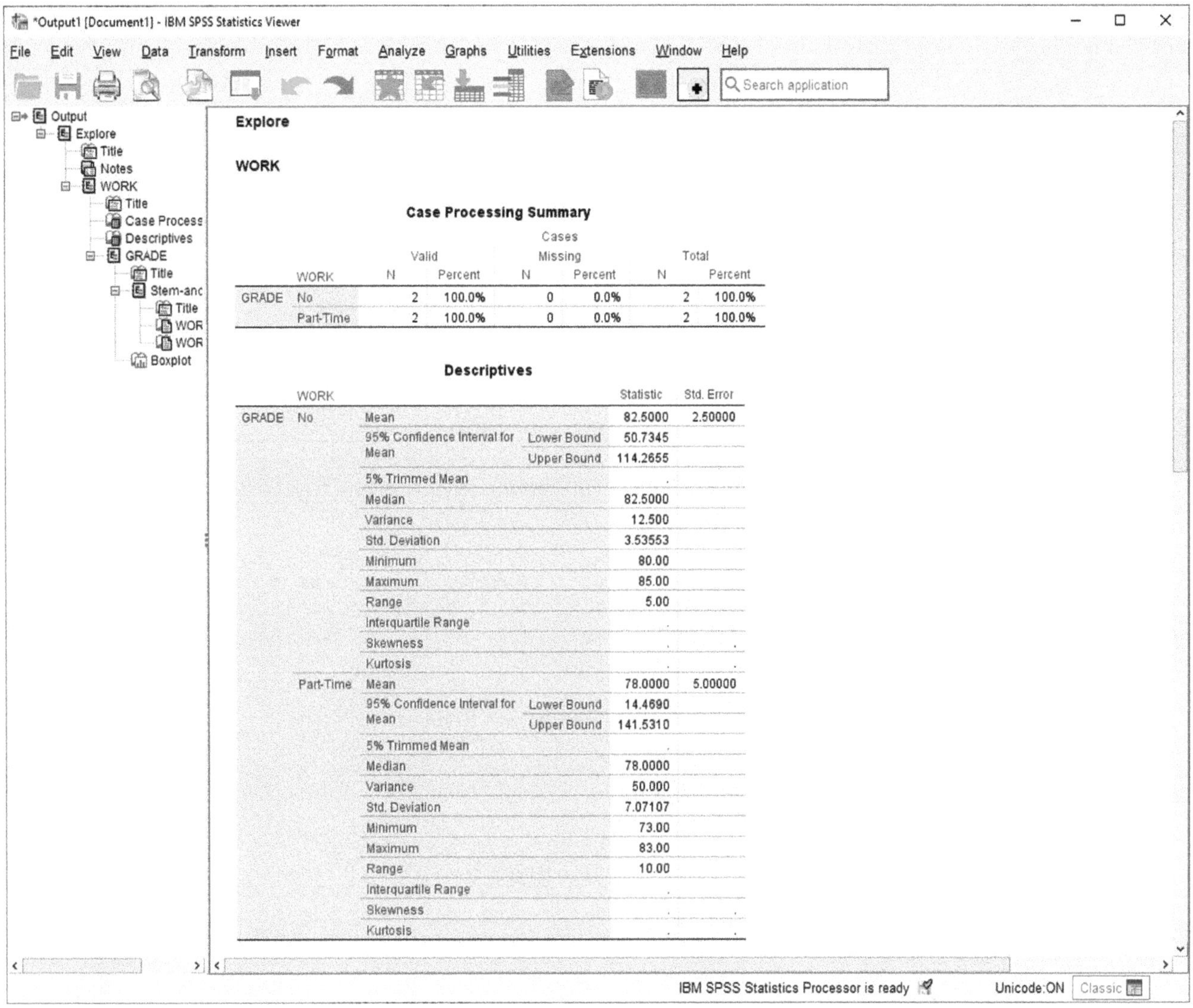

Case Processing Summary

	WORK	Valid N	Valid Percent	Missing N	Missing Percent	Total N	Total Percent
GRADE	No	2	100.0%	0	0.0%	2	100.0%
	Part-Time	2	100.0%	0	0.0%	2	100.0%

Descriptives

	WORK			Statistic	Std. Error
GRADE	No	Mean		82.5000	2.50000
		95% Confidence Interval for Mean	Lower Bound	50.7345	
			Upper Bound	114.2655	
		5% Trimmed Mean		.	
		Median		82.5000	
		Variance		12.500	
		Std. Deviation		3.53553	
		Minimum		80.00	
		Maximum		85.00	
		Range		5.00	
		Interquartile Range		.	
		Skewness		.	.
		Kurtosis		.	.
	Part-Time	Mean		78.0000	5.00000
		95% Confidence Interval for Mean	Lower Bound	14.4690	
			Upper Bound	141.5310	
		5% Trimmed Mean		.	
		Median		78.0000	
		Variance		50.000	
		Std. Deviation		7.07107	
		Minimum		73.00	
		Maximum		83.00	
		Range		10.00	
		Interquartile Range		.	
		Skewness		.	.
		Kurtosis		.	.

For example, the **Mean** GRADE for those who do not work is 82.5 with a **standard deviation** of 3.54. For those who work Part-Time the **Mean** is 78.0 with a **Standard Deviation** of 7.07.

Under that section of output will be a Stem and Leaf plot as well as a figure illustrating the **Mean** and **Confidence Interval** for each variable. Here, the darker line represents the **Mean** and the blue bars represent the **Confidence Intervals**.

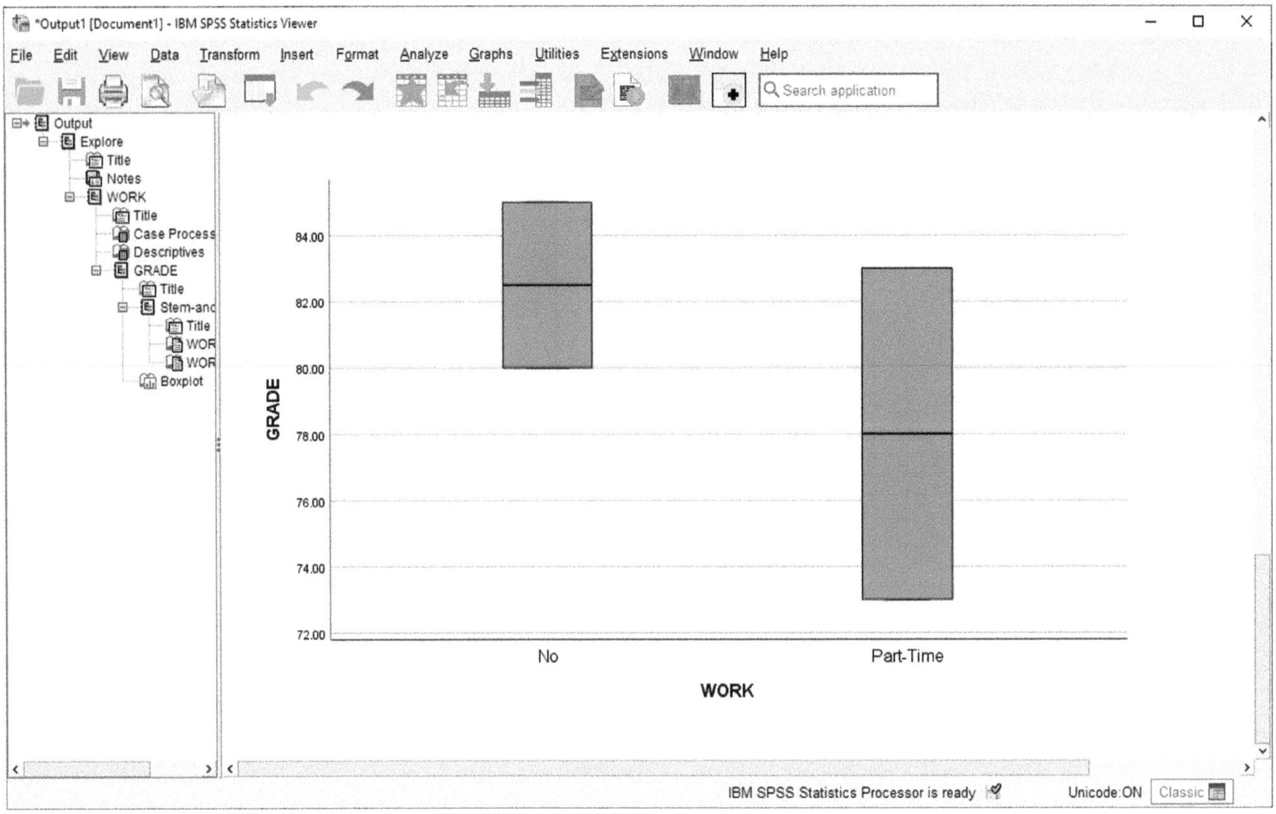
*Output1 [Document1] - IBM SPSS Statistics Viewer
File Edit View Data Transform Insert Format Analyze Graphs Utilities Extensions Window Help
Search application
Output
Explore
Title
Notes
WORK
Title
Case Process
Descriptives
GRADE
Title
Stem-and
Title
Boxplot
GRADE
84.00
82.00
80.00
78.00
76.00
74.00
72.00
No
Part-Time
WORK
IBM SPSS Statistics Processor is ready
Unicode:ON
Classic

Chapter 4
Graphing Data

Section 4.1 Graphing Basics

In addition to the frequency distributions, measures of central tendency, and measures of dispersion discussed in Chapter 3, graphing is a useful way to summarize, organize, and reduce your data. It has been said that a picture is worth a thousand words. In the case of complicated datasets, this is certainly true.

With SPSS, it is possible to make publication-quality graphs. One important advantage of using SPSS to create your graphs instead of using other software (e.g., Excel or SigmaPlot) is that the data have already been entered. Thus, duplication is eliminated, and the chance of making a transcription error is reduced.

Section 4.2 Bar Charts, Pie Charts, and Histograms

Description

Bar charts, pie charts, and histograms represent—through the varying heights of bars or sizes of pie pieces—the number of times each score occurs. They are graphical representations of the frequency distributions discussed in Chapter 3.

Drawing Conclusions

The *Frequencies* command produces output that indicates both the number of cases in the sample with a particular value and the percentage of cases with that value. Thus, conclusions drawn should relate only to describing the numbers or percentages for the sample. If the data are at least **ordinal** in nature, conclusions regarding the cumulative percentages and/or **percentiles** can also be drawn.

SPSS Data Format

You need only one variable to use this command.

Dataset

For the graphing examples, we will use a new set of data. Enter the data that follow by defining the three subject variables in the *Variable View* window: HEIGHT (in inches), WEIGHT (in pounds), and SEX (1 = Male, 2 = Female). When you create the variables, designate HEIGHT and WEIGHT as *Scale* measures and SEX as a *Nominal* measure (in the far-right column of the *Variable View*). Switch to the *Data View* to enter the data values for the 16 participants. Now use the *Save As* command to save the file, naming it HEIGHT.sav.

DOI: 10.4324/9781003450467-4

HEIGHT	WEIGHT	SEX
66	150	1
69	155	1
73	160	1
72	160	1
68	150	1
63	140	1
74	165	1
70	150	1
66	110	2
64	100	2
60	95	2
67	110	2
64	105	2
63	100	2
67	110	2
65	105	2

Make sure you have entered the data correctly by calculating a **mean** for each of the three variables (click *Analyze*, then *Descriptive Statistics*, then *Descriptives*). Compare your results with those in the table below.

Descriptive Statistics

	N	Minimum	Maximum	Mean	Std. Deviation
HEIGHT	16	60.00	74.00	66.9375	3.90672
WEIGHT	16	95.00	165.00	129.0625	26.34507
SEX	16	1.00	2.00	1.5000	.51640
Valid N (listwise)	16				

Running the Command

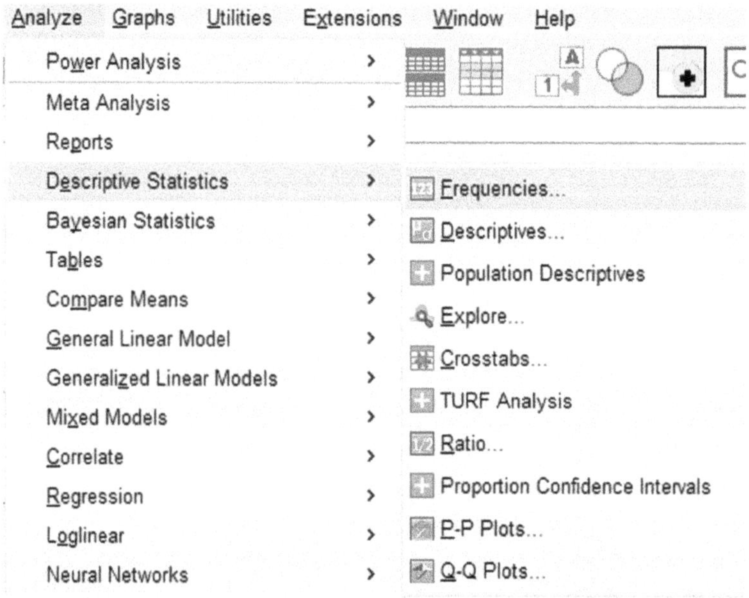

The *Frequencies* command will produce graphical frequency distributions. Click *Analyze*, then *Descriptive Statistics*, then *Frequencies*. You will be presented with the main **dialog box** for the *Frequencies* command, where you can enter the variables for which you would like to create graphs or charts (HEIGHT in this example). See Chapter 3 for other options with this command.

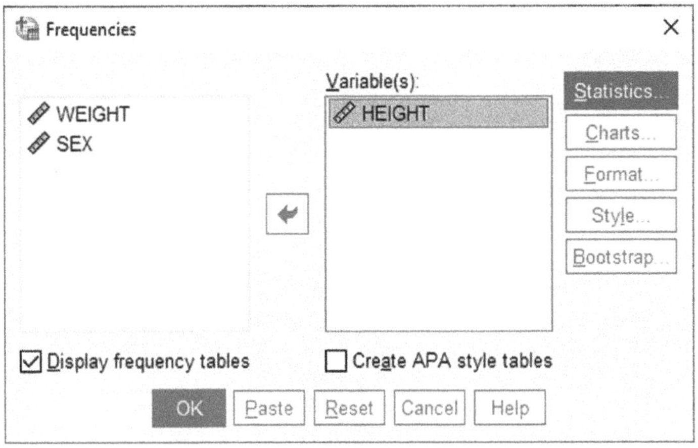

Click the *Charts* button on the right side of the *Frequencies* command to bring up the Charts **dialog box**.

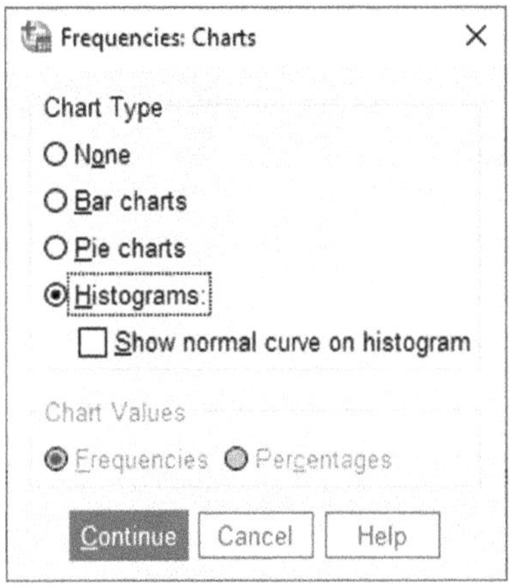

There are three types of charts available with this command: *Bar charts*, *Pie charts*, and *Histograms*. For each type, the *Y*-axis can be either a frequency count or a percentage (selected with the *Chart Values* option). Select *Histograms* and click *Continue*. Next, click *OK*.

You will receive the charts for any variables selected in the main *Frequencies* command **dialog box**.

Output

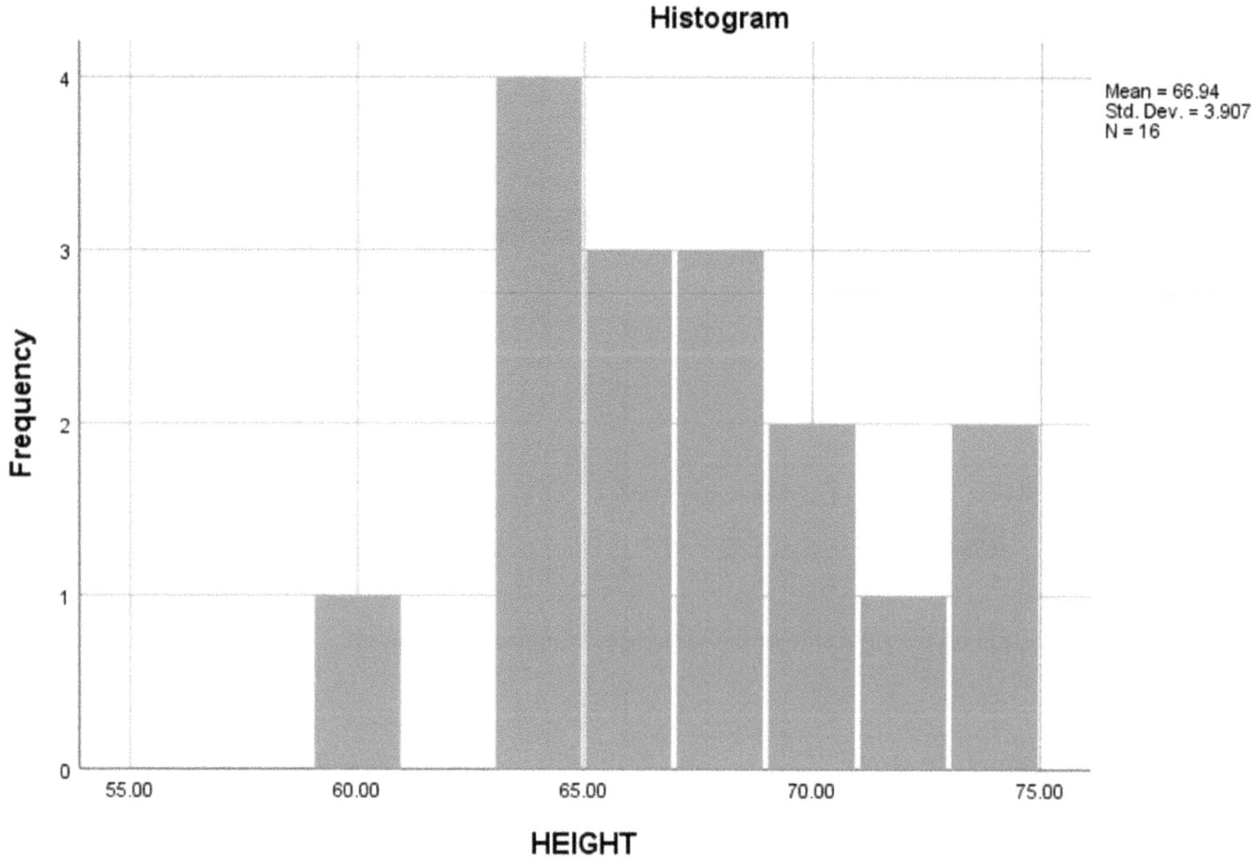

The bar chart consists of a Y-axis, representing the frequency, and an X-axis, representing each score (or groups of scores).

If selected for output, the pie chart shows the percentage of the whole that is represented by each value.

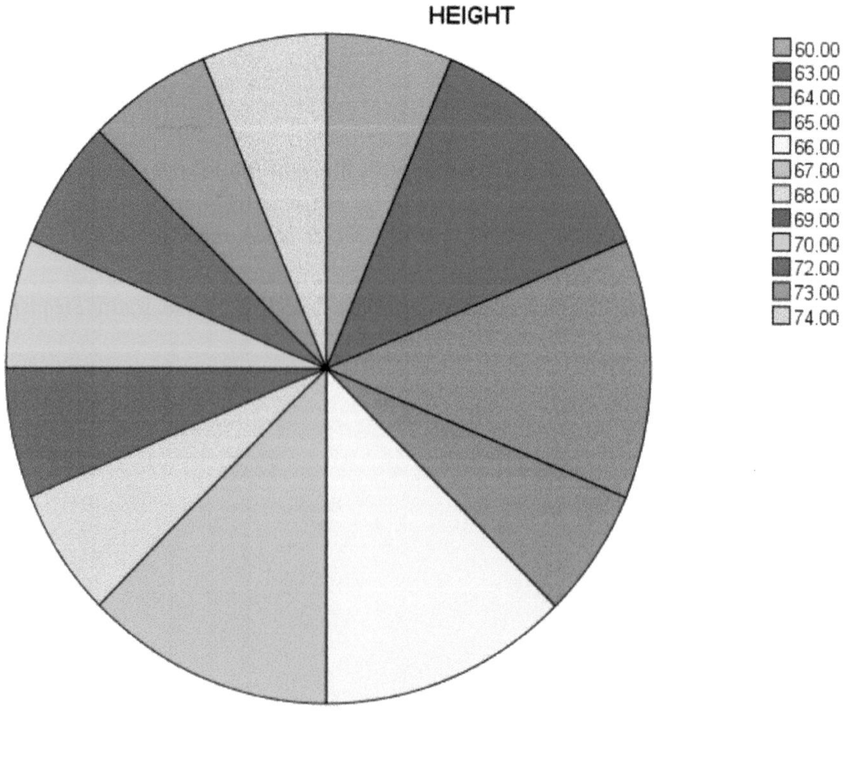

The *Histogram* command creates a grouped frequency distribution. The **range** of scores is split into evenly spaced groups. The midpoint of each group is plotted on the *X*-axis, and the *Y*-axis represents the number of scores for each group.

If you select *Show normal curve on histogram*, a normal curve will be superimposed over the distribution. This is very useful in determining if the distribution you have is approximately normal.

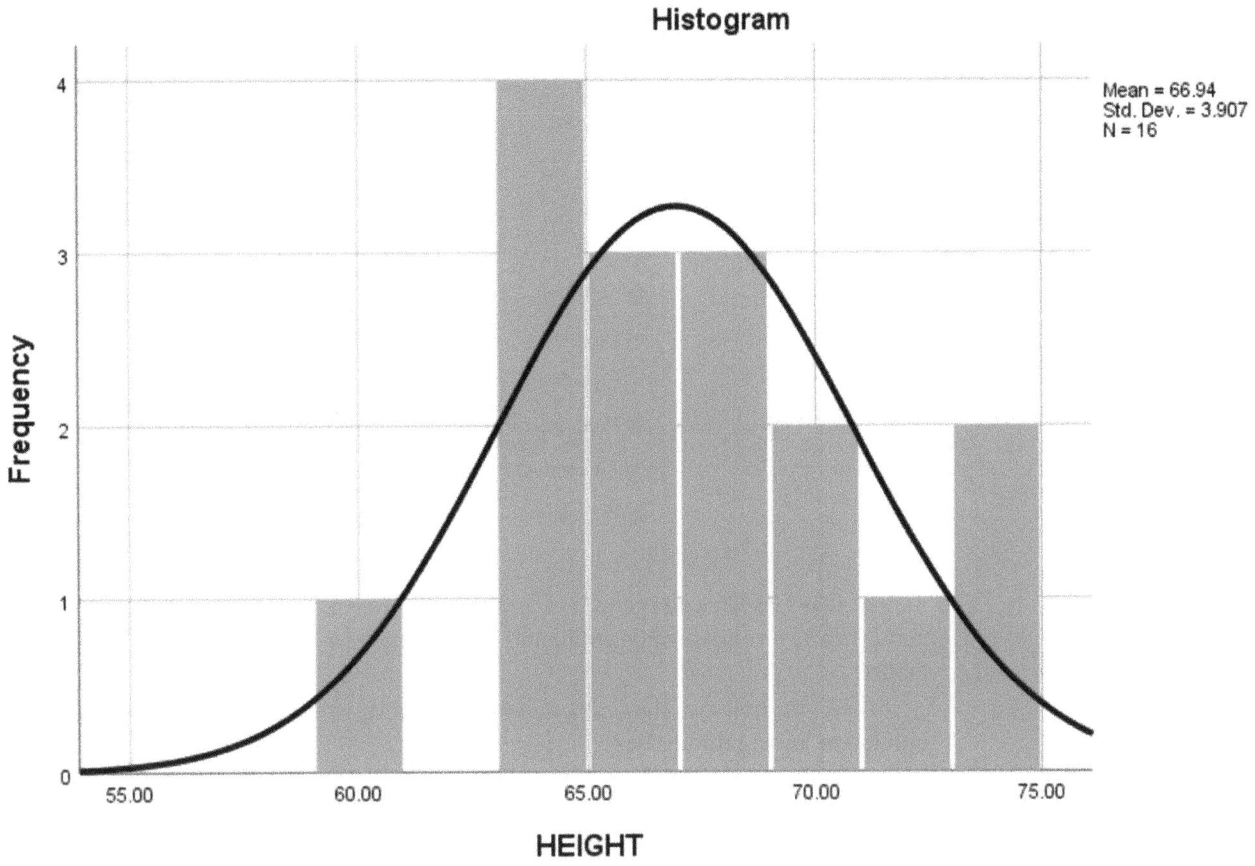

Practice Exercise

Use Practice Dataset 1 in Appendix B. After you have entered the data, first construct a histogram that represents the mathematics skills scores and displays a normal curve, and then construct a bar chart that represents the frequencies for the variable AGE.

Section 4.3 The SPSS Chart Builder

Chart Builder Basics

Make sure that the HEIGHT.sav data file you created in Section 4.2 is open. In order to use the chart builder, you must have a data file open.

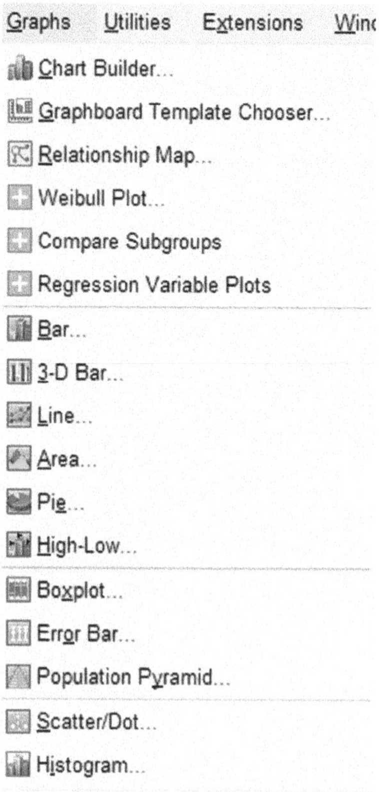

The *Chart Builder* command is accessed using *Graphs*, then *Chart Builder* in the submenu. This is a very versatile command that can make a variety of graphs of excellent quality.

When you first run the *Chart Builder* command, you will probably be presented with the following **dialog box**:

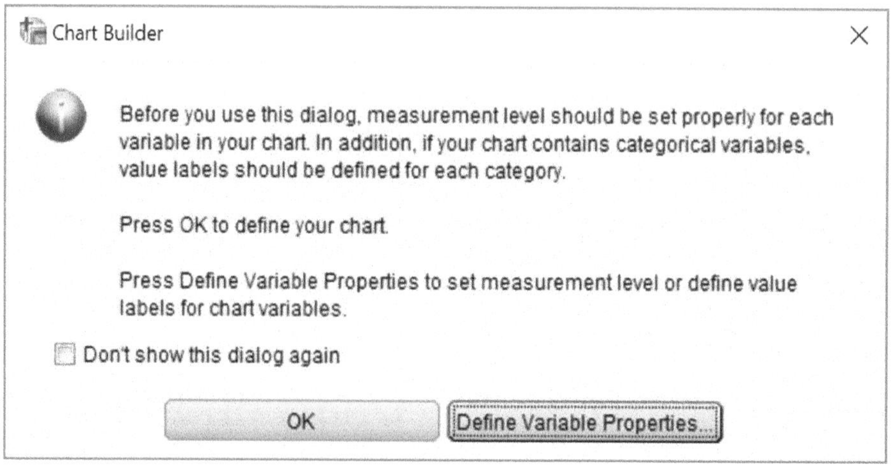

This **dialog box** is asking you to ensure that your variables are properly defined. Refer to Sections 1.3 and 2.1 if you had difficulty defining the variables used in creating the dataset for this example or to refresh your knowledge of this topic. Click *OK*.

The Chart Builder allows you to make any kind of graph that is normally used in publication or presentation, and much of it is beyond the scope of this text. However, this text will go over the basics of the Chart Builder so that you can understand its mechanics.

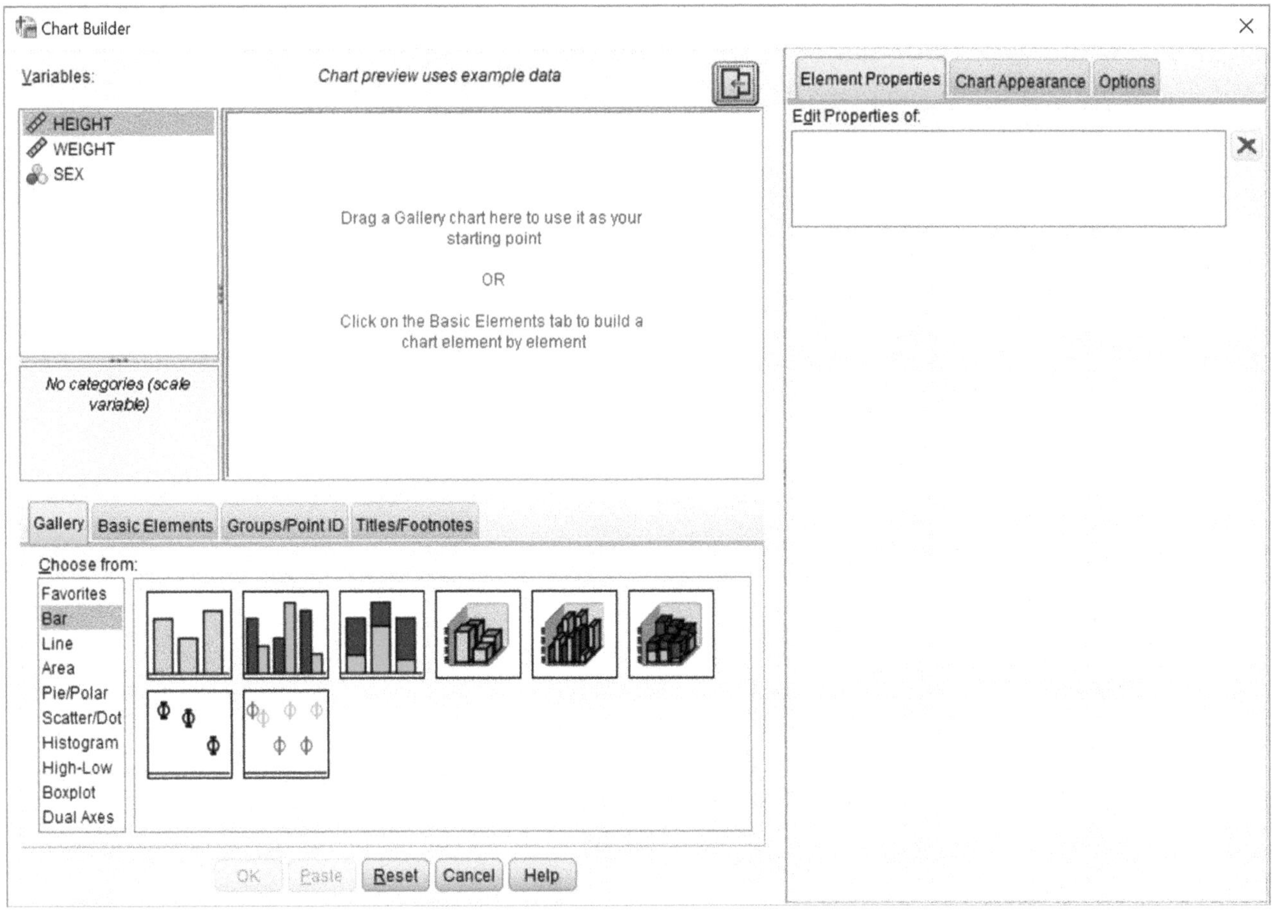

Near the middle of the **dialog box**, there are four main tabs that let you control the graphs you are making. The first one is the *Gallery* tab. The *Gallery* tab allows you to choose the basic format of your graph.

For instance, the screenshot above shows the different kinds of bar charts that the Chart Builder can create.

After you have selected the basic form of graph that you want using the *Gallery* tab, you simply drag the image from the bottom right of the window up to the main window at the top (where it reads, "Drag a Gallery chart here to use it as your starting point").

Alternatively, you can use the *Basic Elements* tab to drag a coordinate system (labeled *Choose Axes*) to the top window, then drag variables and elements into the window.

The other tabs (*Groups/Point ID* and *Titles/Footnotes*) can be used for adding other standard elements to your graphs.

The examples in this text will cover some of the basic types of graphs you can make with the Chart Builder. After a little experimentation on your own, and once you have mastered the examples in this chapter, you will soon gain a full understanding of the Chart Builder.

Section 4.4 Scatterplots

Description

Scatterplots (also called scattergrams or scatter diagrams) display two values for each case with a mark on the graph. The *X*-axis represents the value for one variable. The *Y*-axis represents the value for the second variable.

Assumptions

Both variables should be **interval** or **ratio scales**. If **nominal** or **ordinal** data are used, be cautious about your interpretation of the scattergram.

SPSS Data Format

You need two variables to perform this command.

Running the Command

You can produce scatterplots by clicking *Graphs*, then *Chart Builder*. In *Gallery Choose from* select *Scatter/Dot*. Then drag the *Simple Scatter* icon (top-left) up to the main chart area as shown in the screenshot below. Disregard the *Element Properties* window that pops up by choosing *Close* at the bottom of that window.

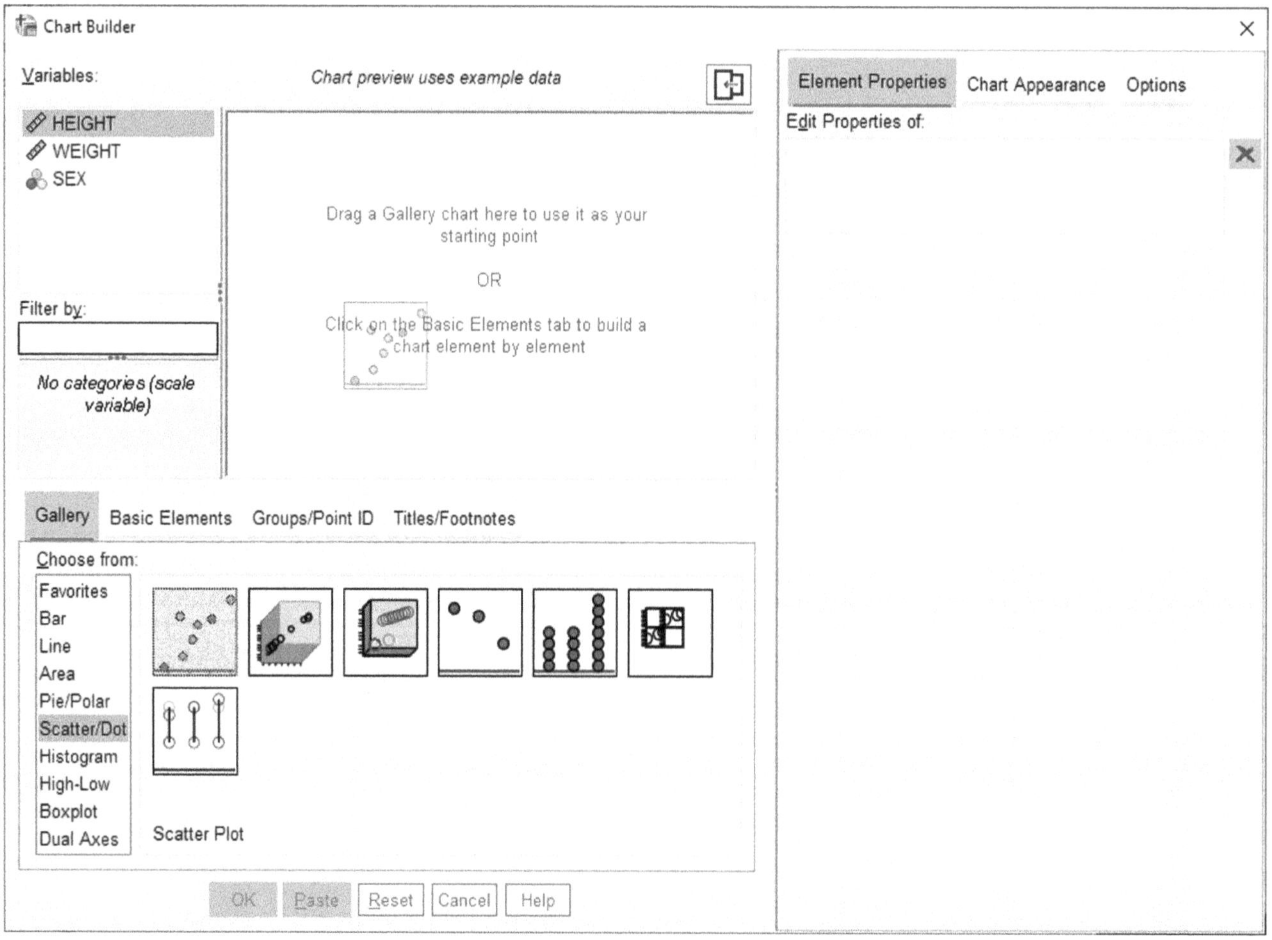

Next, drag the HEIGHT variable to the *X-Axis* area and drag the WEIGHT variable to the *Y-Axis* area. (Remember that standard graphing conventions indicate that **dependent variables** should be *Y* and **independent variables** should be *X*. This would mean that we are trying to predict weights from heights.) At this point, your screen should resemble the example shown on the next page. Note that your actual data are *not* shown—just a set of dummy values, which may look different from the pattern shown here.

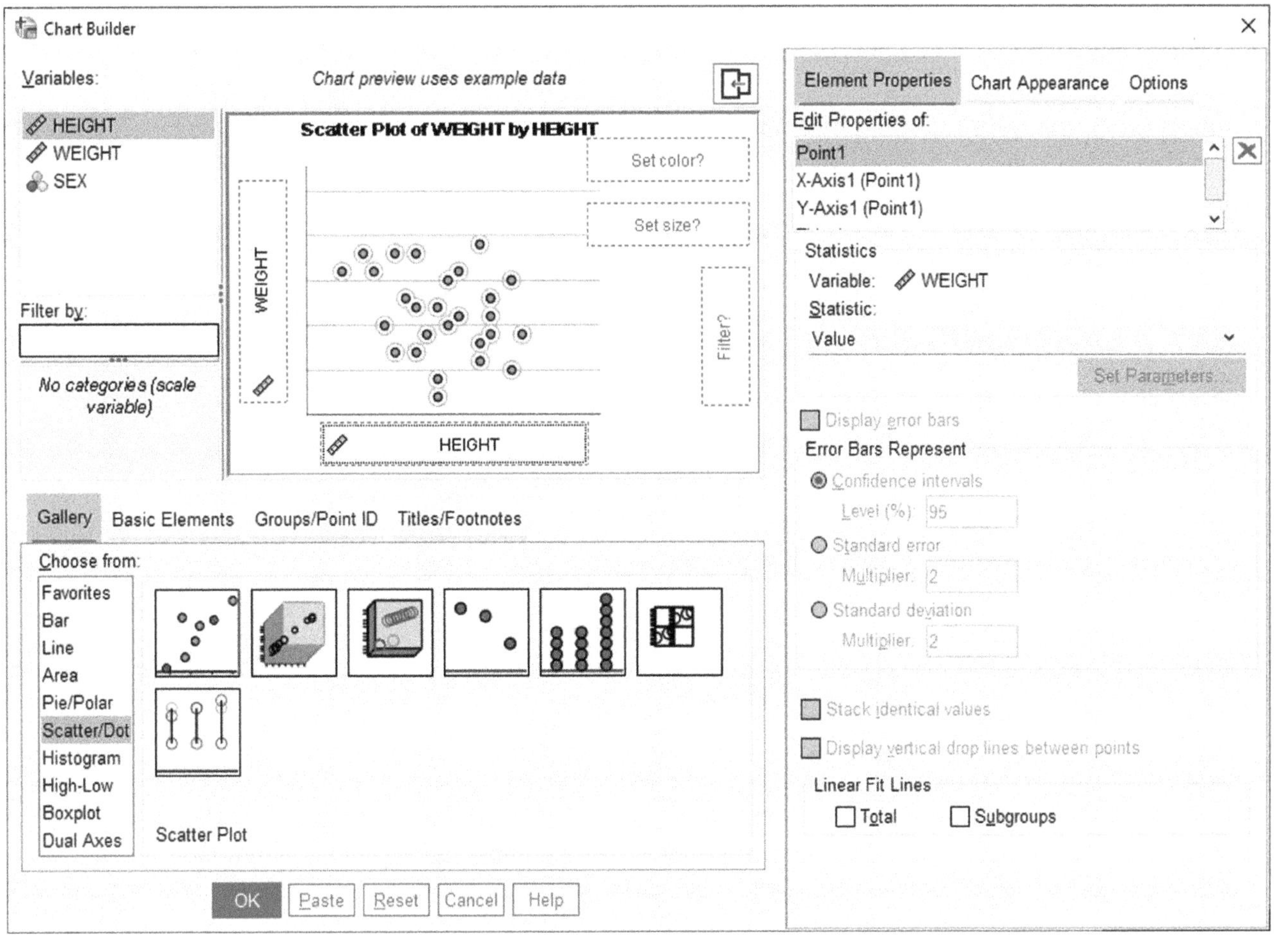

Click *OK*. You should get your new graph as Output.

Output

The output will consist of a mark for each participant at the appropriate *X* and *Y* levels.

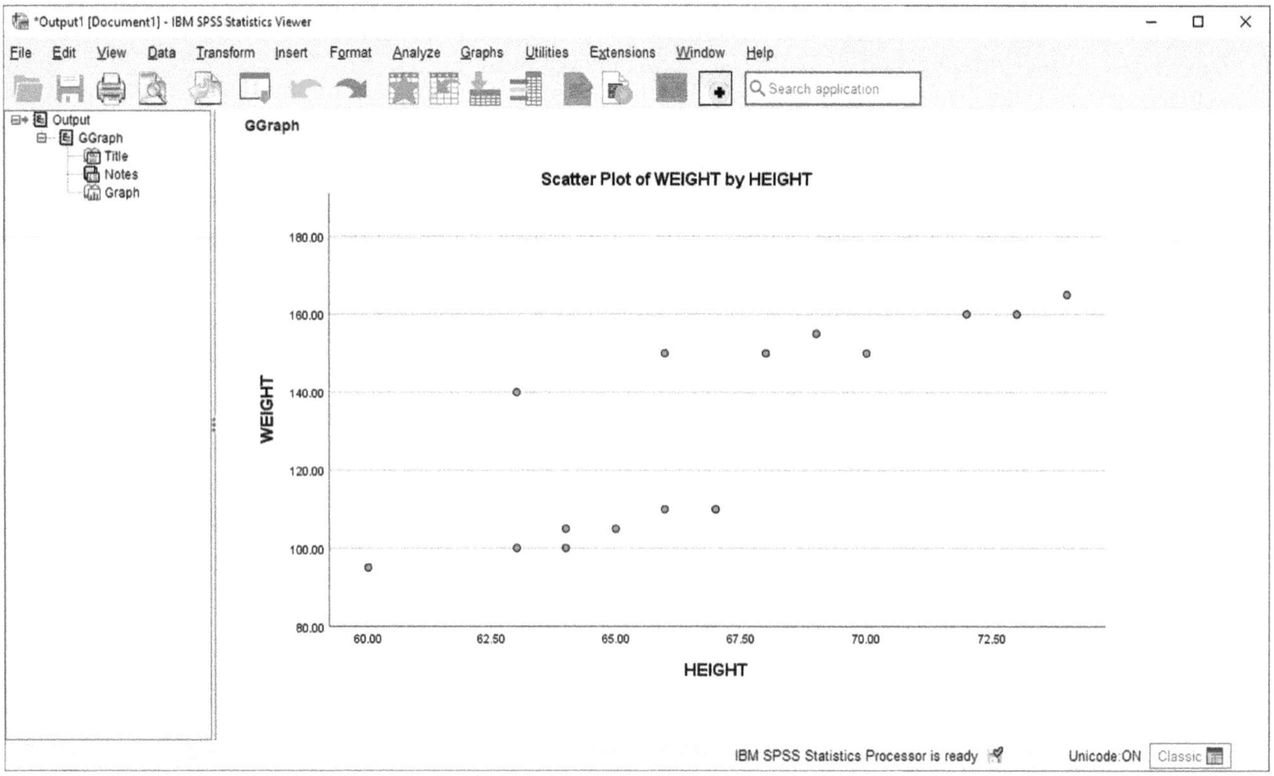

Adding a Third Variable

Even though the scatterplot is a two-dimensional graph, it can plot a third variable. To make it do so, select the *Groups/Point ID* tab in the Chart Builder. Click the *Grouping/ stacking variable* option. Again, disregard the *Element Properties* window that pops up. Next, drag the variable SEX into the upper-right corner where it indicates *Set color*. When this is done, your screen should look like the image shown below. If you are not able to drag the variable SEX, it may be because it is not identified as **nominal** or **ordinal** in the *Variable View* window.

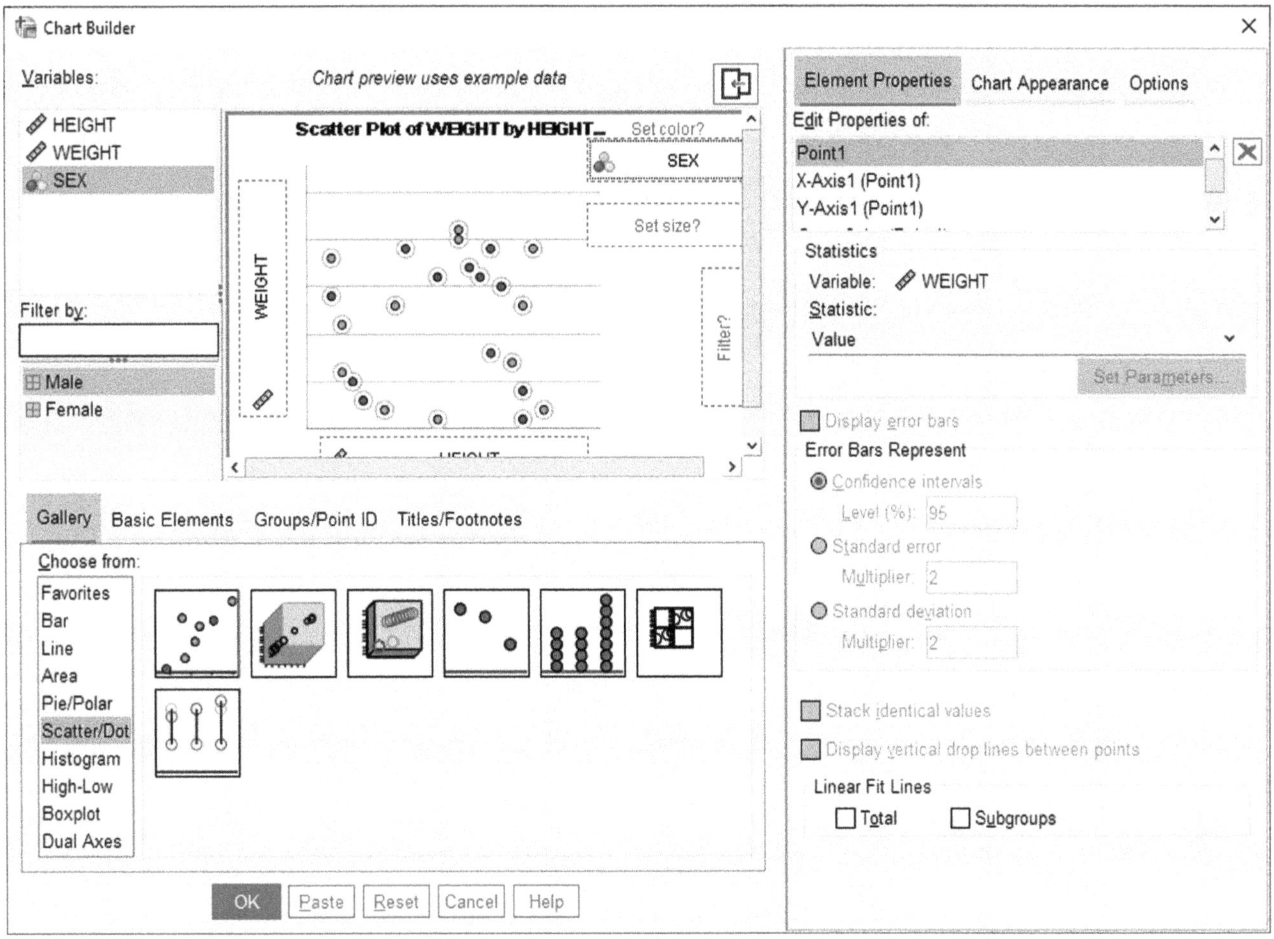

Click *OK* to have SPSS produce the graph.

Now our output will have two different sets of marks. One set represents the male participants, and the second set represents the female participants. These two sets will appear in two different colors on your screen. You can use the SPSS *Chart Editor* (see Section 4.6) to make them different shapes.

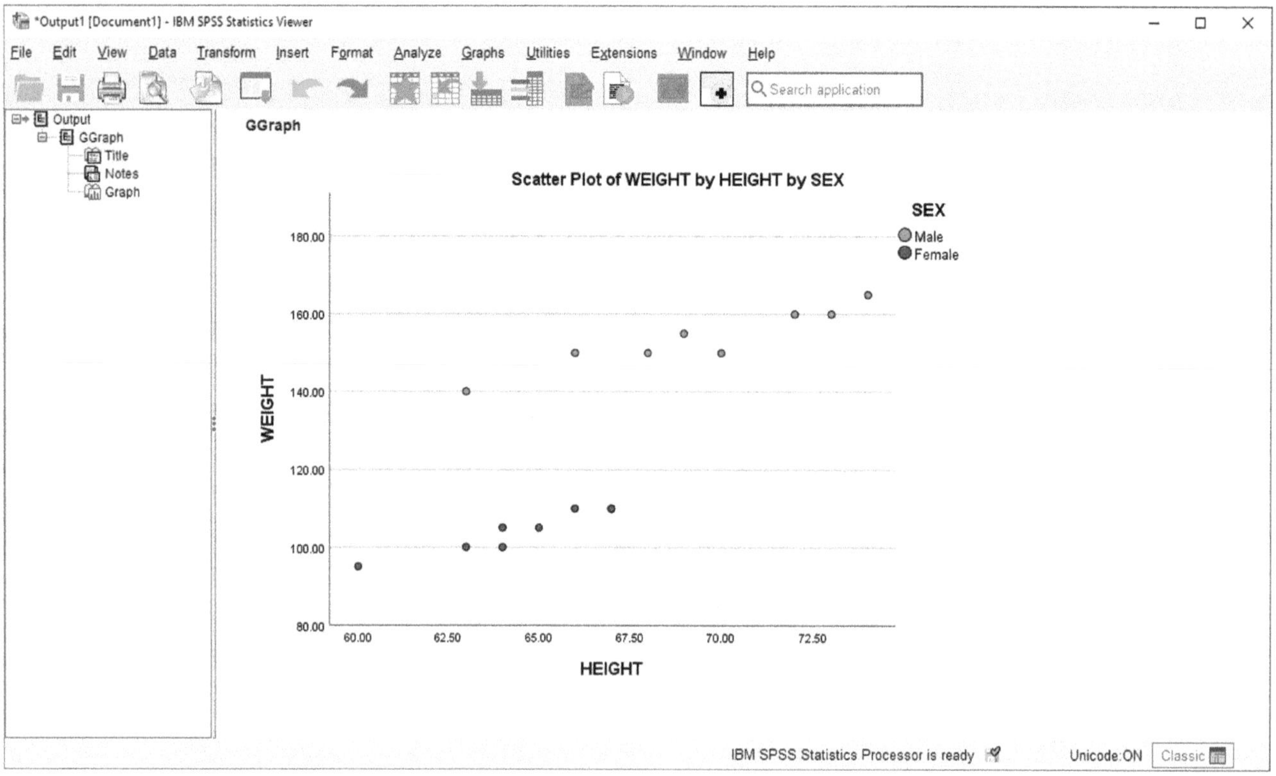

Practice Exercise

Use Practice Dataset 2 in Appendix B. Construct a scatterplot to examine the relationship between SALARY and EDUCATION.

Section 4.5 Advanced Bar Charts

Description

Bar charts can be produced with the *Frequencies* command (see Section 4.2). Sometimes, however, we are interested in a bar chart where the *Y*-axis is not a frequency. To produce such a chart, we need to use the *Bar charts* command.

SPSS Data Format

You need at least two variables to perform this command. There are two basic kinds of bar charts—those for between-subjects designs and those for repeated-measures designs. Use the between-subjects method if one variable is the **independent variable** and the other is the **dependent variable** (representing one score for different groups of individuals). Use the repeated-measures method if you have a **dependent variable** for each value of the **independent variable** (e.g., you would have three variables for a design with three values of the **independent variable**). This normally occurs when you make multiple observations over time.

This example uses the GRADES.sav data file, which will be created in Chapter 6. Please see Section 6.4 for the data if you would like to follow along.

Running the Command

Open the Chart Builder by clicking *Graphs*, then *Chart Builder*. In the *Gallery* tab, select *Bar*. If you had only one **independent variable**, you would select the Simple Bar Chart example (top-left corner). If you have more than one **independent variable** (as in this example), select the Clustered Bar Chart example from the top row.

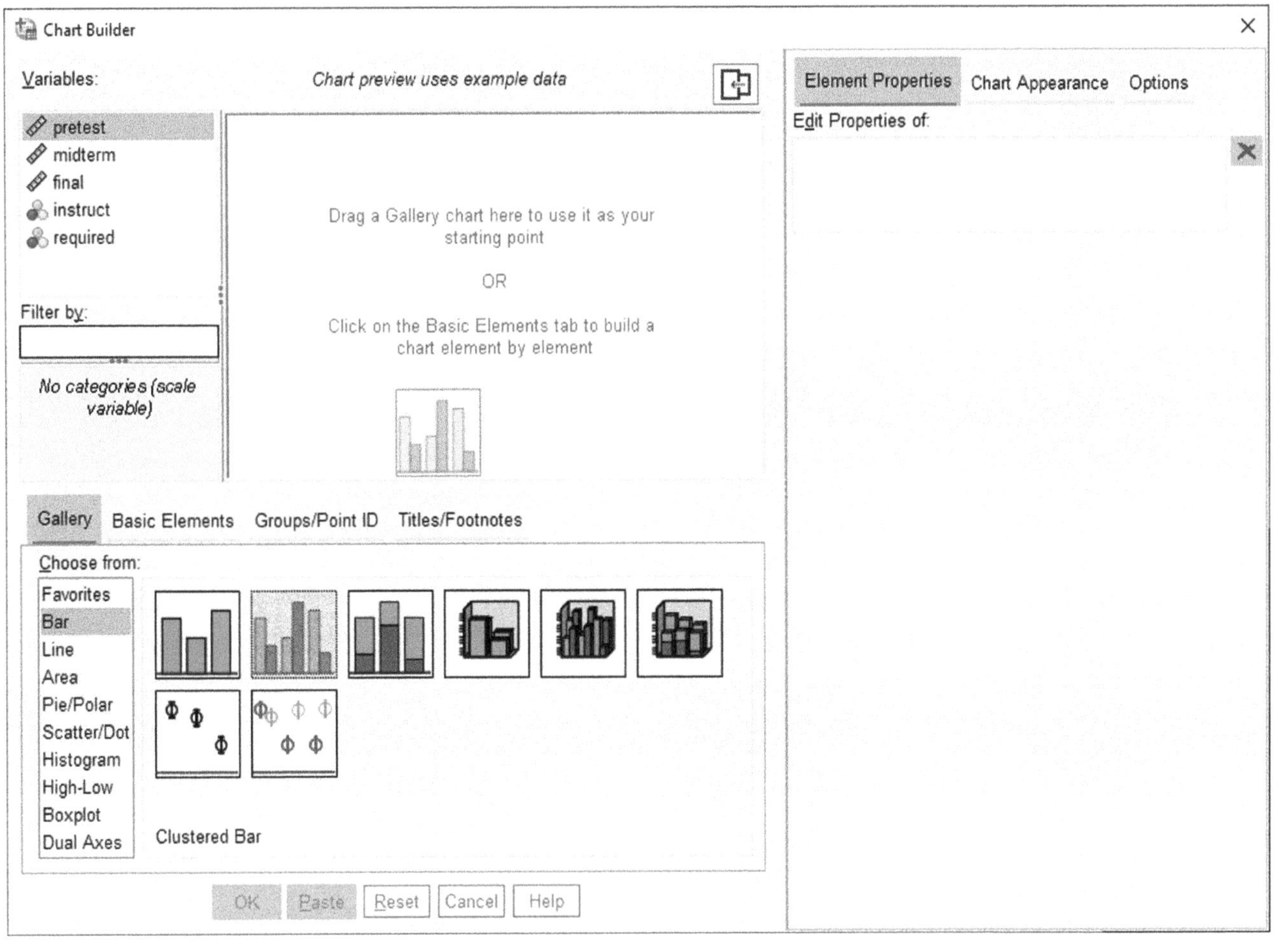

Drag the example to the top working area. Once you do, the working area should look like the screenshot below. (Note that you will need to open the data file you would like to graph in order to run this command.)

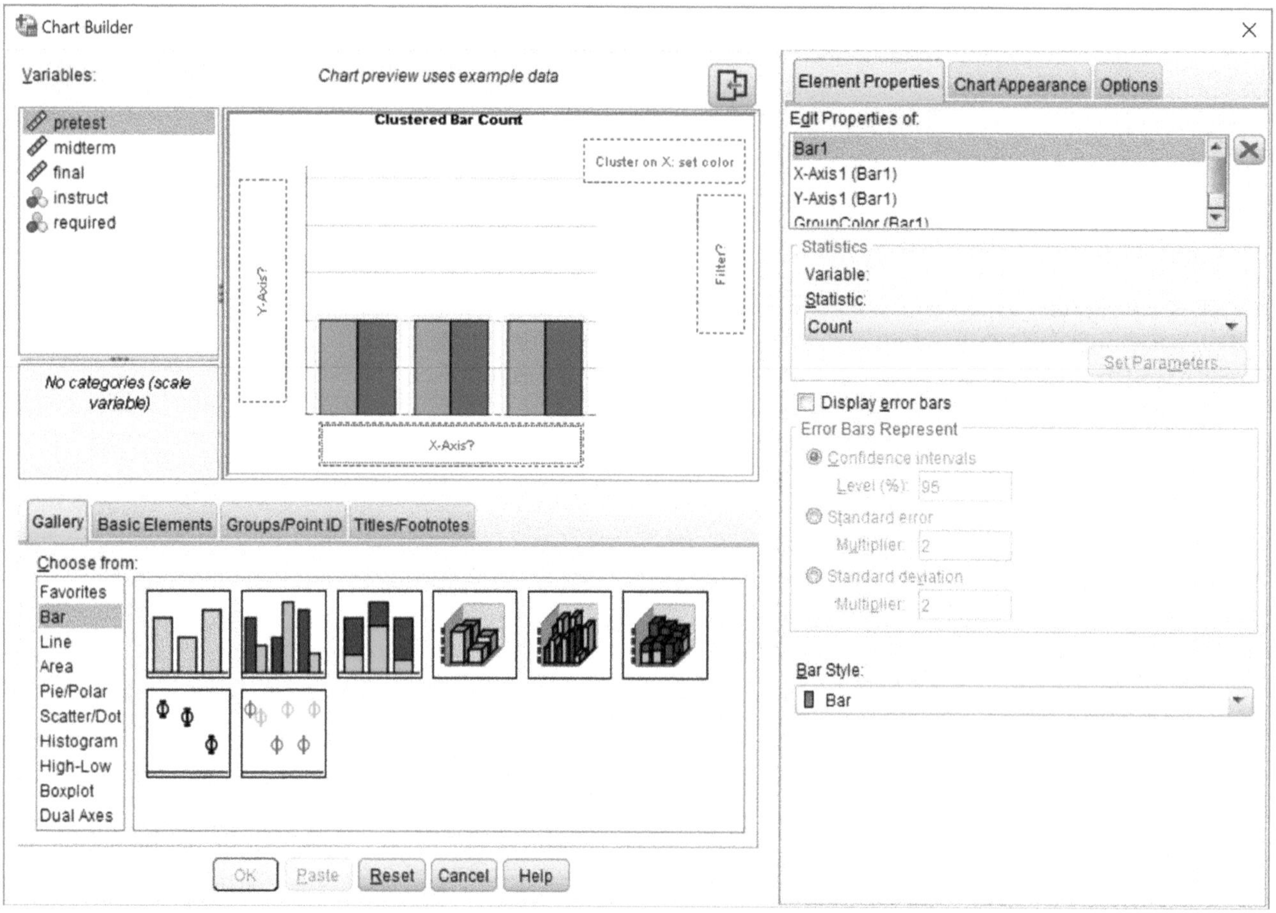

If you are using a repeated-measures design like our example here using GRADES.sav from Chapter 6 (three different variables representing the *Y* values that we want), you need to select all three variables (you can <Ctrl>-click them to select multiple variables) and then drag all three variable names to the *Y-Axis* area. When you do, you will be given the confirmation message shown. Click *OK*.

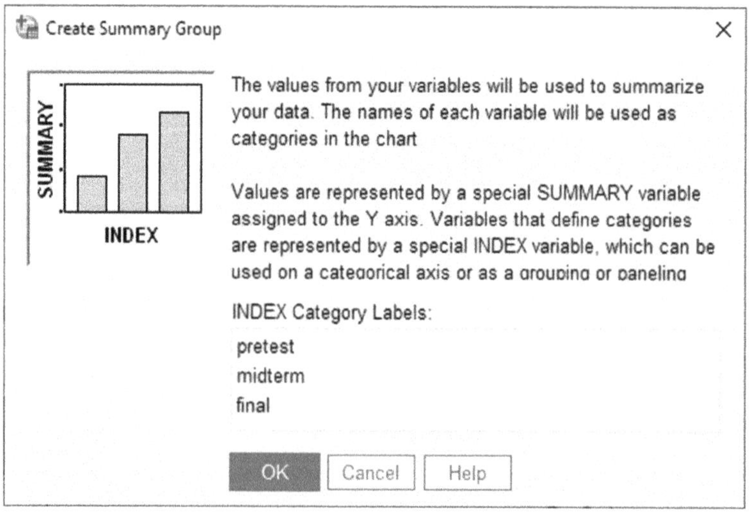

Next, you will need to drag the INSTRUCT variable to the top-right in the *Cluster on X: set color* area. Note: The Chart Builder pays attention to the types of variables that you ask it to graph. If you are getting error messages or unusual results, be sure that your variables being used as category labels are properly designated as *Nominal* in the *Variable View* tab (see Chapter 2, Section 2.1). You can easily change the type of variable by right-clicking on the variable name.

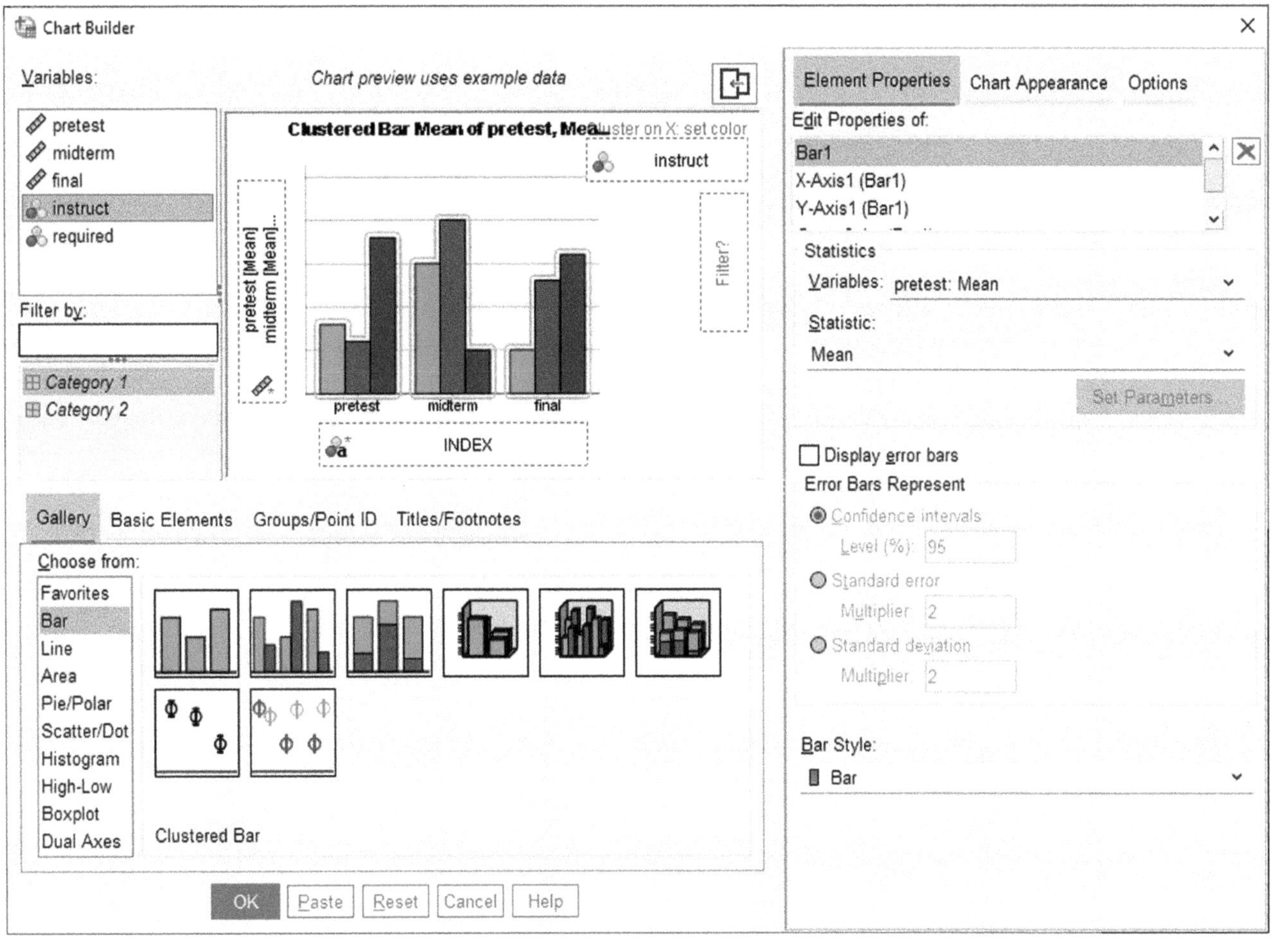
Chart Builder
Variables:
pretest
midterm
final
instruct
required
Filter by:
Category 1
Category 2
Chart preview uses example data
Clustered Bar Mean of pretest, Mea...
Cluster on X: set color
instruct
pretest [Mean] midterm [Mean]...
Filter?
pretest
midterm
final
INDEX
Gallery
Basic Elements
Groups/Point ID
Titles/Footnotes
Choose from:
Favorites
Bar
Line
Area
Pie/Polar
Scatter/Dot
Histogram
High-Low
Boxplot
Dual Axes
Clustered Bar
OK
Paste
Reset
Cancel
Help
Element Properties
Chart Appearance
Options
Edit Properties of:
Bar1
X-Axis1 (Bar1)
Y-Axis1 (Bar1)
Statistics
Variables: pretest: Mean
Statistic:
Mean
Set Parameters...
Display error bars
Error Bars Represent
Confidence intervals
Level (%): 95
Standard error
Multiplier: 2
Standard deviation
Multiplier: 2
Bar Style:
Bar

Output

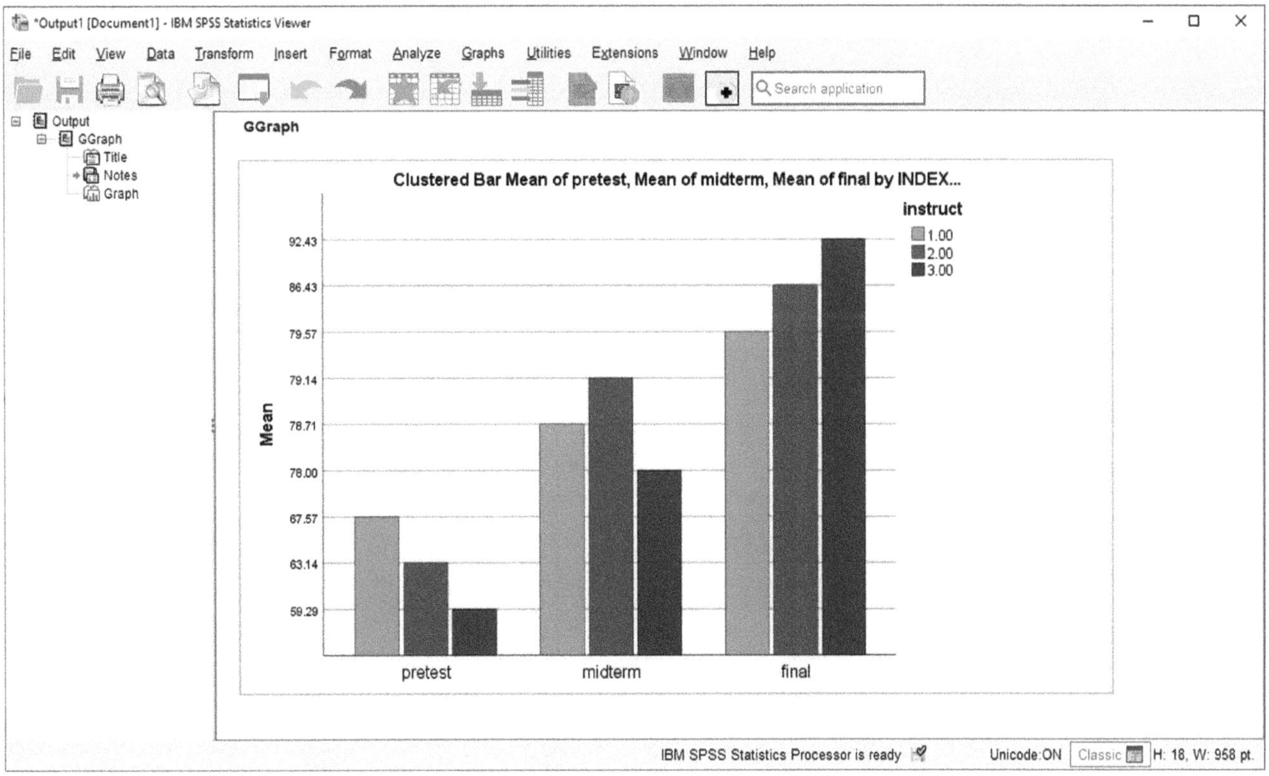
*Output1 [Document1] - IBM SPSS Statistics Viewer
File Edit View Data Transform Insert Format Analyze Graphs Utilities Extensions Window Help
Search application
Output
GGraph
Title
Notes
Graph
GGraph
Clustered Bar Mean of pretest, Mean of midterm, Mean of final by INDEX...
instruct
1.00
2.00
3.00
Mean
92.43
86.43
79.57
79.14
78.71
78.00
67.57
63.14
59.29
pretest
midterm
final
IBM SPSS Statistics Processor is ready
Unicode:ON
Classic
H: 18, W: 958 pt

Practice Exercise

Using Practice Dataset 1 in Appendix B, construct a clustered bar graph examining the relationship between MATHEMATICS SKILLS scores (as the **dependent variable)** and MARITAL STATUS and SEX (as **independent variables**). Make sure you classify both SEX and MARITAL STATUS as nominal variables.

Section 4.6 Editing SPSS Graphs

Whatever command you use to create your graph, you will probably want to do some editing to make it appear exactly as you want it to look. In SPSS, you do this in much the same way that you edit graphs in other software programs (e.g., Excel). After your graph is made, in the **output window**, select your graph (this will create handles around the outside of the entire object) and right-click. Then, click *Edit Content* and click *In Separate Window*. Alternatively, you can double-click on the graph to open it for editing.

When you open the graph, the *Chart Editor* window will appear. (Note: This example uses the histogram created in Section 4.2.)

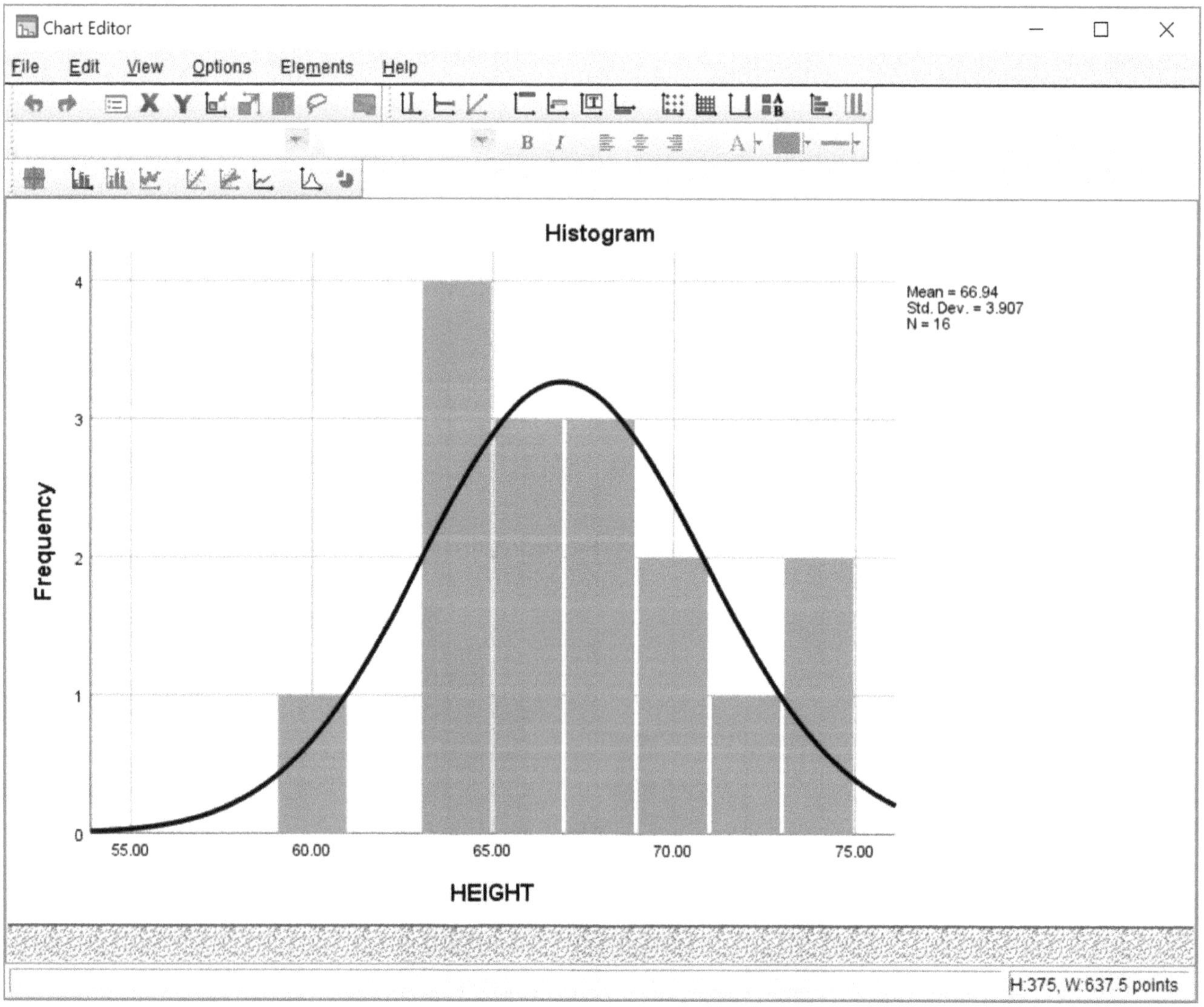

Once *Chart Editor* is open, you can easily edit each element of the graph. To select an element, click on the relevant spot on the graph. For instance, if you have added a title to your graph ("Histogram" in the example shown above, which was created earlier in this chapter and saved as HEIGHT.sav), you may select the element representing the title of the graph by clicking anywhere on the title.

Once you have selected an element, you can tell whether the correct element is selected because it will have handles around it.

If the item you have selected is a text element (e.g., the title of the graph), a cursor will be present and you can edit the text as you would in a word processing program. If you would like to change another attribute of the element (e.g., the color or font size), use the Properties box.

With a little practice, you can make excellent graphs using SPSS. Once your graph is formatted the way you want it, simply select *File*, then *Close*.

Chapter 5

Prediction and Association

Section 5.1 Pearson Correlation Coefficient

Description

The Pearson correlation coefficient (sometimes called the *Pearson product-moment correlation coefficient* or simply the *Pearson r*) determines the strength of the linear relationship between two variables.

Assumptions

Both variables should be measured on **interval** or **ratio scales** (or as a **dichotomous** nominal variable). If a relationship exists between them, that relationship should be linear. Because the Pearson correlation coefficient is computed with *z*-scores, both variables should also be normally distributed. If your data do not meet these assumptions, consider using the Spearman *rho* correlation coefficient instead.

SPSS Data Format

Two variables are required in your SPSS data file. Each subject must have data for both variables.

DOI: 10.4324/9781003450467-5

Running the Command

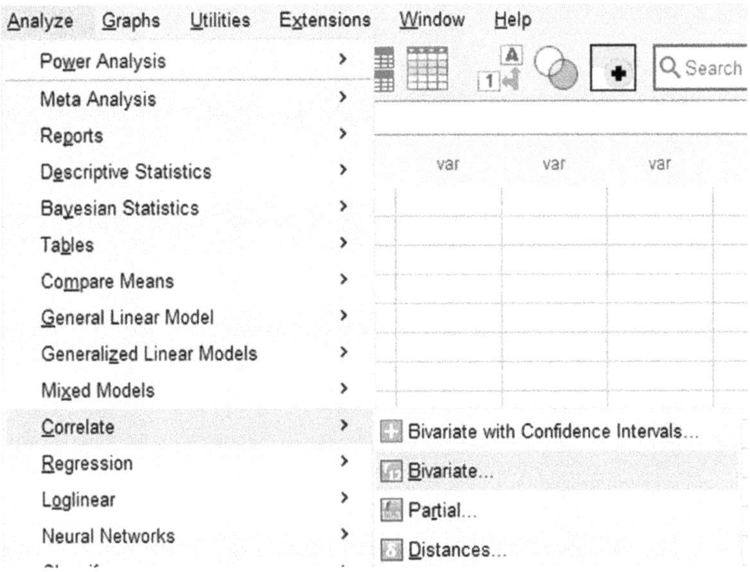

To select the Pearson correlation coefficient, click *Analyze*, then *Correlate*, then *Bivariate* (bivariate refers to two variables). This will bring up the Bivariate Correlations **dialog box**. This example uses the HEIGHT.sav data file entered at the start of Chapter 4.

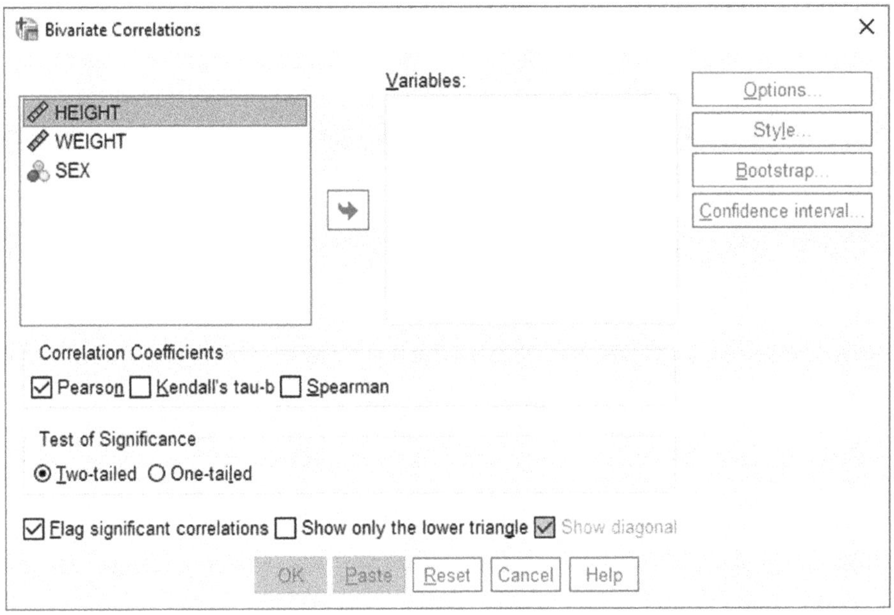

Move at least two variables from the box at the left into the box at the right by using the transfer arrow (or by double-clicking each variable). Make sure that a check is in the *Pearson* box under *Correlation Coefficients*. It is acceptable to move more than two variables.

For our example, we will move all three variables over and click *OK*.

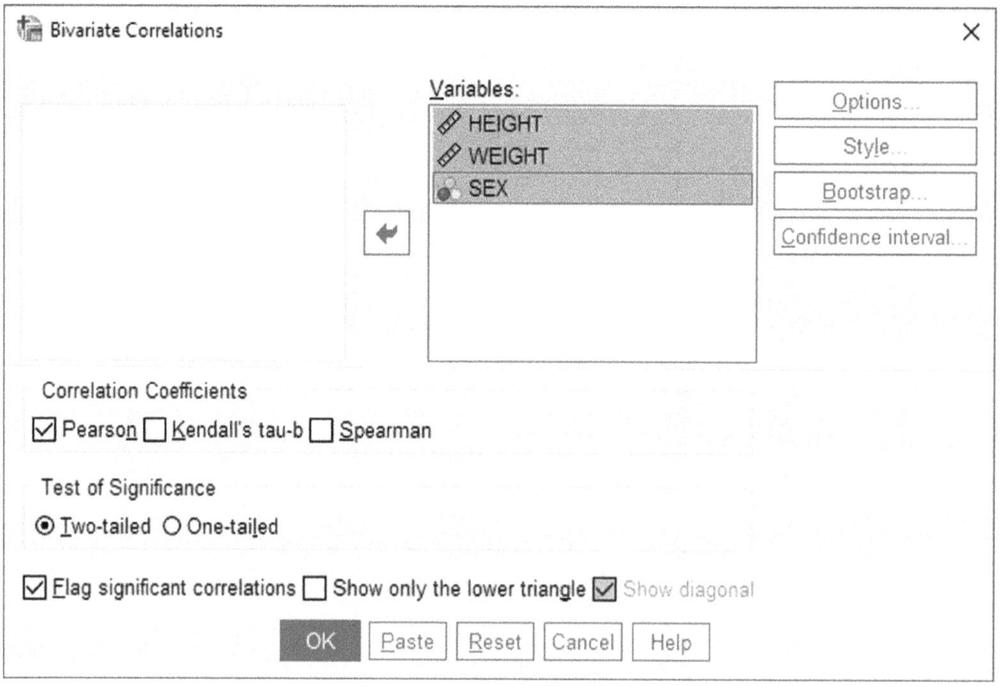

Reading the Output

The output consists of a **correlation matrix**. Every variable you entered in the command is represented as both a row and a column. We entered three variables in our command. Therefore, we have a 3 × 3 table. There are also three rows in each cell—the correlation, the **significance** level, and the *N*. If a correlation is significant at less than the .05 level, a single * will appear next to the correlation. If it is significant at the .01 level or lower, ** will appear next to the correlation. For instance, all of the correlations in the output above have a **significance** level of < .01, so they are flagged with ** to indicate that they are less than .01.

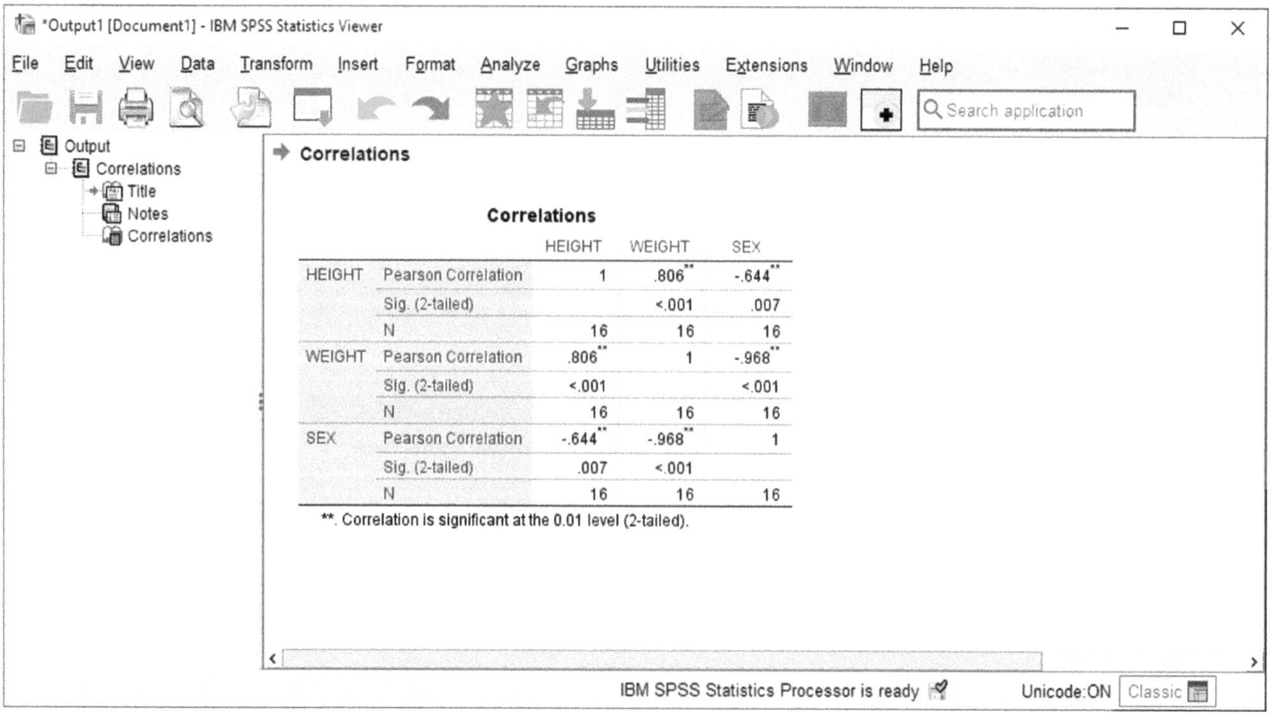

Correlations

		HEIGHT	WEIGHT	SEX
HEIGHT	Pearson Correlation	1	.806**	-.644**
	Sig. (2-tailed)		<.001	.007
	N	16	16	16
WEIGHT	Pearson Correlation	.806**	1	-.968**
	Sig. (2-tailed)	<.001		<.001
	N	16	16	16
SEX	Pearson Correlation	-.644**	-.968**	1
	Sig. (2-tailed)	.007	<.001	
	N	16	16	16

**. Correlation is significant at the 0.01 level (2-tailed).

To read the correlations, select a row and a column. For instance, the correlation between height and weight is determined through selection of the WEIGHT row and the HEIGHT column (.806). We get the same answer by selecting the HEIGHT row and the WEIGHT column. The correlation between a variable and itself is always 1, so there is a diagonal set of 1s.

Drawing Conclusions

The correlation coefficient will be between –1.0 and +1.0. Coefficients close to 0.0 represent a weak relationship. Coefficients close to 1.0 or –1.0 represent a strong relationship. Generally, correlations with an absolute value greater than 0.7 are considered strong. Correlations with an absolute value less than 0.3 are considered weak. Correlations with an absolute value between 0.3 and 0.7 are considered moderate. Positive correlations indicate that as one variable gets larger, the other variable also gets larger. Negative correlations indicate that as one variable gets larger, the other variable gets smaller.

Significant correlations are flagged with asterisks. A significant correlation indicates a reliable relationship, but not necessarily a strong correlation. With enough participants, a very small correlation can be significant. See Appendix A for a discussion of **effect sizes** for correlations.

Phrasing Results That Are Significant

In the preceding example, we obtained a correlation of .806 between HEIGHT and WEIGHT. A correlation of .806 is a strong positive correlation, and it is significant at the .001 level. Thus, we could state the following in a results section:

> A Pearson correlation coefficient was calculated for the relationship between participants' height and weight. A strong positive correlation was found (r (14) = .806, $p < .001$), indicating a significant linear relationship between the two variables. Taller participants tend to weigh more.

The conclusion states the direction (positive), strength (strong), value (.806), degrees of freedom (14), and **significance** level (< .001) of the correlation. In addition, a statement of direction is included (taller is heavier).

Note that the degrees of freedom given in parentheses is 14. The output indicates an N of 16. While most SPSS procedures give degrees of freedom, the *Correlation* command gives only the N (the number of pairs). For a correlation, the degrees of freedom is $N - 2$.

Phrasing Results That Are Not Significant

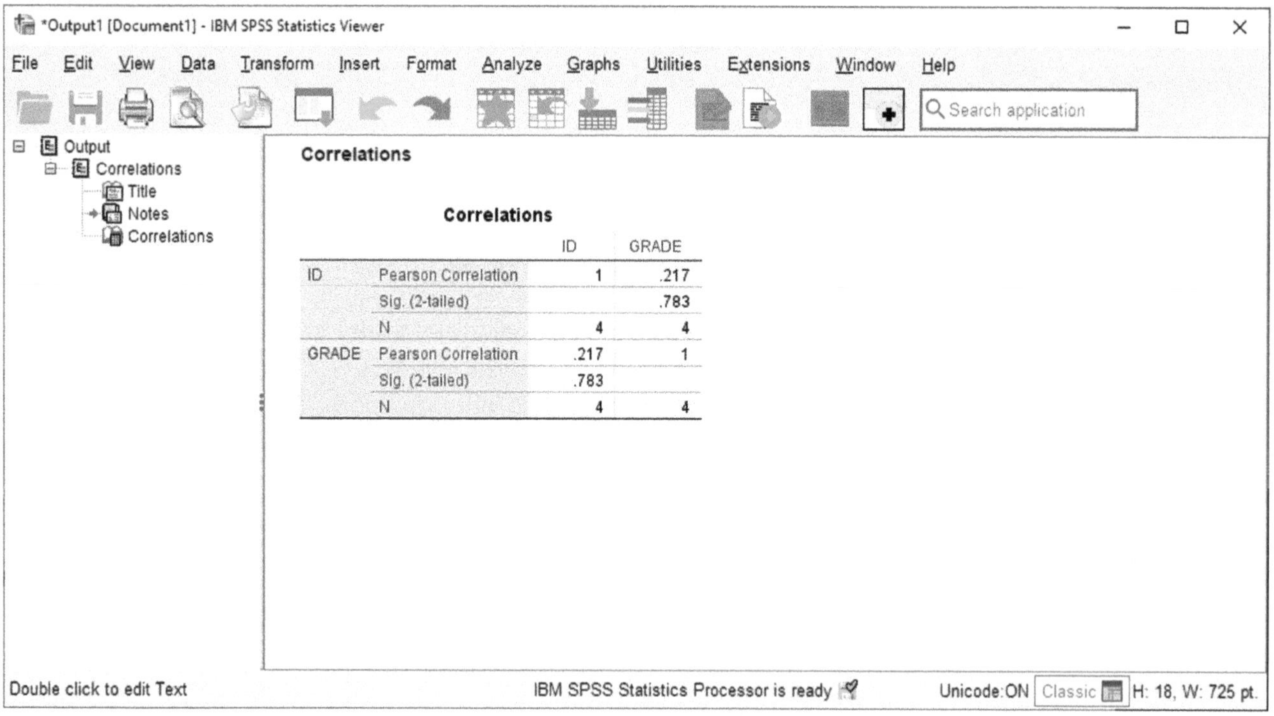

Correlations

		ID	GRADE
ID	Pearson Correlation	1	.217
	Sig. (2-tailed)		.783
	N	4	4
GRADE	Pearson Correlation	.217	1
	Sig. (2-tailed)	.783	
	N	4	4

Using our SAMPLE.sav dataset from the previous chapters, we can calculate a correlation between ID and GRADE. If we do so, we get the output shown here. The correlation has a **significance** level of .783. Thus, we could write the following in a results section (note that the degrees of freedom is $N - 2$):

> A Pearson correlation was calculated examining the relationship between participants' ID numbers and grades. A weak correlation that was not significant was found (r (2) = .217, $p > .05$). ID number is not related to grade in the course.

Practice Exercise

Use Practice Dataset 2 in Appendix B. Determine the value of the Pearson correlation coefficient for the relationship between SALARY and YEARS OF EDUCATION, and phrase your results.

Section 5.2 Spearman Correlation Coefficient

Description

The Spearman correlation coefficient determines the strength of the relationship between two variables. It is a nonparametric procedure. Therefore, it is weaker than the Pearson correlation coefficient, but it can be used in more situations.

Assumptions

Because the Spearman correlation coefficient functions on the basis of the ranks of data, it requires **ordinal** (or **interval** or **ratio)** data for both variables. They do not need to be normally distributed.

SPSS Data Format

Two variables are required in your SPSS data file. Each subject must provide data for both variables.

Running the Command

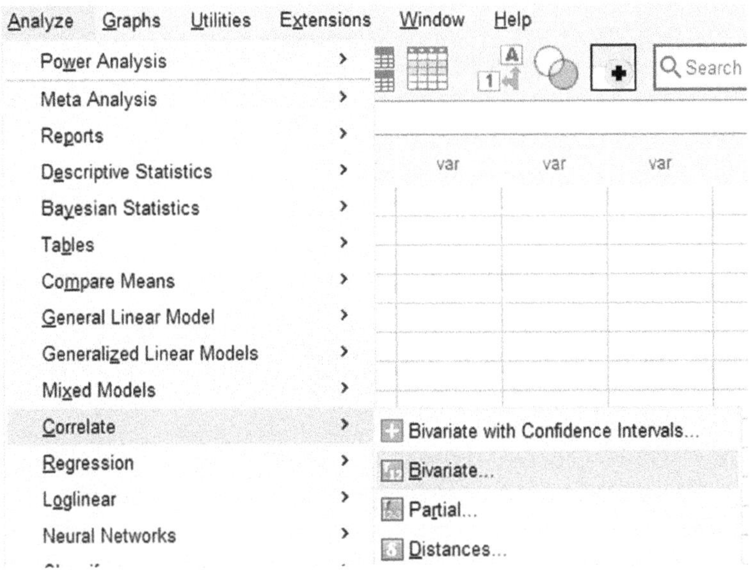

Using the HEIGHT.sav file created in Chapter 4, click *Analyze*, then *Correlate*, then *Bivariate*. This will bring up the main **dialog box** for Bivariate Correlations (just like the Pearson correlation). About halfway down the **dialog box**, there is a section for indicating the type of correlation you will compute. You can select as many correlations as you want. For our example, remove the check in the *Pearson* box (by clicking on it) and click on the *Spearman* box.

Use the variables HEIGHT and WEIGHT. This is also one of the few commands that allows you to choose a one-tailed test, if desired.

Reading the Output

The output is essentially the same as for the Pearson correlation. Each pair of variables has its correlation coefficient indicated twice. The Spearman *rho* can range from –1.0 to +1.0, just like the Pearson *r*.

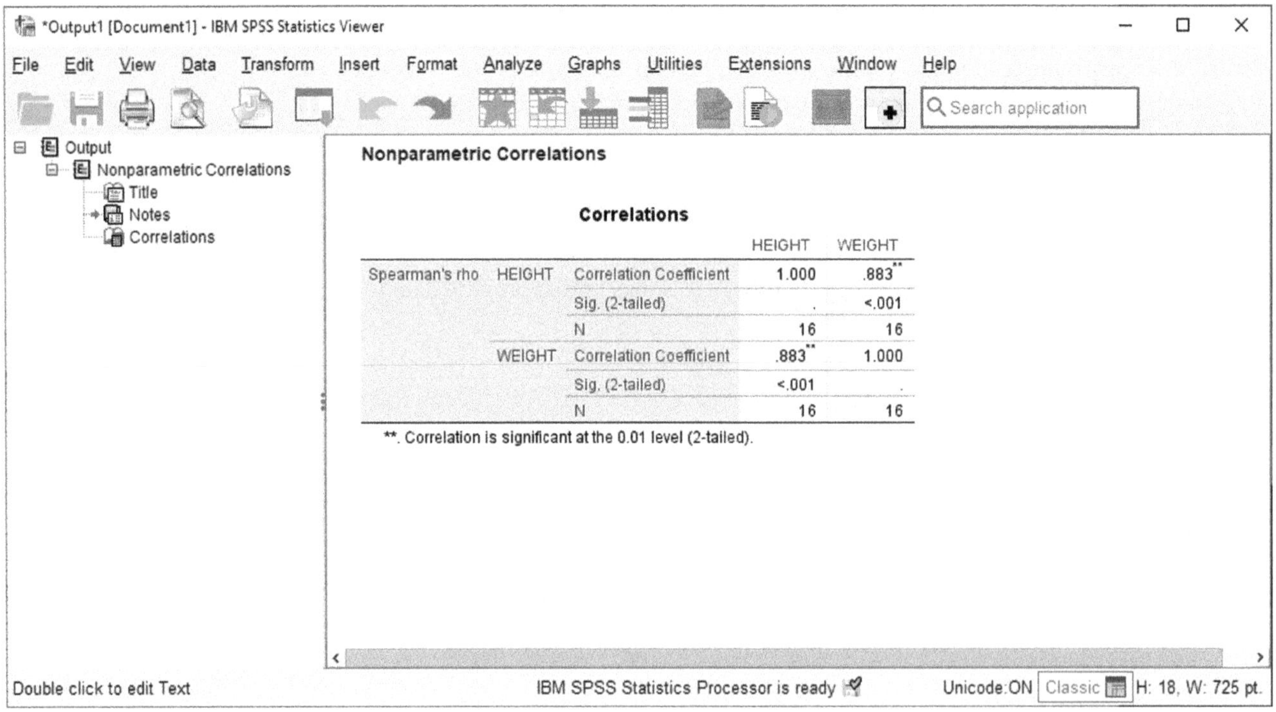

Correlations

			HEIGHT	WEIGHT
Spearman's rho	HEIGHT	Correlation Coefficient	1.000	.883**
		Sig. (2-tailed)	.	<.001
		N	16	16
	WEIGHT	Correlation Coefficient	.883**	1.000
		Sig. (2-tailed)	<.001	.
		N	16	16

**. Correlation is significant at the 0.01 level (2-tailed).

The output listed above indicates a correlation of .883 between HEIGHT and WEIGHT.

Drawing Conclusions

The correlation will be between –1.0 and +1.0. Scores close to 0.0 represent a weak relationship. Scores close to 1.0 or –1.0 represent a strong relationship. Significant correlations are flagged with asterisks. A significant correlation indicates a reliable relationship, but not necessarily a strong correlation. With enough participants, a very small correlation can be significant. Generally, correlations greater than 0.7 are considered strong. Correlations less than 0.3 are considered weak. Correlations between 0.3 and 0.7 are considered moderate.

Phrasing Results That Are Significant

In the prior example, we obtained a correlation of .883 between HEIGHT and WEIGHT. A correlation of .883 is a strong positive correlation, and it is significant at the .001 level. Thus, we could state the following in a results section:

> A Spearman *rho* correlation coefficient was calculated for the relationship between participants' height and weight. A strong positive correlation was found (*rho* (14) = .883, $p < .001$), indicating a significant relationship between the two variables. Taller participants tend to weigh more.

The conclusion states the direction (positive), strength (strong), value (.883), degrees of freedom (14), and **significance** level (< .001) of the correlation. In addition, a statement of direction is included (taller is heavier). Note that the degrees of freedom given in parentheses is 14. The output indicates an *N* of 16. For a correlation, the degrees of freedom is $N - 2$.

Phrasing Results That Are Not Significant

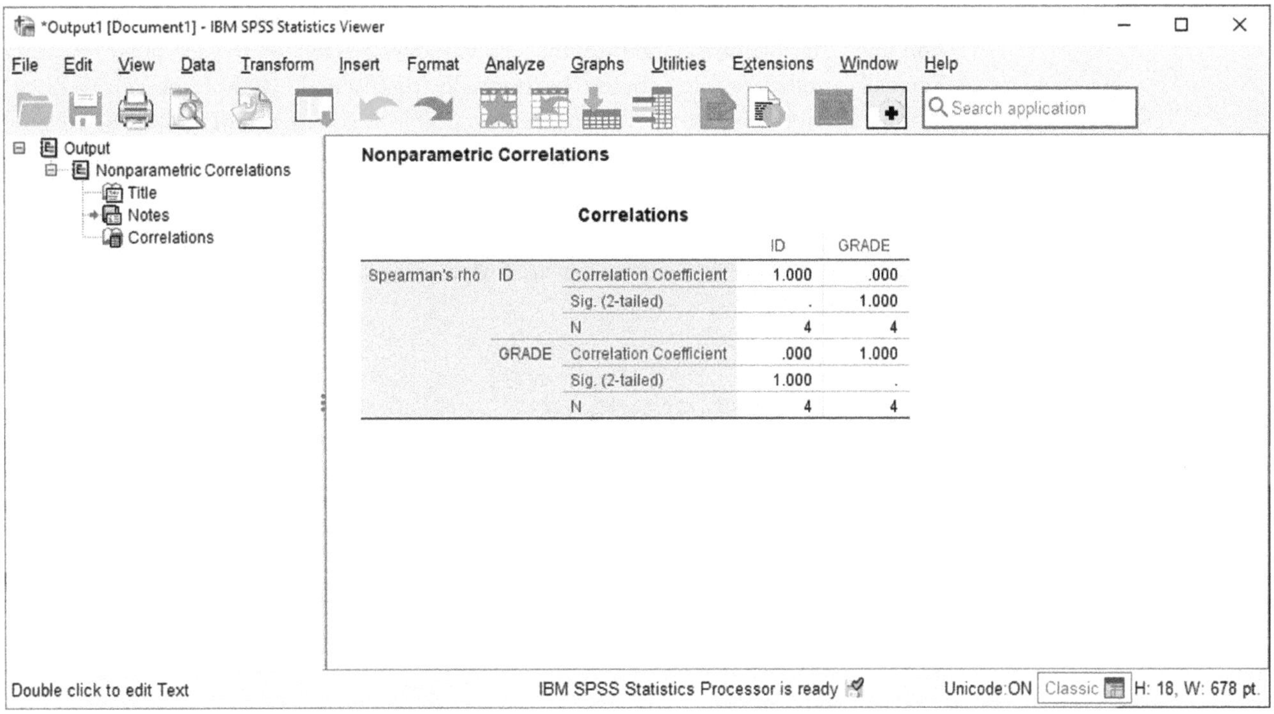

Correlations

			ID	GRADE
Spearman's rho	ID	Correlation Coefficient	1.000	.000
		Sig. (2-tailed)	.	1.000
		N	4	4
	GRADE	Correlation Coefficient	.000	1.000
		Sig. (2-tailed)	1.000	.
		N	4	4

Using our SAMPLE.sav dataset from the previous chapters, we could calculate a Spearman *rho* correlation between ID and GRADE. If we did so, we would get the output shown here. The correlation coefficient equals .000 and has a **significance** level of 1.000. Note that though this value is rounded up and is not, in fact, exactly 1.000, we could state the following in a results section:

> A Spearman *rho* correlation coefficient was calculated for the relationship between a subject's ID number and grade. An extremely weak correlation that was not significant was found (r (2) = .000, $p > .05$). ID number is not related to grade in the course.

Practice Exercise

Use Practice Dataset 2 in Appendix B. Determine the strength of the relationship between salary and job classification by calculating the Spearman *rho* correlation.

Section 5.3 Simple Linear Regression

Description

Simple linear regression allows for the prediction of one variable from another.

Assumptions

Simple linear regression assumes that both variables are **interval-** or **ratio-scaled**. In addition, the **dependent variable** should be normally distributed around the prediction line. This, of course, assumes that the variables are related to each other linearly. Typically, both variables should be normally distributed. **Dichotomous variables** (variables with only two **levels**) are also acceptable as **independent variables**.

SPSS Data Format

Two variables are required in the SPSS data file. Each subject must contribute to both values.

Running the Command

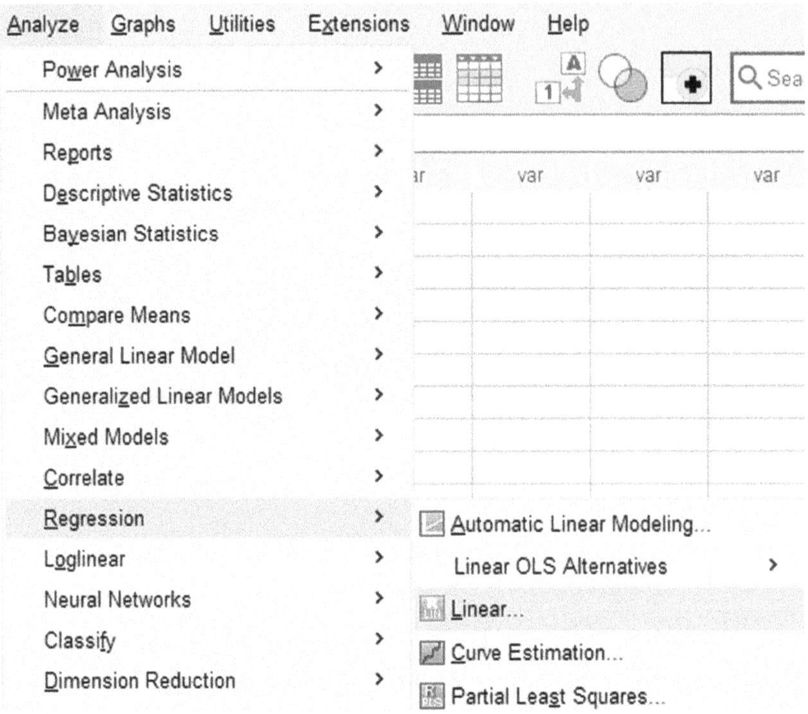

Returning to the HEIGHT.sav data file, click *Analyze*, then *Regression*, then *Linear*. This will bring up the main **dialog box** for Linear Regression, shown below. On the left side of the **dialog box** is a list of the variables in your data file. On the right are blocks for the **dependent variable** (the variable we are trying to predict) and the **independent variable** (the variable from which we are predicting).

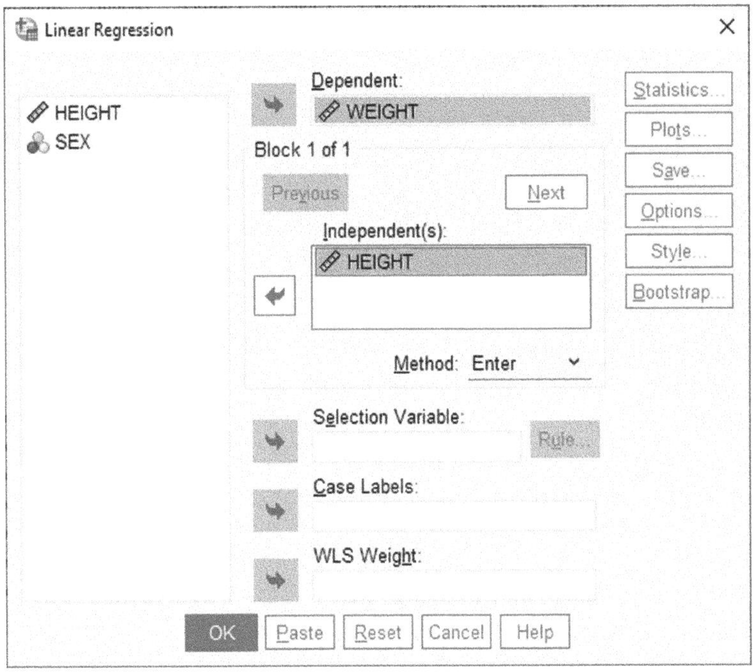

We are interested in predicting someone's weight on the basis of his or her height. Thus, we should place the variable WEIGHT in the **dependent variable** block and the variable HEIGHT in the **independent variable** block. Then, we can click *OK* to run the analysis.

Reading the Output

For simple linear regressions, we are interested in three components of the output. The first is called the Model Summary, and it occurs after the Variables Entered/Removed section. For our example, you should see this output. *R* Square (called the **coefficient of determination**) gives you the proportion of the **variance** of your **dependent variable** (WEIGHT) that can be explained by variation in your **independent variable** (HEIGHT). Thus, 64.9 percent of the variation in weight can be explained by differences in height (taller individuals weigh more). For information on R Square as a measure of effect size please refer to Appendix A.

The **standard error of estimate** gives you a measure of dispersion for your prediction equation. When the prediction equation is used, 68 percent of the data will fall within one **standard error of estimate** (predicted) value. Just over 95 percent will fall within two standard errors. Thus, in the previous example, 95 percent of the time, our estimated weight will be within 32.296 pounds of being correct (i.e., $2 \times 16.148 = 32.296$).

The second part of the output that we are interested in is the ANOVA summary table. The important number here is the **significance** level in the rightmost column. If that value is less than .05, then we have a significant linear regression. If it is larger than .05, we do not. For more information on reading ANOVA tables, refer to the sections on ANOVA in Chapter 7.

The final section of the output is the table of coefficients. This is where the actual prediction equation can be found.

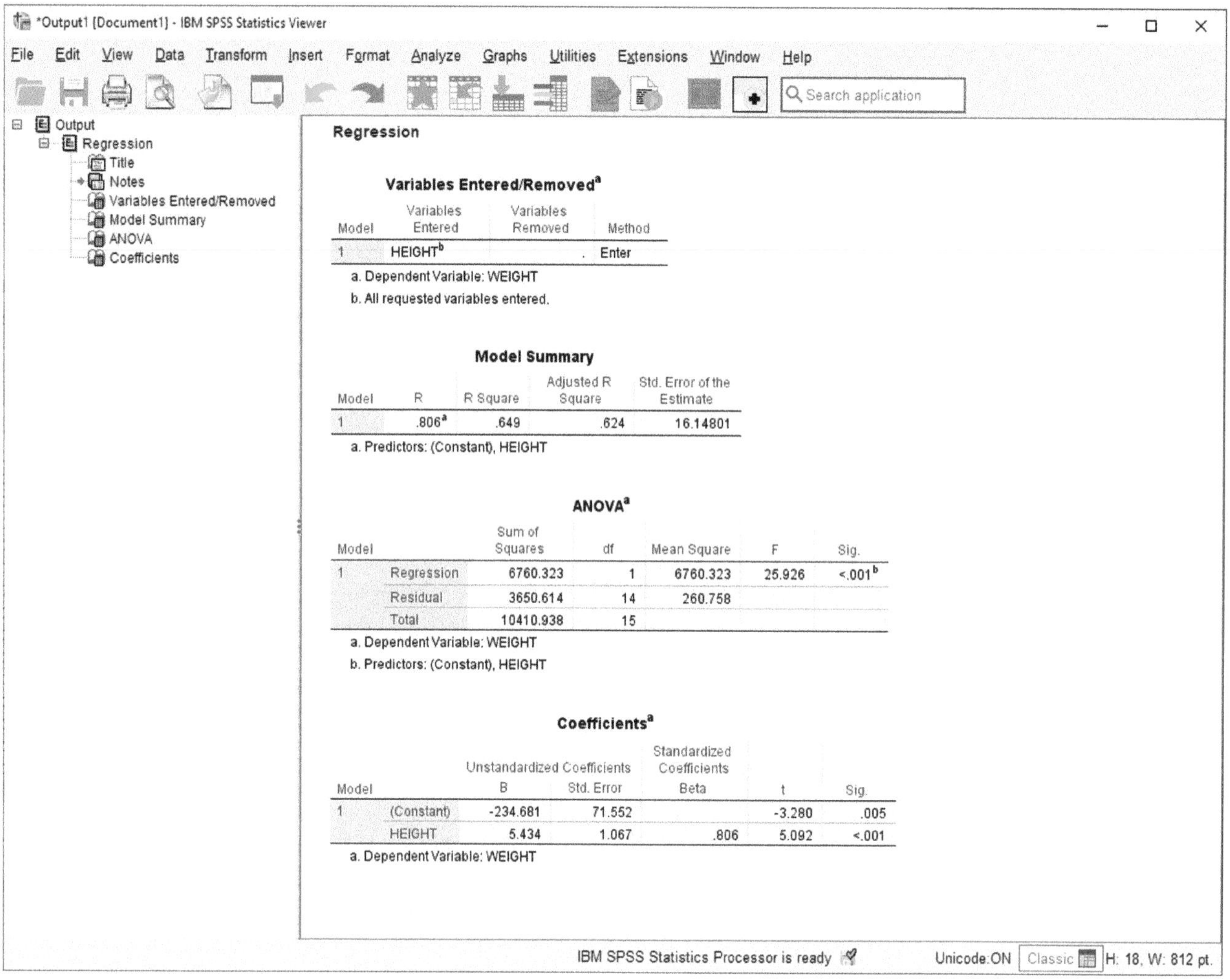

Regression

Variables Entered/Removed[a]

Model	Variables Entered	Variables Removed	Method
1	HEIGHT[b]	.	Enter

a. Dependent Variable: WEIGHT
b. All requested variables entered.

Model Summary

Model	R	R Square	Adjusted R Square	Std. Error of the Estimate
1	.806[a]	.649	.624	16.14801

a. Predictors: (Constant), HEIGHT

ANOVA[a]

Model		Sum of Squares	df	Mean Square	F	Sig.
1	Regression	6760.323	1	6760.323	25.926	<.001[b]
	Residual	3650.614	14	260.758		
	Total	10410.938	15			

a. Dependent Variable: WEIGHT
b. Predictors: (Constant), HEIGHT

Coefficients[a]

Model		Unstandardized Coefficients		Standardized Coefficients		
		B	Std. Error	Beta	t	Sig.
1	(Constant)	-234.681	71.552		-3.280	.005
	HEIGHT	5.434	1.067	.806	5.092	<.001

a. Dependent Variable: WEIGHT

In most texts, you learn that $Y' = a + bX$ is the regression equation. Y' (pronounced "Y prime") is your **dependent variable** (primes are normally predicted values or dependent variables), and X is your **independent variable**. In SPSS output, the values of both a and b are found in the B column. The first value, –234.681, is the value of a (labeled Constant). The second value, 5.434, is the value of b (labeled with the name of the **independent variable**). Thus, our prediction equation for the example above is WEIGHT′ = –234.681 + 5.434(HEIGHT). In other words, the average subject who is an inch taller than another subject weighs 5.434 pounds more. A person who is 60 inches tall should weigh –234.681 + 5.434(60) = 91.359 pounds. Given our earlier discussion of **standard error of estimate**, 95 percent of individuals who are 60 inches tall will weigh between 59.063 (91.359–32.296 = 59.063) and 123.655 (91.359 + 32.296 = 123.655) pounds. This range (59.063 to 123655) is the confidence interval for your prediction.

Drawing Conclusions

Conclusions from regression analyses indicate (a) whether or not a significant prediction equation was obtained, (b) the direction of the relationship, and (c) the equation itself. The direction of the relationship is obtained from the sign of the Beta coefficient. Positive coefficients indicate positive relationships. Negative coefficients indicate negative relationships.

Phrasing Results That Are Significant

In the earlier example, we obtained an *R* Square of .649 and a regression equation of WEIGHT′ = –234.681 + 5.434(HEIGHT). The ANOVA resulted in $F = 25.926$ with 1 and 14 degrees of freedom. The *F* is significant at the less than .001 level. Thus, we could state the following in a results section:

> A simple linear regression was calculated to predict participants' weight based on their height. A significant regression equation was found ($F(1,14) = 25.926$, $p < .001$), with an R^2 of .649. Participants' predicted weight is equal to –234.68 + 5.43(HEIGHT) pounds when height is measured in inches. Participants' average weight increased 5.43 pounds for each inch of height.

The conclusion states the direction (increase), strength (.649), value (25.926), degrees of freedom (1,14), and **significance** level (< .001) of the regression. In addition, a statement of the equation itself is included.

Phrasing Results That Are Not Significant

If the ANOVA is not significant (this example attempts to predict GRADE from TRAINING using the SAMPLE.sav dataset), the section of the output labeled *Sig.* for the ANOVA will be greater than .05, and the regression equation will not be significant. A results section might include the following statement:

> A simple linear regression was calculated to predict participants' grades from whether or not they received training. The regression equation was not significant ($F(1,2) = 4.245$, $p > .05$) with an R^2 of .68. Training is not a significant predictor of grades.

Note that for results that are not significant, the ANOVA results and R^2 results are given, but the regression equation is not.

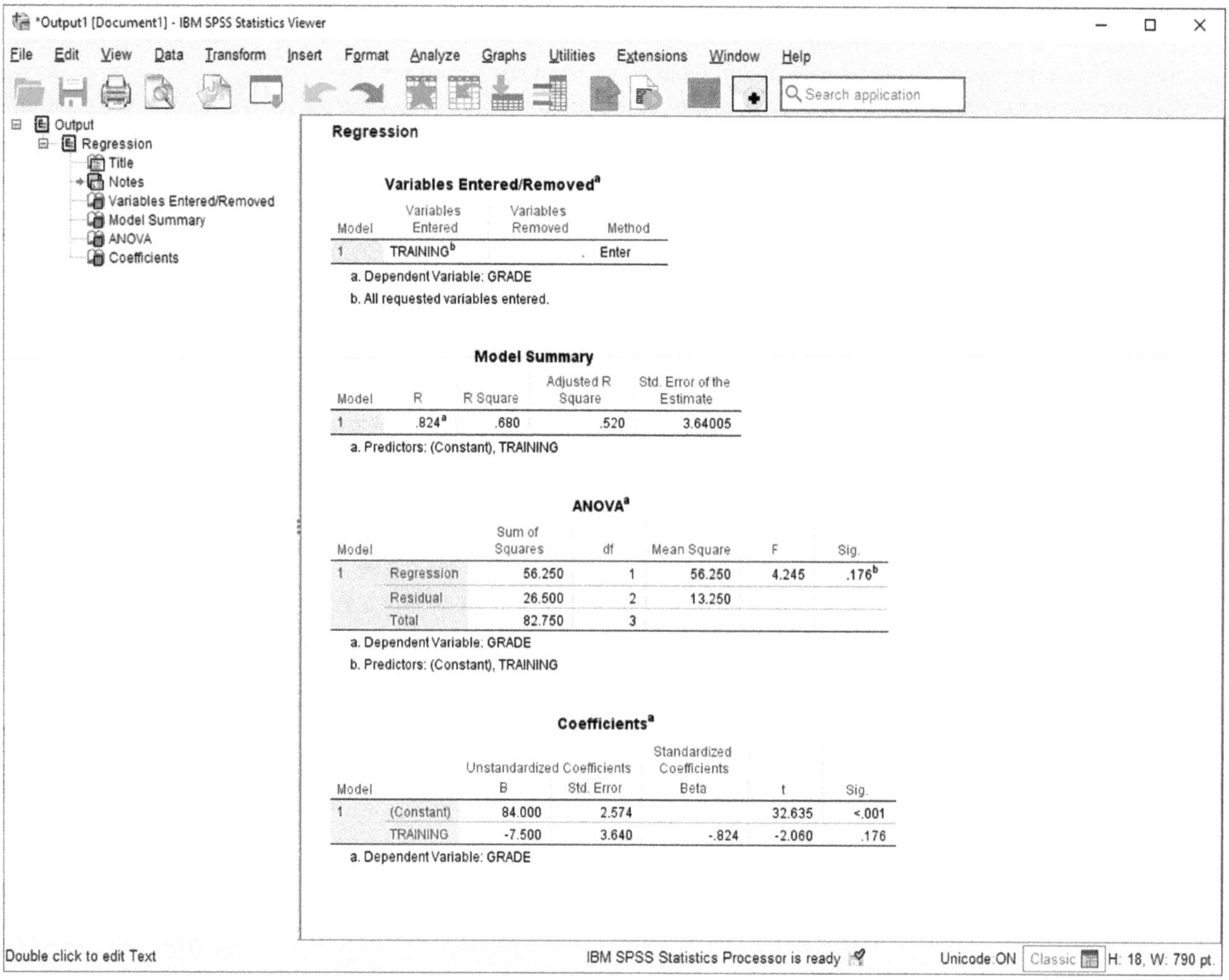

Regression

Variables Entered/Removed[a]

Model	Variables Entered	Variables Removed	Method
1	TRAINING[b]	.	Enter

a. Dependent Variable: GRADE
b. All requested variables entered.

Model Summary

Model	R	R Square	Adjusted R Square	Std. Error of the Estimate
1	.824[a]	.680	.520	3.64005

a. Predictors: (Constant), TRAINING

ANOVA[a]

Model		Sum of Squares	df	Mean Square	F	Sig.
1	Regression	56.250	1	56.250	4.245	.176[b]
	Residual	26.500	2	13.250		
	Total	82.750	3			

a. Dependent Variable: GRADE
b. Predictors: (Constant), TRAINING

Coefficients[a]

Model		Unstandardized Coefficients B	Std. Error	Standardized Coefficients Beta	t	Sig.
1	(Constant)	84.000	2.574		32.635	<.001
	TRAINING	-7.500	3.640	-.824	-2.060	.176

a. Dependent Variable: GRADE

Practice Exercise

Use Practice Dataset 2 in Appendix B. If we want to predict salary from years of education, what salary would you predict for someone with 12 years of education? What salary would you predict for someone with a college education (16 years)?

Section 5.4 Multiple Linear Regression

Description

The multiple linear regression analysis allows the prediction of one variable from several other variables.

Assumptions

Multiple linear regression assumes that all variables are **interval-** or **ratio-scaled**. In addition, the **dependent variable** should be normally distributed around the prediction line. This, of course, assumes that the variables are related to each other linearly. All variables should be normally distributed. **Dichotomous variables** (e.g., SEX in the example below) are also acceptable as **independent variables**.

SPSS Data Format

At least three variables are required in the SPSS data file. Each subject must provide data for all variables.

Running the Command

Click *Analyze*, then *Regression*, then *Linear*. This will bring up the main **dialog box** for Linear Regression. On the left side of the **dialog box** is a list of the variables in your data file. (We are using the HEIGHT.sav data file from the start of this chapter.) On the right side of the **dialog box** are blanks for the **dependent variable** (the variable you are trying to predict) and the **independent variables** (the variables from which you are predicting).

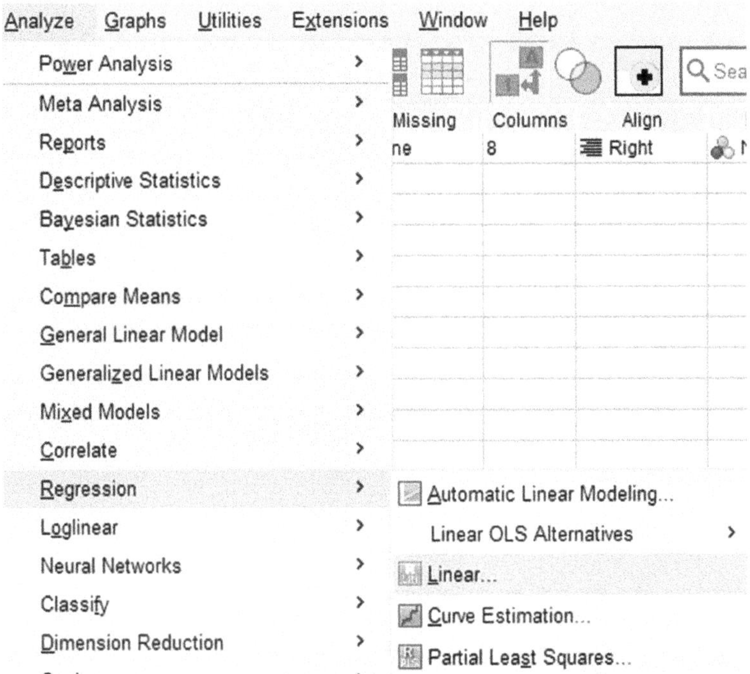

We are interested in predicting someone's weight based on his or her height and sex. We believe that both sex and height are related to weight. Thus, we should place the **dependent variable** WEIGHT in the *Dependent* block and the **independent variables** HEIGHT and SEX in the *Independent(s)* block.

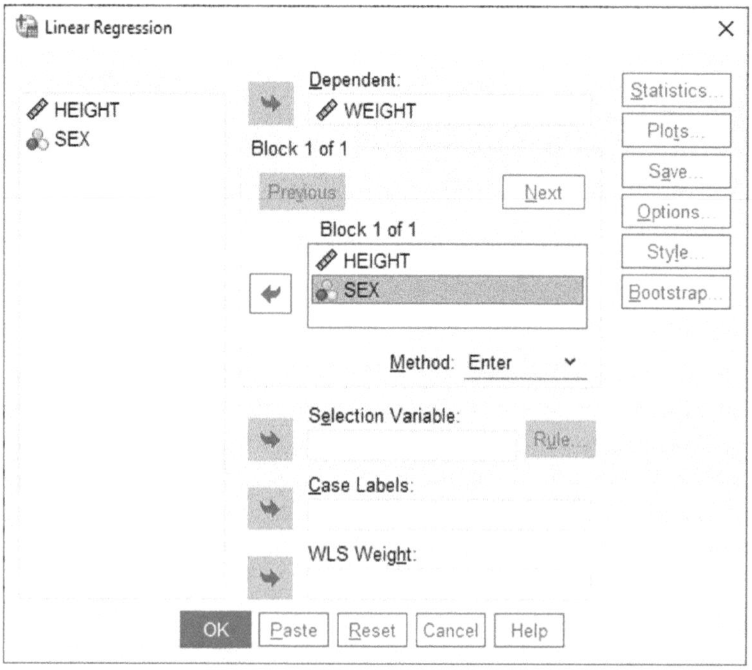

This will perform an analysis to determine if WEIGHT can be predicted from SEX and/or HEIGHT. There are several methods SPSS can use to conduct this analysis. These can be selected with the *Method* box. The most widely used is *Enter* (the default), which puts all the variables in the equation, whether or not they are significant. Other methods can be selected by clicking on the down arrow next to the word *Enter.* The other methods use various means to enter only those variables that are significant predictors. In this case, we will use the default. Click *OK* to run the analysis.

Reading the Output

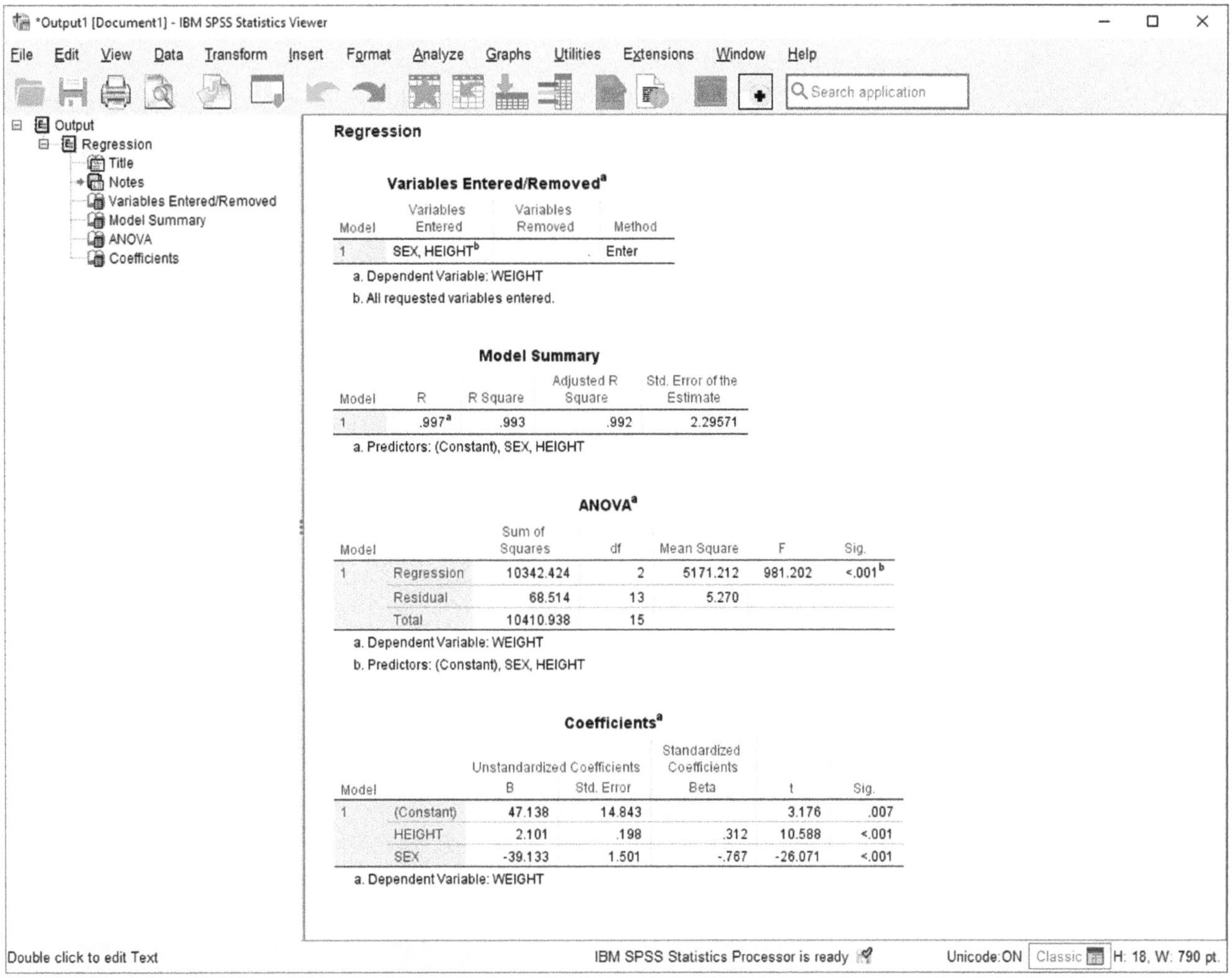

Regression

Variables Entered/Removed[a]

Model	Variables Entered	Variables Removed	Method
1	SEX, HEIGHT[b]	.	Enter

a. Dependent Variable: WEIGHT
b. All requested variables entered.

Model Summary

Model	R	R Square	Adjusted R Square	Std. Error of the Estimate
1	.997[a]	.993	.992	2.29571

a. Predictors: (Constant), SEX, HEIGHT

ANOVA[a]

Model		Sum of Squares	df	Mean Square	F	Sig.
1	Regression	10342.424	2	5171.212	981.202	<.001[b]
	Residual	68.514	13	5.270		
	Total	10410.938	15			

a. Dependent Variable: WEIGHT
b. Predictors: (Constant), SEX, HEIGHT

Coefficients[a]

Model		Unstandardized Coefficients B	Unstandardized Coefficients Std. Error	Standardized Coefficients Beta	t	Sig.
1	(Constant)	47.138	14.843		3.176	.007
	HEIGHT	2.101	.198	.312	10.588	<.001
	SEX	-39.133	1.501	-.767	-26.071	<.001

a. Dependent Variable: WEIGHT

For multiple linear regression, there are three components of the output in which we are interested. The first is called the Model Summary, which is found after the Variables Entered/Removed section. For our example, you should get the output above.

R Square (called the **coefficient of determination**) tells you the proportion of the **variance** in the **dependent variable** (WEIGHT) that can be explained by variation in the **independent variables** (HEIGHT and SEX, in this case). Thus, 99.3 percent of the variation in weight can be explained by differences in height and sex (taller individuals weigh more, and men weigh more). Note that when a second variable is added, our *R* Square goes up from .649 to .993. The .649 was obtained using the Simple Linear Regression example in Section 5.3. For information on R Square as a measure of effect size please refer to Appendix A.

The **standard error of the estimate** gives you a margin of error for the prediction equation. Using the prediction equation, 68 percent of the data will fall within one **standard error of estimate** (predicted) value. Just over 95 percent will fall within two **standard errors of estimate**. Thus, in the example above, 95 percent of the time our estimated weight will be within 4.591 (2.29571 × 2) pounds of being correct. In our Simple Linear Regression example in Section 5.3, this number was 32.296. Note the higher degree of accuracy.

The second part of the output that we are interested in is the ANOVA summary table. For more information on reading ANOVA tables, refer to the sections on ANOVA in Chapter 7. For now, the important number is the **significance** in the rightmost column. If that value is less than .05, we have a significant linear regression. If it is larger than .05, we do not.

The final section of output we are interested in is the table of coefficients. This is where the actual prediction equation can be found.

In most texts, you learn that $Y' = a + bX$ is the regression equation. For multiple regression, our equation changes to $Y' = B_0 + B_1 X_1 + B_2 X_2 + \ldots + B_z X_z$ (where z is the number of independent variables). Y' is your **dependent variable**, and the Xs are your **independent variables**. The Bs are listed in a column. Thus, our prediction equation for the example above is WEIGHT′ = 47.138–39.133(SEX) + 2.101(HEIGHT) (where SEX is coded as 1 = Male, 2 = Female, and HEIGHT is in inches). In other words, the average difference in weight for participants who differ by one inch in height is 2.101 pounds. Males tend to weigh 39.133 pounds more than females. (Males have a value of SEX that is one less than females. Therefore, their predicted weight will be (–1)(–39.133) = 39.133 different.) A female who is 60 inches tall should weigh 47.138–39.133(2) + 2.101(60) = 94.932 pounds. Given our earlier discussion of the **standard error of estimate**, 95 percent of females who are 60 inches tall will weigh between 90.341 (94.932–4.591 = 90.341) and 99.523 (94.932 + 4.591 = 99.523) pounds.

Drawing Conclusions

Conclusions from regression analyses indicate (a) whether or not a significant prediction equation was obtained, (b) the direction of the relationship, and (c) the equation itself. Multiple regression is generally much more powerful than simple linear regression. Compare our two examples.

With multiple regression, you must also consider the **significance** level of each **independent variable**. In the example above, the **significance** level of both **independent variables** is less than .001.

Phrasing Results That Are Significant

In our example, we obtained an R Square of .993 and a regression equation of WEIGHT′ = 47.138–39.133(SEX) + 2.101(HEIGHT). The ANOVA resulted in $F = 981.202$ with 2 and 13 degrees of freedom. F is significant at the less than .001 level. Thus, we could state the following in a results section:

> A multiple linear regression was calculated to predict participants' weight based on their height and sex. A significant regression equation was found (F (2,13) = 981.202, $p < .001$), with an R^2 of .993. Participants' predicted weight is equal to 47.138–39.133(SEX) + 2.101(HEIGHT), where SEX is coded as 1 = Male, 2 = Female, and HEIGHT is measured in inches. Participants increased 2.101 pounds for each inch of height, and males weighed 39.133 pounds more than females. Both SEX and HEIGHT were significant predictors.

The conclusion states the direction (increase), strength (.993), value (981.20), degrees of freedom (2,13), and **significance** level (< .001) of the regression. In addition, a statement of the equation itself is included. Because there are multiple **independent variables**, we have noted whether or not each is significant.

Phrasing Results That Are Not Significant

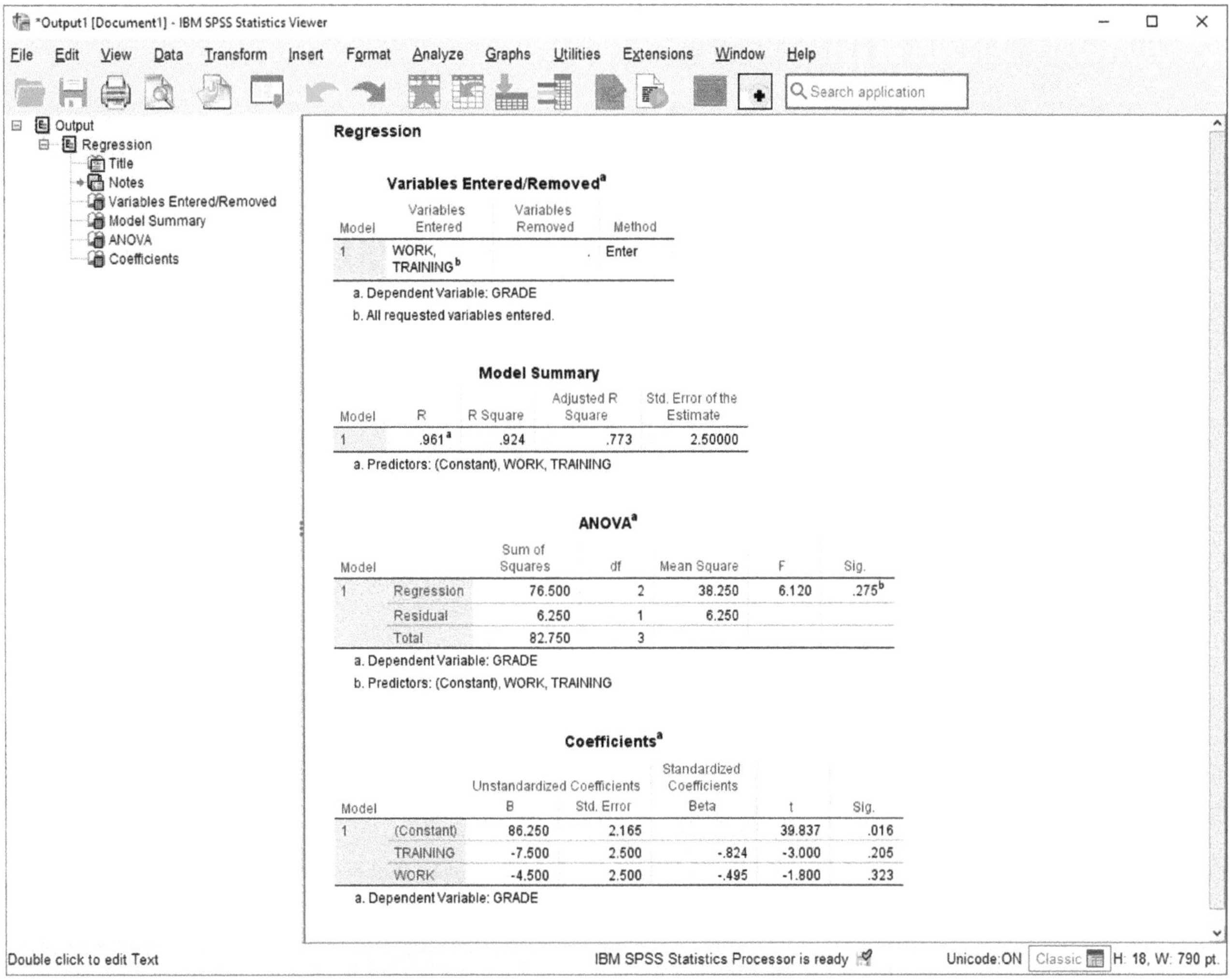

Regression

Variables Entered/Removed[a]

Model	Variables Entered	Variables Removed	Method
1	WORK, TRAINING[b]	.	Enter

a. Dependent Variable: GRADE
b. All requested variables entered.

Model Summary

Model	R	R Square	Adjusted R Square	Std. Error of the Estimate
1	.961[a]	.924	.773	2.50000

a. Predictors: (Constant), WORK, TRAINING

ANOVA[a]

Model		Sum of Squares	df	Mean Square	F	Sig.
1	Regression	76.500	2	38.250	6.120	.275[b]
	Residual	6.250	1	6.250		
	Total	82.750	3			

a. Dependent Variable: GRADE
b. Predictors: (Constant), WORK, TRAINING

Coefficients[a]

Model		Unstandardized Coefficients B	Unstandardized Coefficients Std. Error	Standardized Coefficients Beta	t	Sig.
1	(Constant)	86.250	2.165		39.837	.016
	TRAINING	-7.500	2.500	-.824	-3.000	.205
	WORK	-4.500	2.500	-.495	-1.800	.323

a. Dependent Variable: GRADE

If the ANOVA does not find a significant relationship, the *Sig.* section of the output will be greater than .05, and the regression equation will not be significant. A results section for the output shown here (based on SAMPLE.sav) might include the following statement:

> A multiple linear regression was calculated predicting participants' grades based on whether or not they received training and the amount they work. The regression equation was not significant (F (2,1) = 6.12, $p > .05$) with an R^2 of .924. Neither training nor working is a significant predictor of grades.

Note that for results that are not significant, the ANOVA results and R^2 results are given, but the regression equation is not.

Practice Exercise

Use Practice Dataset 2 in Appendix B. Determine the regression equation for predicting salary based on education, years of service, and sex. Which variables are significant predictors? If you believe that men were paid more than women, what would you conclude after conducting this analysis?

Chapter 6

Basic Parametric Inferential Statistics and *t* Tests

Parametric **inferential statistical** procedures allow you to draw inferences about populations based on samples of those populations. To make these inferences, you must be able to make certain assumptions about the shape of the distributions of the population samples.

Section 6.1 Review of Basic Hypothesis Testing

The Null Hypothesis

In hypothesis testing, we create two hypotheses that are **mutually exclusive** (i.e., both cannot be true at the same time) and **all inclusive** (i.e., one of them must be true). We refer to those two hypotheses as the **null hypothesis** and the **alternative hypothesis**. The **null hypothesis** generally states that any difference we observe is caused by random error. The **alternative hypothesis** generally states that any difference we observe is caused by a systematic difference between groups.

Type I and Type II Errors

All hypothesis testing attempts to draw conclusions about the real world based on the results of a test (a statistical test, in this case). There are four possible combinations of results.

DOI: 10.4324/9781003450467-6

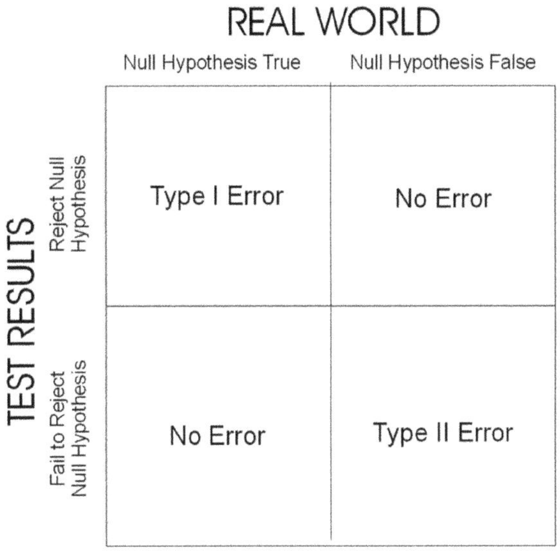

REAL WORLD
Null Hypothesis True
Null Hypothesis False
TEST RESULTS
Reject Null Hypothesis
Fail to Reject Null Hypothesis
Type I Error
No Error
No Error
Type II Error

Two of the possible results are correct test results. The other two results are errors. A **Type I error** occurs when we reject a **null hypothesis** that is, in fact, true, while a **Type II error** occurs when we fail to reject a **null hypothesis** that is, in fact, false.

Significance tests determine the probability of making a **Type I error** (often called *alpha* <α>, *Sig.*, or *p*). In other words, after performing a series of calculations, we obtain a probability of obtaining a value as far from expected as we got assuming the **null hypothesis** is true. If there is a low probability, such as 5 or less in 100 (.05), by convention, we reject the **null hypothesis**. In other words, we typically use the .05 level (or less) as the maximum **Type I error** rate we are willing to accept.

When there is a low probability of a **Type I error**, such as .05, we can state that the **significance** test has led us to "reject the **null hypothesis**." This is synonymous with saying that a difference is "statistically significant." For instance, on a reading test, suppose you found that a random sample of girls from a school district scored higher than a random sample of boys. This result may have been obtained merely because the chance errors associated with random sampling created the observed difference (this is what the **null hypothesis** asserts). If there is a sufficiently low probability that random errors were the cause (as determined by a **significance** test), we can state that the difference between boys and girls is statistically significant.

Significance Levels Versus Critical Values

Most statistics textbooks present hypothesis testing by using the concept of a critical value. With such an approach, we obtain a value for a test statistic and compare it to a critical value we look up in a table. If the obtained value is larger than the critical value, we reject the **null hypothesis** and conclude that we have found a significant difference (or relationship). If the obtained value is less than the critical value, we fail to reject the **null hypothesis** and conclude that there is not a significant difference.

The critical-value approach is well suited to hand calculations. Tables that give critical values for alpha levels of .001, .01, .05, and so on can be created. It is not practical to create a table for every possible alpha level.

On the other hand, SPSS can determine the exact alpha level associated with any value of a test statistic. Thus, looking up a critical value in a table is not necessary. This, however, does change the basic procedure for determining whether or not to reject the **null hypothesis**.

The section of SPSS output labeled *Sig.* (sometimes *p* or *alpha* <α>) indicates the likelihood of making a **Type I error** if we reject the **null hypothesis**. A value of .05 or less indicates that we should reject the **null hypothesis** (assuming an alpha level of .05). A value greater than .05 indicates that we should fail to reject the **null hypothesis**.

In other words, when using SPSS, we normally reject the **null hypothesis** if the output value under *Sig.* is equal to or smaller than .05, and we fail to reject the **null hypothesis** if the output value is larger than .05.

One-Tailed Versus Two-Tailed Tests

SPSS output generally includes a two-tailed alpha level (normally labeled *Sig.* or *p* in the output). A two-tailed hypothesis attempts to determine whether any difference (either positive or negative) exists. Thus, you have an opportunity to make a **Type I error** on either of the two tails of the **normal distribution**.

A one-tailed test examines a difference in a specific direction. Thus, we can make a **Type I error** on only one side (tail) of the distribution. If we have a one-tailed hypothesis, but our SPSS output gives a two-tailed **significance** result, we can take the **significance** level in the output and divide it by two. Thus, if our difference is in the right direction, and if our output indicates a **significance** level of .084 (two-tailed), but we have a one-tailed hypothesis, we can report a **significance** level of .042 (one-tailed). Note that in newer versions of SPSS many procedures now indicate both one and two-tailed **significance** levels (as shown here One-Sided *p* and Two-Sided *p*).

Significance	
One-Sided p	Two-Sided p
.237	.474

Phrasing Results

Results of hypothesis testing can be stated in different ways, depending on the conventions specified by your institution. The following examples illustrate some of these differences.

Degrees of Freedom

Sometimes the degrees of freedom are given in parentheses immediately after the symbol representing the test, as in this example:

$$t(3) = 7.00, p < .01$$

Other times, the degrees of freedom are given within the statement of results, as in this example:

$$t = 7.00, df = 3, p < .01$$

Significance Level

When you obtain results that are significant, they can be described in different ways. For instance, if you obtained a **significance** level of .006 on a *t* test, you could describe it in any of the following three ways:

$$t(3) = 7.00, p < .05$$

$$t(3) = 7.00, p < .01$$

$$t(3) = 7.00, p = .006$$

Note that because the exact probability is .006, both .05 and .01 are also correct. There are also various ways of describing results that are not significant. For instance, if you obtained a **significance** level of .505, any of the following three statements could be used:

$$t(2) = 0.805, ns$$

$$t(2) = 0.805, p > .05$$

$$t(2) = 0.805, p = .505$$

Statement of Results

Sometimes the results will be stated in terms of the **null hypothesis**, as in the following example:

> The null hypothesis was rejected ($t = 7.00$, $df = 3$, $p = .006$).

Other times, the results are stated in terms of their level of **significance**, as in the following example:

> A statistically significant difference was found: $t\ (3) = 7.00$, $p < .01$.

Statistical Symbols

Generally, statistical symbols are presented in *italics.* Prior to the widespread use of computers and desktop publishing, statistical symbols were underlined. Underlining is a signal to a printer that the underlined text should be set in italics. Institutions vary on their requirements for student work, so you are advised to consult your instructor about this.

Section 6.2 Single-Sample *t* Test

Description

The single-sample *t* test compares the **mean** of a single sample to a known population mean. It is useful for determining if the current set of data has changed from a long-term value (e.g., comparing the current year's temperatures to a historical average to determine if global warming is occurring).

Assumptions

The distributions from which the scores are taken should be normally distributed. However, the t test is **robust** and can handle violations of the assumption of a **normal distribution**. The **dependent variable** must be measured on an **interval** or **ratio scale**.

SPSS Data Format

The SPSS data file for the single-sample t test requires a single variable in SPSS. That variable represents the set of scores in the sample that we will compare to the population mean.

Running the Command

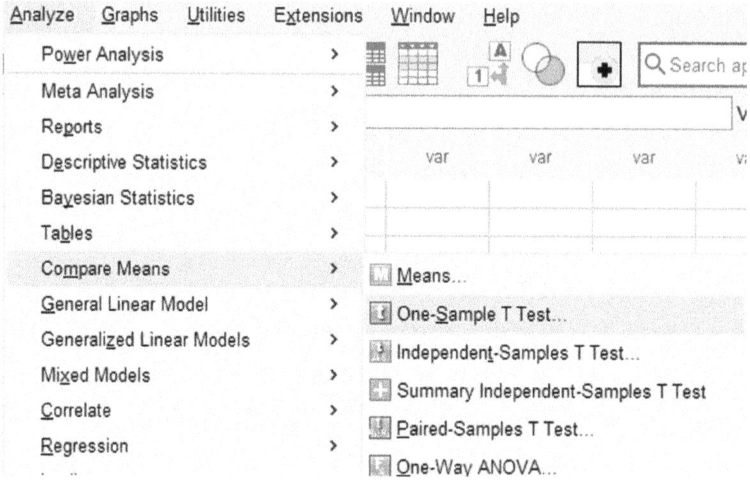

This example uses the HEIGHT.sav data file created in Chapter 4. The one sample (also called "single-sample") t test is located in the *Compare Means* submenu, under the *Analyze* menu.

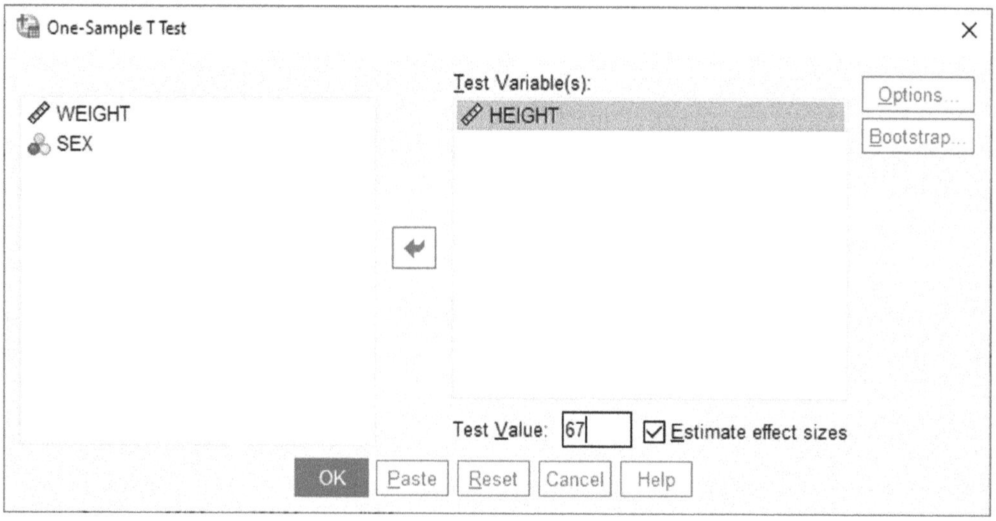

The **dialog box** for the single-sample t test requires that we transfer the variable representing the current set of scores to the *Test Variable(s)* section. We must also enter the population average in the *Test Value* blank. The example presented here is testing the variable HEIGHT against a population **mean** of 67. In other words, we are testing whether the sample came from a population with a **mean** of 67. Click *OK* to run the analysis.

Reading the Output

The output for the single-sample *t* test consists of two sections. The first section lists the sample variable and some basic **descriptive statistics** (*N*, **mean**, **standard deviation**, and standard error).

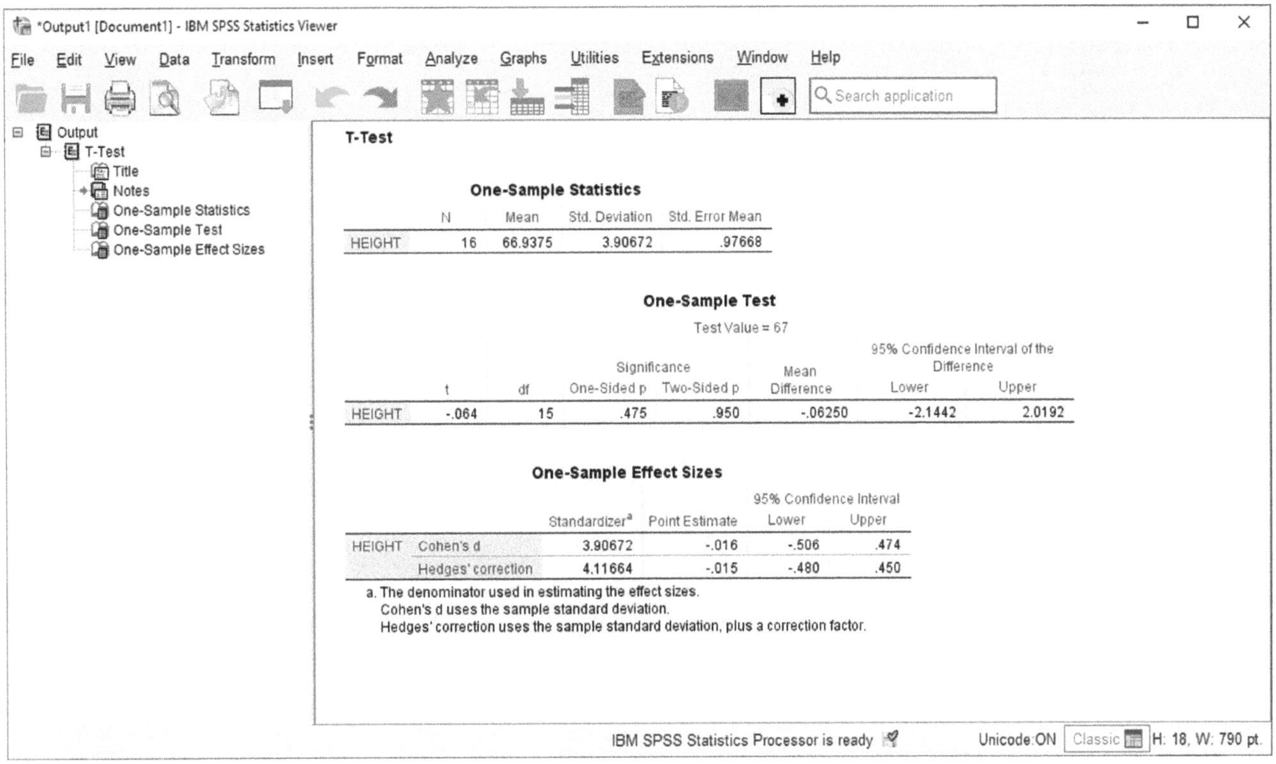

One-Sample Statistics

	N	Mean	Std. Deviation	Std. Error Mean
HEIGHT	16	66.9375	3.90672	.97668

One-Sample Test

Test Value = 67

			Significance			95% Confidence Interval of the Difference	
	t	df	One-Sided p	Two-Sided p	Mean Difference	Lower	Upper
HEIGHT	-.064	15	.475	.950	-.06250	-2.1442	2.0192

One-Sample Effect Sizes

				95% Confidence Interval	
		Standardizer[a]	Point Estimate	Lower	Upper
HEIGHT	Cohen's d	3.90672	-.016	-.506	.474
	Hedges' correction	4.11664	-.015	-.480	.450

a. The denominator used in estimating the effect sizes.
Cohen's d uses the sample standard deviation.
Hedges' correction uses the sample standard deviation, plus a correction factor.

The second section of output contains the results of the *t* test. The example presented above indicates a *t* value of –.064, with 15 degrees of freedom and a **significance** level of .95. The **mean** difference of –.0625 is the difference between the sample average (66.9375) and the population average we entered in the **dialog box** to conduct the test (67).

If you selected *Estimate effect sizes* in the main **dialog box**, then your output will also give effect sizes. Here, Cohen's *d* is equal to .016, and we use the standard cut-offs of .2 for small, .5 for moderate, and .8 for large. Our result of .016 is a small **effect size**.

Drawing Conclusions

The assumption when conducting a *t* test is that the means are equal. Therefore, a significant result indicates that the sample **mean** is not equivalent to the population **mean** (hence the term "significantly different"). A result that is not significant means that there is not a significant difference between the means. It does not mean that they are equal. Refer to your statistics text for the section on failure to reject the **null hypothesis**.

Phrasing Results That Are Significant

The above example did not find a significant difference between the population **mean** and the sample **mean**. If, however, we had used a population **mean** of 64 instead of 67, we would have obtained the following output:

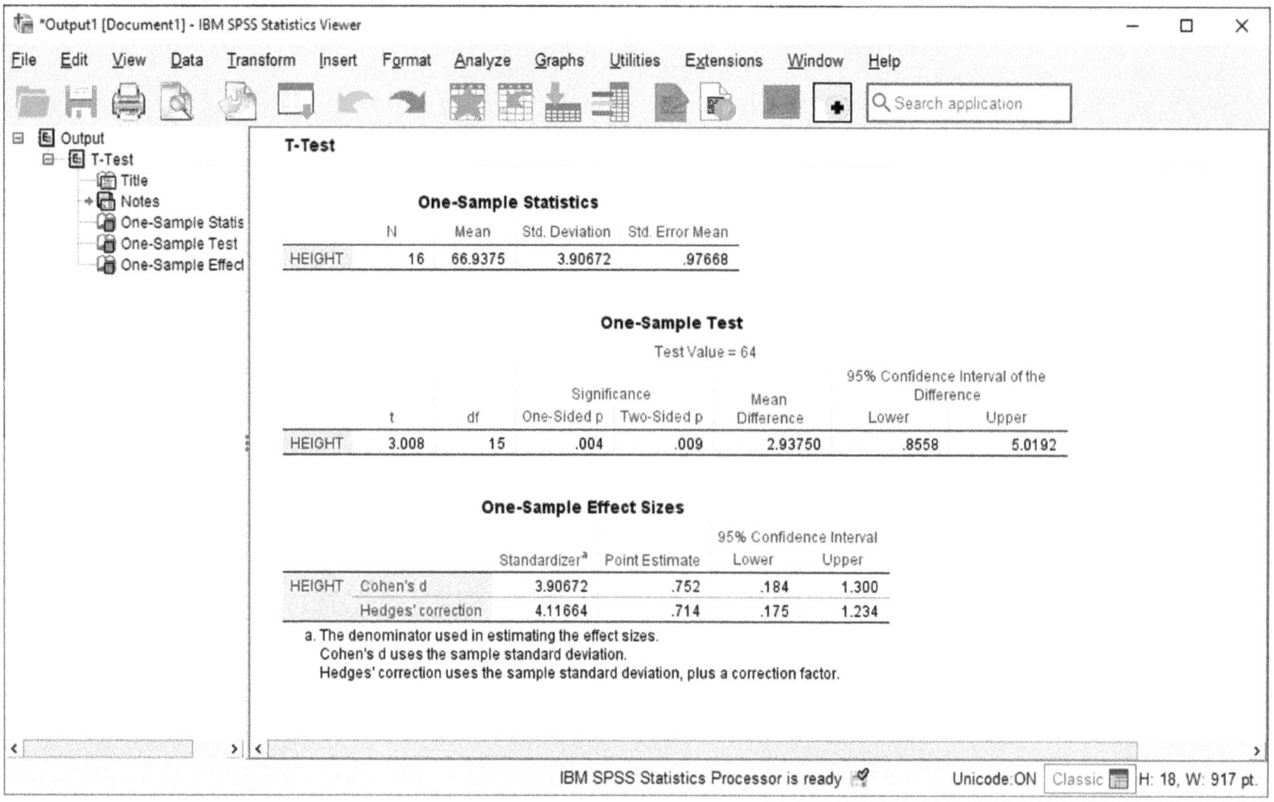

T-Test

One-Sample Statistics

	N	Mean	Std. Deviation	Std. Error Mean
HEIGHT	16	66.9375	3.90672	.97668

One-Sample Test

	Test Value = 64						
			Significance			95% Confidence Interval of the Difference	
	t	df	One-Sided p	Two-Sided p	Mean Difference	Lower	Upper
HEIGHT	3.008	15	.004	.009	2.93750	.8558	5.0192

One-Sample Effect Sizes

				95% Confidence Interval	
		Standardizer[a]	Point Estimate	Lower	Upper
HEIGHT	Cohen's d	3.90672	.752	.184	1.300
	Hedges' correction	4.11664	.714	.175	1.234

a. The denominator used in estimating the effect sizes.
Cohen's d uses the sample standard deviation.
Hedges' correction uses the sample standard deviation, plus a correction factor.

In this case, we could state the following:

> A single-sample *t* test that compared the mean height of the sample to a population value of 64 was conducted. A significant difference was found (t (15) = 3.008, p = .009). The sample mean of 66.9375 (sd = 3.907) was significantly greater than the population mean. The effect size was moderate (d=.752).

Phrasing Results That Are Not Significant

In our first example, the **significance** level was greater than .05. Thus, we would state the following:

> A single-sample *t* test that compared the mean height of the sample to a population value of 67 was conducted. No significant difference was found (t (15) = –.064, p = .95). The sample mean of 66.9375 (sd = 3.907) was not significantly greater than the population mean. The effect size was small (d=.016).

Practice Exercise

The **mean** salary in the United States is a hypothetical average of $25,000. Determine if the **mean** salary of the participants in Practice Dataset 2 in Appendix B is significantly greater than this value. Note that this is a one-tailed hypothesis.

Section 6.3 Independent-Samples *t* Test

Description

The independent-samples *t* test compares the means of two independent samples. Samples are independent if there is no relationship between them (generally meaning each participant provides data for only one sample).

Assumptions

The two groups being compared should be independent of each other. Observations are independent when information about one is unrelated to the other. Normally, this means that one group of participants provides data for one sample and a different group of participants provides data for the other sample (and individuals in one group are not matched with individuals in the other group). One way to accomplish this is through using **random assignment** to form two groups.

The scores should be normally distributed, but the *t* test is **robust** and can handle violations of the assumption of a **normal distribution**. The two samples should, however, have the same variance.

The **dependent variable** must be measured on an **interval** or **ratio scale**. The **independent variable** should have only two **discrete** levels.

SPSS Data Format

The SPSS data file for the independent *t* test requires two variables. One variable, the **grouping variable**, represents the value of the **independent variable**. The **grouping variable** should have two distinct values (e.g., 0 for a control group and 1 for an experimental group). The second variable represents the **dependent variable**, such as scores on a test.

Conducting an Independent-Samples *t* Test

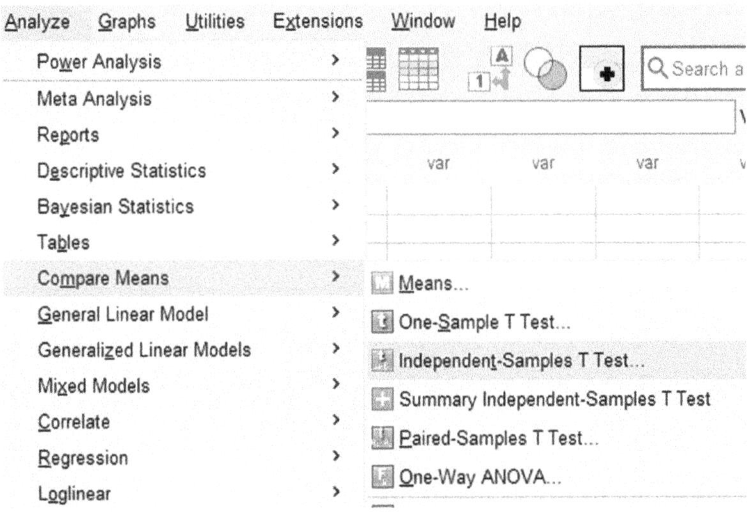

For our example, we will use the SAMPLE.sav data file. Click *Analyze*, then *Compare Means*, then *Independent-Samples T Test*. This will bring up the main **dialog box**. Transfer the **dependent variable(s)** into the *Test Variable(s)* blank. For our example, we will use the variable GRADE.

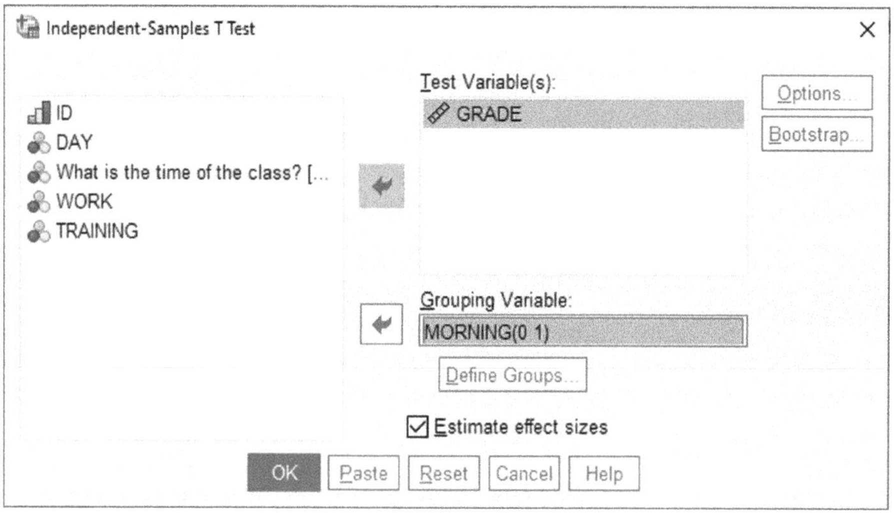

Transfer the **independent variable** into the *Grouping Variable* section. For our example, we will use the variable MORNING.

Next, click *Define Groups* and enter the values of the two **levels** of the **independent variable** (0 for nonmorning and 1 for morning). Independent *t* tests are capable of comparing only two **levels** at a time. Click *Continue*, then click *OK* to run the analysis.

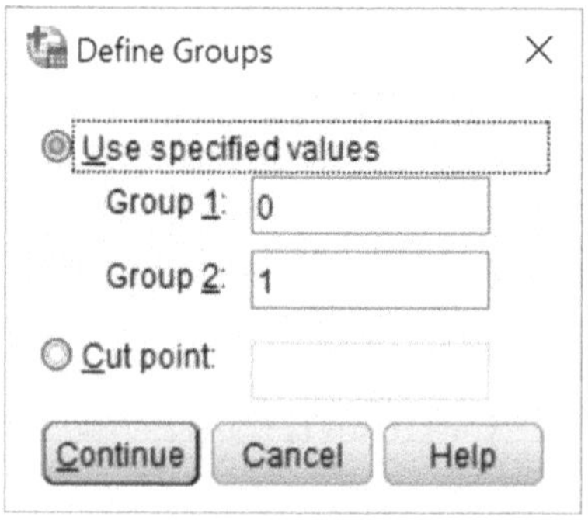

Output from the Independent-Samples *t* Test

The output will have a section labeled "Group Statistics." This section provides the basic **descriptive statistics** for the **dependent variable(s)** for each value of the **independent variable**. Next, there will be a section with the results of the *t* test. The output should look like the image below.

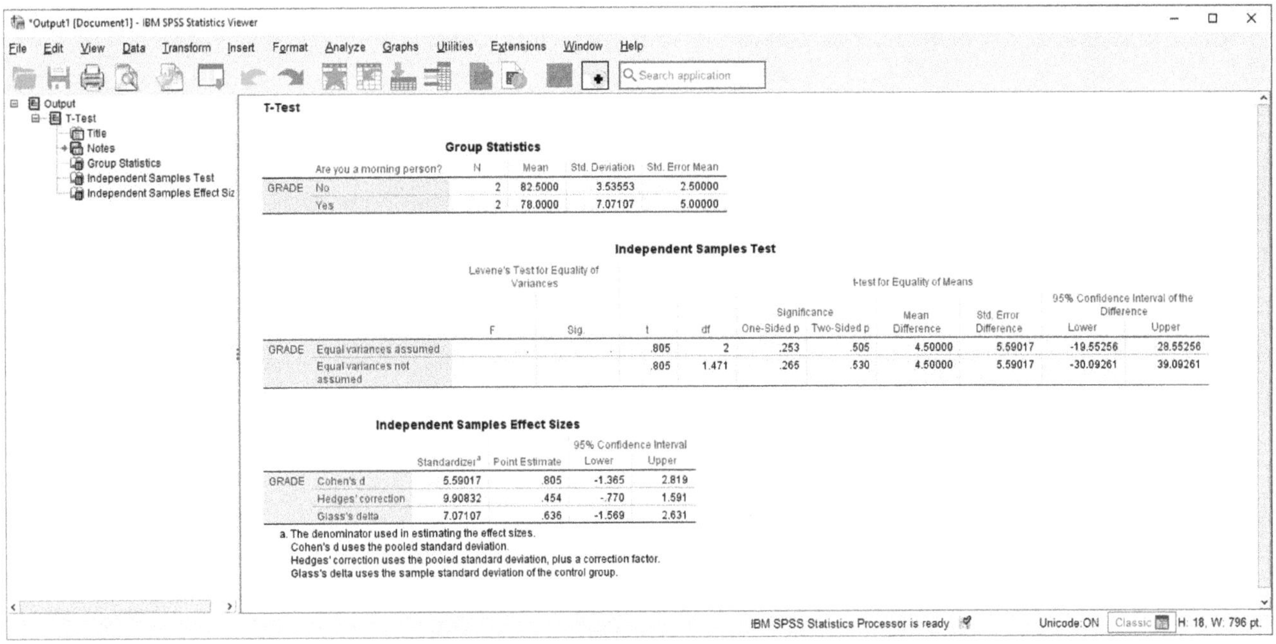

Group Statistics

	Are you a morning person?	N	Mean	Std. Deviation	Std. Error Mean
GRADE	No	2	82.5000	3.53553	2.50000
	Yes	2	78.0000	7.07107	5.00000

Independent Samples Test

		Levene's Test for Equality of Variances		t-test for Equality of Means							
						Significance				95% Confidence Interval of the Difference	
		F	Sig.	t	df	One-Sided p	Two-Sided p	Mean Difference	Std. Error Difference	Lower	Upper
GRADE	Equal variances assumed	.	.	.805	2	.253	.505	4.50000	5.59017	-19.55256	28.55256
	Equal variances not assumed			.805	1.471	.265	.530	4.50000	5.59017	-30.09261	39.09261

Independent Samples Effect Sizes

		Standardizer[a]	Point Estimate	95% Confidence Interval Lower	95% Confidence Interval Upper
GRADE	Cohen's d	5.59017	.805	-1.365	2.819
	Hedges' correction	9.90832	.454	-.770	1.591
	Glass's delta	7.07107	.636	-1.569	2.631

a. The denominator used in estimating the effect sizes.
Cohen's d uses the pooled standard deviation.
Hedges' correction uses the pooled standard deviation, plus a correction factor.
Glass's delta uses the sample standard deviation of the control group.

The columns labeled *t*, *df*, and *Sig. (2-tailed)* provide the standard answer for the *t* test. They provide the value of *t*, the degrees of freedom (number of participants minus 2, in this case), and the **significance** level (often called *p*). Normally, we use the "Equal variances assumed" row. This will provide you with an answer equivalent to what you would calculate by hand.

If you selected *Estimate effect sizes* in the main dialog box, the output will also show effect sizes. As discussed in Appendix A, Cohen's *d* is the most appropriate effect size for most problems, and we use .2, .5, and .8 as cut-offs for small, moderate, and large effect sizes. Here we have an effect size of .805 indicating a large effect size.

Drawing Conclusions

Recall from the previous section that the *t* test assumes an equality of **means**. Therefore, a significant result indicates that the **means** are not equivalent. When drawing conclusions about a *t* test, you must state the direction of the difference (i.e., which **mean** was larger than the other). You should also include information about the value of *t*, the degrees of freedom, the **significance** level, the **means**, and **standard deviations** for the two groups.

Phrasing Results That Are Significant

For a significant *t* test (e.g., the output below), you might state the following:

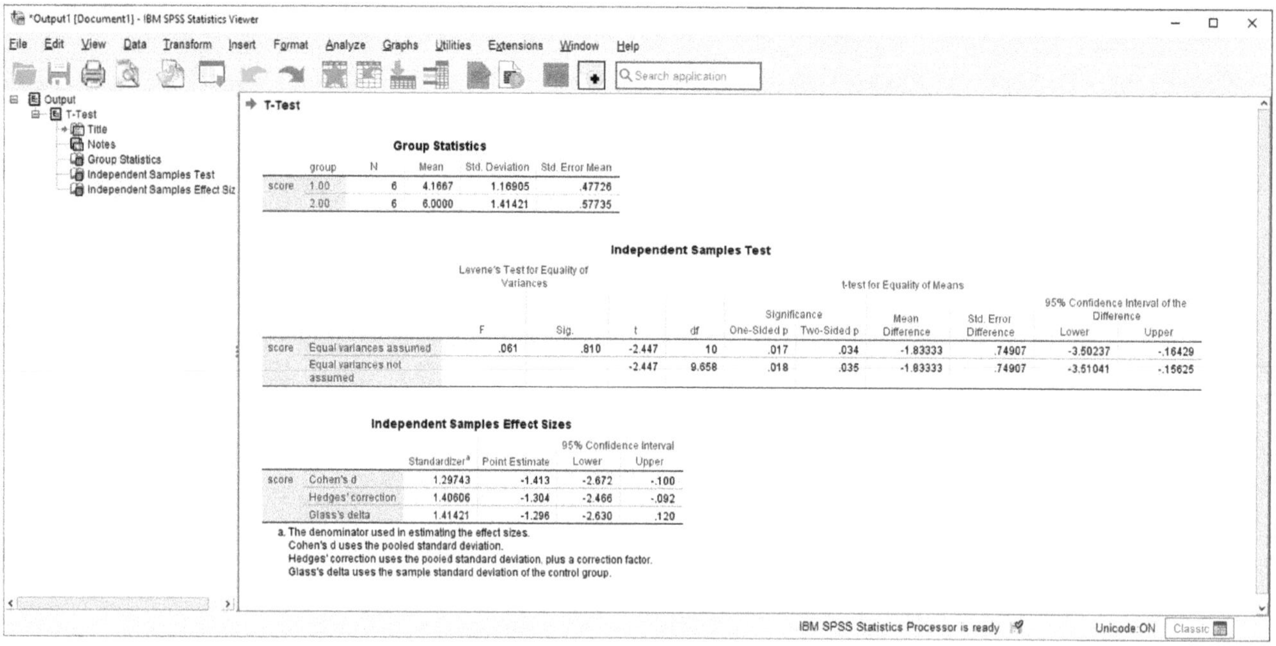

Group Statistics

	group	N	Mean	Std. Deviation	Std. Error Mean
score	1.00	6	4.1667	1.16905	.47726
	2.00	6	6.0000	1.41421	.57735

Independent Samples Test

		Levene's Test for Equality of Variances		t-test for Equality of Means							
						Significance				95% Confidence Interval of the Difference	
		F	Sig.	t	df	One-Sided p	Two-Sided p	Mean Difference	Std. Error Difference	Lower	Upper
score	Equal variances assumed	.061	.810	-2.447	10	.017	.034	-1.83333	.74907	-3.50237	-.16429
	Equal variances not assumed			-2.447	9.658	.018	.035	-1.83333	.74907	-3.51041	-.15625

Independent Samples Effect Sizes

				95% Confidence Interval	
		Standardizer[a]	Point Estimate	Lower	Upper
score	Cohen's d	1.29743	-1.413	-2.672	-.100
	Hedges' correction	1.40606	-1.304	-2.466	-.092
	Glass's delta	1.41421	-1.296	-2.630	.120

a. The denominator used in estimating the effect sizes.
Cohen's d uses the pooled standard deviation.
Hedges' correction uses the pooled standard deviation, plus a correction factor.
Glass's delta uses the sample standard deviation of the control group.

> An independent-samples *t* test comparing the mean scores of the experimental and control groups found a significant difference between the means of the two groups (t (5) = 2.447, $p < .05$). The mean of the experimental group was significantly lower ($M = 4.17$, $sd = 1.17$) than the mean of the control group ($M = 6.00$, $sd = 1.41$). The effect size was large (d=1.413).

Phrasing Results That Are Not Significant

In our example at the start of this section, we compared the scores of the morning people to the scores of the nonmorning people. We did not find a significant difference, so we could state the following:

> An independent-samples *t* test was calculated comparing the mean score of participants who identified themselves as morning people to the mean score of participants who did not identify themselves as morning people. No significant difference was found (t (2) = 0.805, $p > .05$). The mean of the morning people ($M = 78.00$, $sd = 7.07$) was not significantly different from the mean of nonmorning people ($M = 82.50$, $sd = 3.54$). The effect size was large (d=.805).

Practice Exercise

Use Practice Dataset 1 in Appendix B to solve this problem. We believe that young individuals have lower mathematics skills than older individuals. We would test this hypothesis by comparing participants 25 or younger (the "young" group) with participants 26 or older (the "old" group). Hint: You may need to create a new variable that represents each age group. See Chapter 2 for help.

Section 6.4 Paired-Samples *t* Test

Description

The paired-samples *t* test (also called a dependent *t* test) compares the **means** of two scores from related samples. For instance, comparing a pretest and a posttest score for a group of participants would require a paired-samples *t* test.

Assumptions

The paired-samples *t* test assumes that both variables are at the **interval** or **ratio** levels and are normally distributed. The two variables should also be measured with the same scale.

SPSS Data Format

Two variables in the SPSS data file are required. These variables should represent two measurements from each participant.

Running the Command

We will create a new data file containing five variables: PRETEST, MIDTERM, FINAL, INSTRUCT, and REQUIRED. INSTRUCT represents three different instructors for a course. REQUIRED represents whether the course was required or was an elective (0 = elective, 1 = required). The other three variables represent exam scores (100 being the highest score possible).

PRETEST	MIDTERM	FINAL	INSTRUCT	REQUIRED
56	64	69	1	0
79	91	86	1	0
68	77	81	1	0
59	69	71	1	1
64	77	75	1	1
74	88	86	1	1
73	85	86	1	1
47	64	69	2	0
78	98	100	2	0
61	77	85	2	0
68	86	93	2	1
64	77	87	2	1
53	67	76	2	1
71	85	95	2	1
61	79	97	3	0
57	77	89	3	0
49	65	83	3	0
71	93	100	3	1
61	83	94	3	1
58	75	92	3	1
58	74	92	3	1

Enter the data and save them as GRADES.sav. You can check your data entry by computing a **mean** for each instructor using the *Means* command. (See Chapter 3 for more information.) Use INSTRUCT as the **independent variable** and enter PRETEST, MIDTERM, and FINAL as your **dependent variables**.

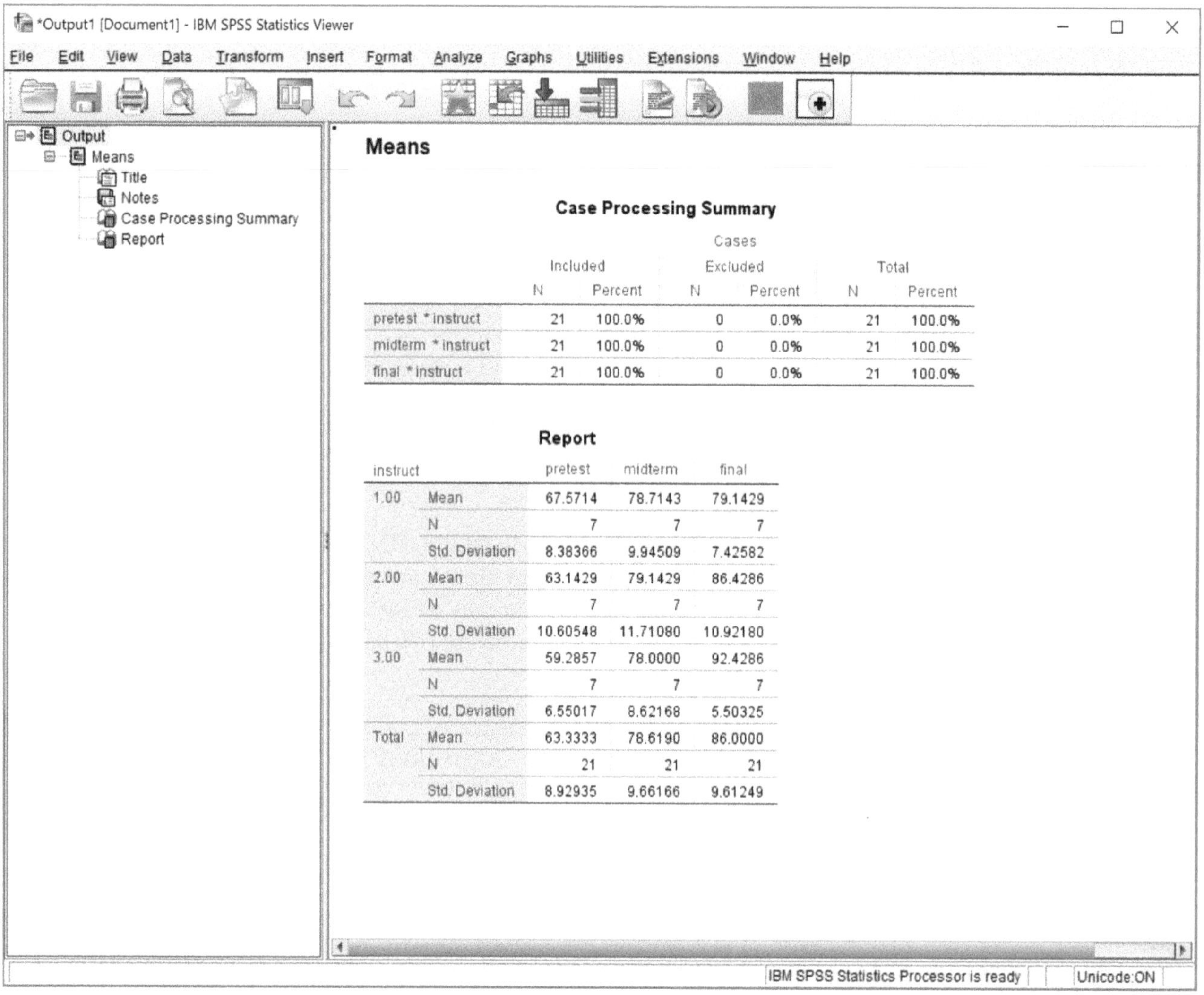

Case Processing Summary

	Cases					
	Included		Excluded		Total	
	N	Percent	N	Percent	N	Percent
pretest * instruct	21	100.0%	0	0.0%	21	100.0%
midterm * instruct	21	100.0%	0	0.0%	21	100.0%
final * instruct	21	100.0%	0	0.0%	21	100.0%

Report

instruct		pretest	midterm	final
1.00	Mean	67.5714	78.7143	79.1429
	N	7	7	7
	Std. Deviation	8.38366	9.94509	7.42582
2.00	Mean	63.1429	79.1429	86.4286
	N	7	7	7
	Std. Deviation	10.60548	11.71080	10.92180
3.00	Mean	59.2857	78.0000	92.4286
	N	7	7	7
	Std. Deviation	6.55017	8.62168	5.50325
Total	Mean	63.3333	78.6190	86.0000
	N	21	21	21
	Std. Deviation	8.92935	9.66166	9.61249

Once you have entered the data, conduct a paired-samples *t* test comparing pretest scores and final scores.

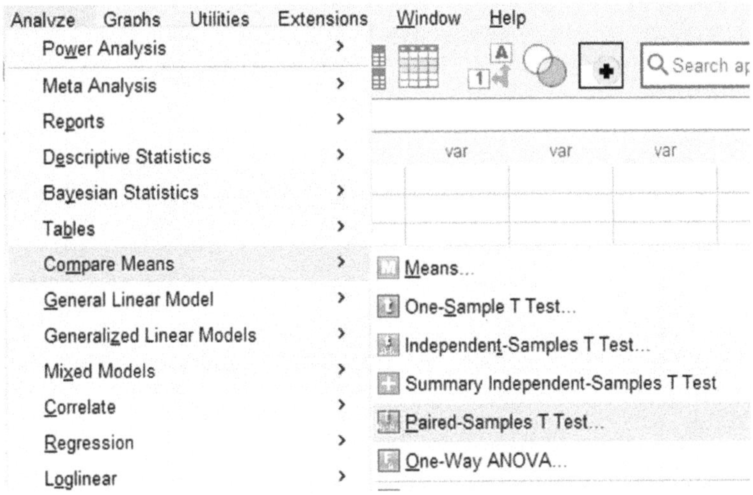

Click *Analyze*, then *Compare Means*, then *Paired-Samples T Test*. This will bring up the main **dialog box**.

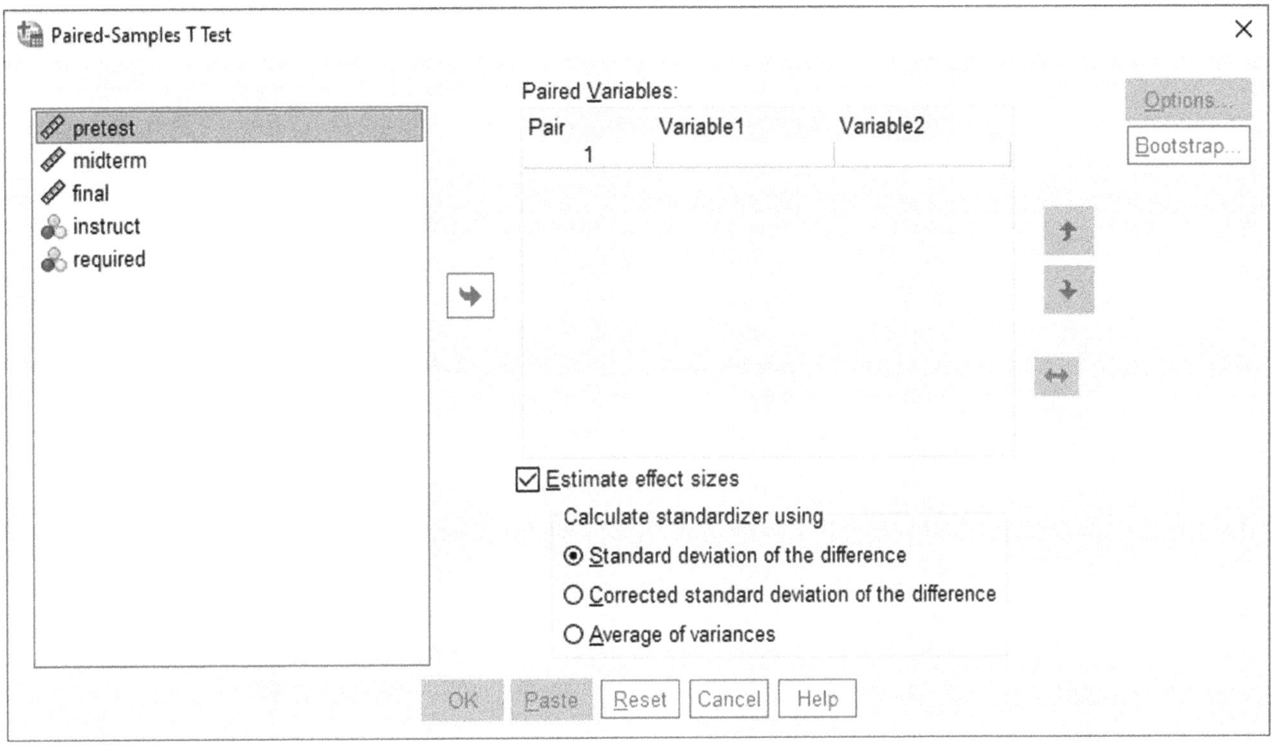

You must select pairs of variables to compare. Click on PRETEST, then click the button. Click on FINAL, then click the button. (Alternatively, you can double-click or control-click them.) The variables will be moved into the *Paired Variables* area. Click *OK* to conduct the test.

Reading the Output

The output for the paired-samples *t* test consists of three components. The first part gives you basic **descriptive statistics** for the pair of variables. The PRETEST average was 63.3, with a **standard deviation** of 8.93. The FINAL average was 86.14, with a **standard deviation** of 9.63.

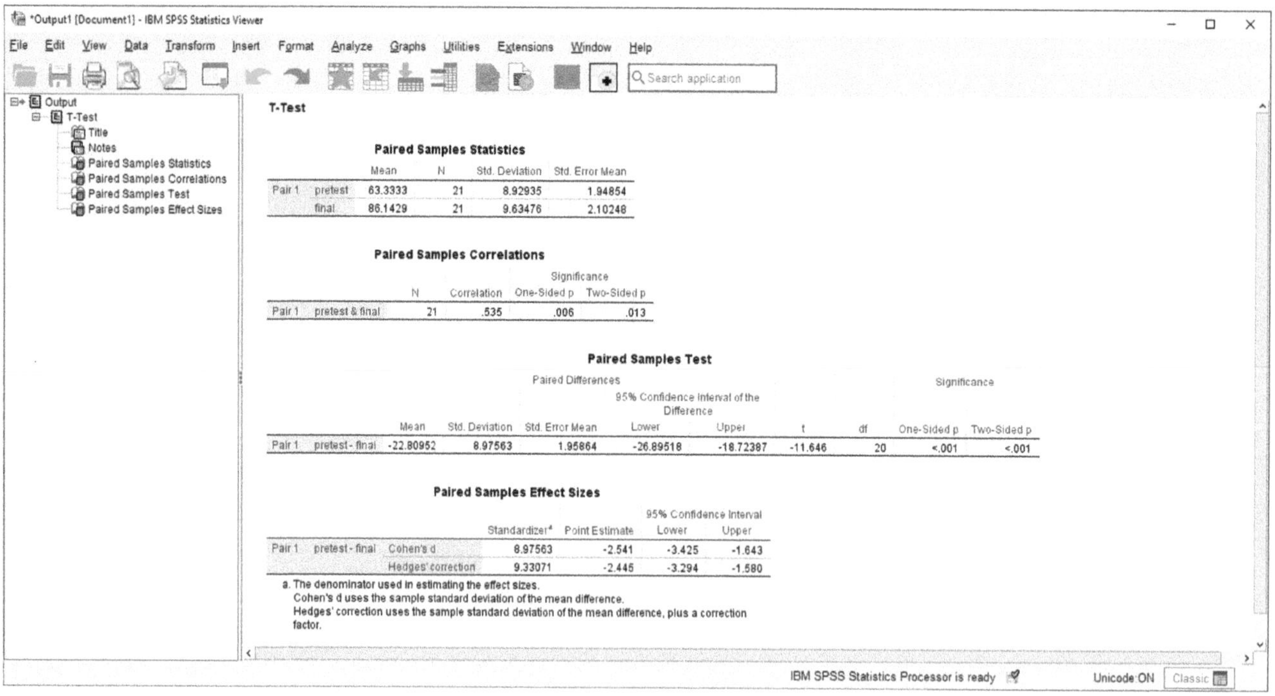

T-Test

Paired Samples Statistics

		Mean	N	Std. Deviation	Std. Error Mean
Pair 1	pretest	63.3333	21	8.92935	1.94854
	final	86.1429	21	9.63476	2.10248

Paired Samples Correlations

		N	Correlation	Significance One-Sided p	Significance Two-Sided p
Pair 1	pretest & final	21	.535	.006	.013

Paired Samples Test

		Paired Differences Mean	Std. Deviation	Std. Error Mean	95% Confidence Interval of the Difference Lower	Upper	t	df	Significance One-Sided p	Two-Sided p
Pair 1	pretest - final	-22.80952	8.97563	1.95864	-26.89518	-18.72387	-11.646	20	<.001	<.001

Paired Samples Effect Sizes

			Standardizer[a]	Point Estimate	95% Confidence Interval Lower	Upper
Pair 1	pretest - final	Cohen's d	8.97563	-2.541	-3.425	-1.643
		Hedges' correction	9.33071	-2.445	-3.294	-1.580

a. The denominator used in estimating the effect sizes.
Cohen's d uses the sample standard deviation of the mean difference.
Hedges' correction uses the sample standard deviation of the mean difference, plus a correction factor.

The second part of the output is a Pearson correlation coefficient for the pair of variables. Within the third part of the output (labeled *Paired Samples Test*), the section called *Paired Differences* contains information about the differences between the two variables. You may have learned in your statistics class that the paired-samples t test is essentially a single-sample t test calculated on the differences between the scores. The final three columns contain the value of t, the degrees of freedom, and the **significance** level. In the example presented here, we obtained a t of –11.646, with 20 degrees of freedom, and a **significance** level of <.001.

If you selected *Estimate effect sizes*, then **Effect Sizes** will be shown at the bottom of the output. Here, Cohen's d is equal to 2.541, a large effect size given our standard cut-offs of .2 for small, .5 for moderate, and .8 for large.

Drawing Conclusions

Paired-samples t tests determine whether or not two means are significantly different from each other. Significant values indicate that the two means are different (technically, that the samples come from populations with different means). Values that are not significant indicate that the scores are not significantly different.

Phrasing Results That Are Significant

When stating the results of a paired-samples t test, you should give the value of t, the degrees of freedom, and the **significance** level. You should also give the **mean** and **standard deviation** for each variable, as well as a statement of results that indicates whether you conducted a one- or two-tailed test. Our example above was significant, so we could state the following:

> A paired-samples t test was calculated to compare the mean pretest score to the mean final exam score. The mean on the pretest was 63.33 (sd = 8.93), and the mean on the final was 86.14 (sd = 9.63). A significant increase from pretest to final was found (t (20) = –11.646, p < .001). There was a large effect size (d=2.541).

Phrasing Results That Are Not Significant

If the **significance** level had been greater than .05 (or greater than .10, if you were conducting a one-tailed test), the result would not have been significant. If you select only the students who have instructor #1 and conduct a test to compare midterm and final scores, you will get the output here.

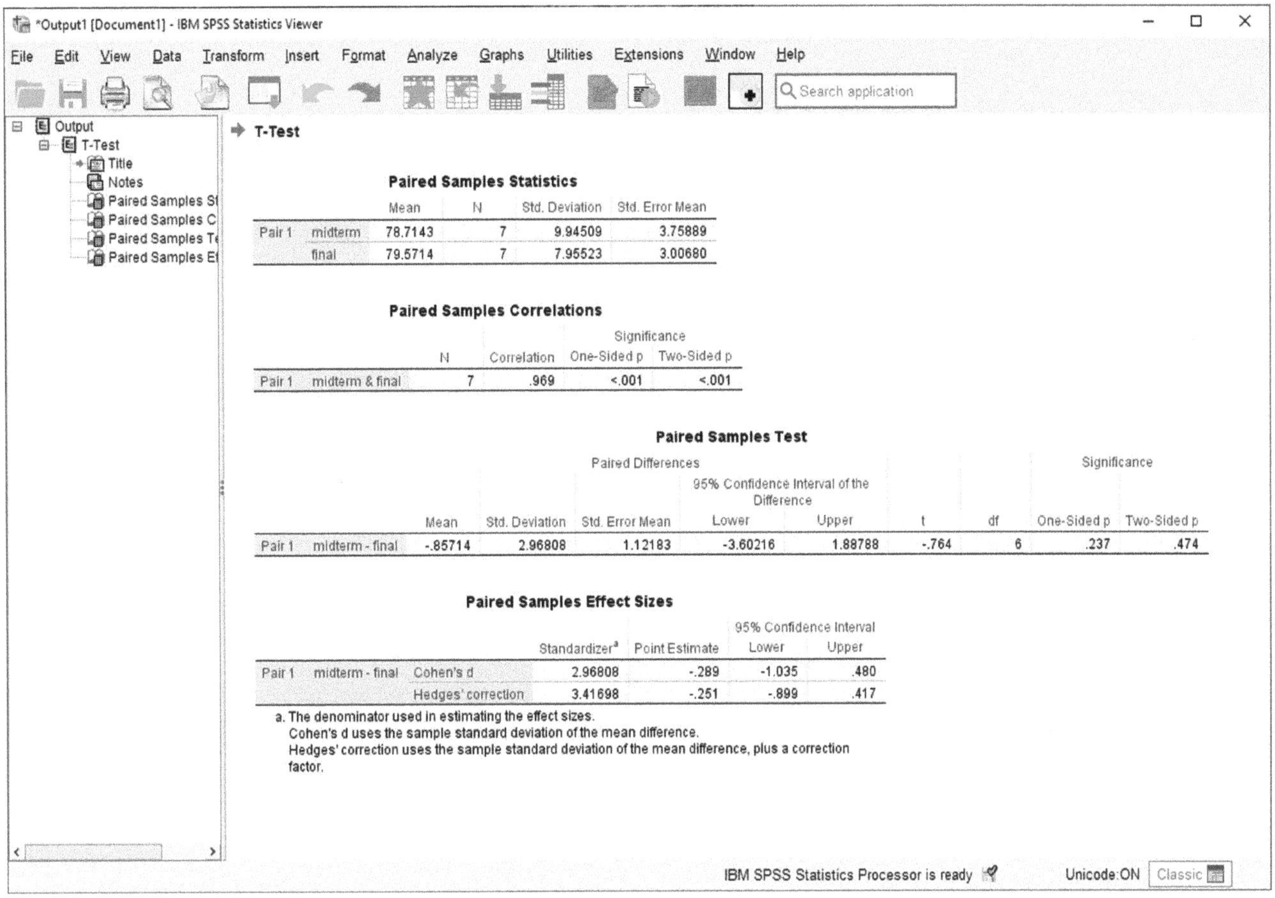

T-Test

Paired Samples Statistics

		Mean	N	Std. Deviation	Std. Error Mean
Pair 1	midterm	78.7143	7	9.94509	3.75889
	final	79.5714	7	7.95523	3.00680

Paired Samples Correlations

		N	Correlation	Significance One-Sided p	Significance Two-Sided p
Pair 1	midterm & final	7	.969	<.001	<.001

Paired Samples Test

		Paired Differences							Significance	
					95% Confidence Interval of the Difference					
		Mean	Std. Deviation	Std. Error Mean	Lower	Upper	t	df	One-Sided p	Two-Sided p
Pair 1	midterm - final	-.85714	2.96808	1.12183	-3.60216	1.88788	-.764	6	.237	.474

Paired Samples Effect Sizes

					95% Confidence Interval	
			Standardizer[a]	Point Estimate	Lower	Upper
Pair 1	midterm - final	Cohen's d	2.96808	-.289	-1.035	.480
		Hedges' correction	3.41698	-.251	-.899	.417

a. The denominator used in estimating the effect sizes.
Cohen's d uses the sample standard deviation of the mean difference.
Hedges' correction uses the sample standard deviation of the mean difference, plus a correction factor.

For this analysis, we could state the following:

> A paired-samples *t* test was calculated to compare midterm and final exam scores for students who had instructor one. The mean on the midterm was 78.71 (*sd* = 9.95), and the mean on the final was 79.57 (*sd* = 7.96). No significant difference from pretest to final was found (*t* (6) = −.764, *p* > .05). There was a moderate effect size (*d*=.289).

Practice Exercise

Use the same GRADES.sav data file and compute a paired-samples *t* test to determine if scores increased from midterm to final.

Chapter 7
ANOVA Models

Section 7.1 ANOVA Models

Analysis of variance (ANOVA) is a procedure that determines the proportion of variability attributed to each of several components. It is one of the most useful and adaptable statistical techniques available.

In general ANOVA procedures in SPSS produce what is called a **Source Table** in which each row represents a source of variability and the columns represent what we know about the variability attributable to each source. In the simplest case total variability is split into that attributable to differences between the groups and the variability attributable to differences within the groups.

Normally there is a column called SS (Sum of Squares), one for df (Degrees of Freedom), one for MS (Mean Squares), and then F—which is the actual answer.

Section 7.2 One-Way ANOVA

Description

The one-way ANOVA compares the means of two or more groups of participants that vary on a single **independent variable** (thus, the one-way designation). When we have three groups, we could use a *t* test to determine differences between the groups, but we would have to conduct three *t* tests (Group 1 compared to Group 2, Group 1 compared to Group 3, and Group 2 compared to Group 3). When we conduct multiple *t* tests, we inflate the **Type I error** rate and increase our chance of drawing an inappropriate conclusion. ANOVA compensates for these multiple comparisons and gives us a single answer that tells us if any of the groups is different from any of the other groups.

Assumptions

The one-way ANOVA requires a single **dependent variable** and a single **independent variable**. Which group participants belong to is determined by the value of the **independent variable**. Groups should be independent of each other. If our participants belong to more than one group each, we will have to conduct a repeated-measures ANOVA. If we have more than one **independent variable**, we should conduct a factorial ANOVA.

ANOVA also assumes that the **dependent variable** is at the **interval** or **ratio** level and is normally distributed and that the variances of the **dependent variable** for each level of the **independent variable** are equal.

SPSS Data Format

Two variables are required in the SPSS data file. One variable serves as the **dependent variable** and the other as the **independent variable**. Each participant should provide only one score for the **dependent variable**.

DOI: 10.4324/9781003450467-7

Running the Command

For this example, we will use the GRADES.sav data file we created in the previous section.

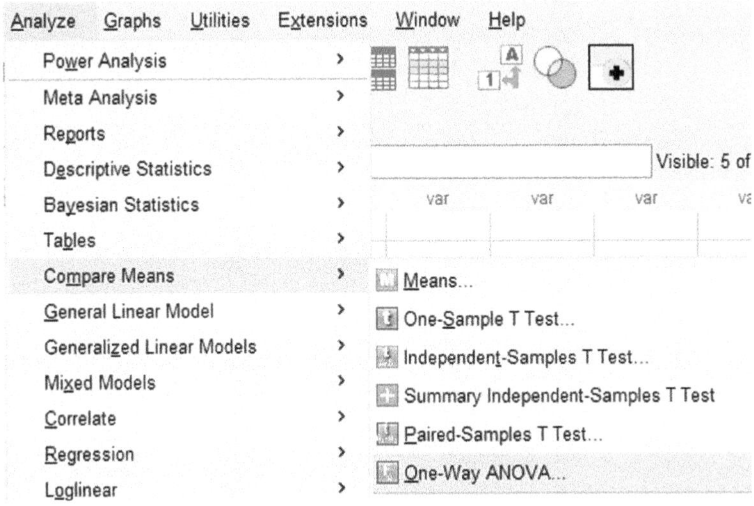

For you to conduct a one-way ANOVA, click *Analyze*, then *Compare Means*, then *One-Way ANOVA*. This will bring up the main **dialog box** for the *One-Way ANOVA* command.

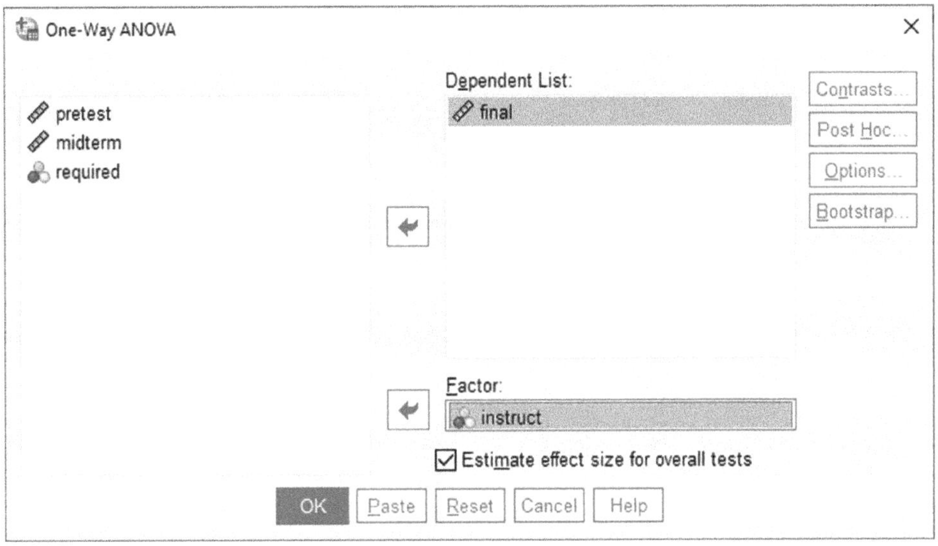

You should place the **independent variable** in the *Factor* box. For our example, INSTRUCT represents three different instructors, and it will be used as our **independent variable**. Our **dependent variable** will be FINAL. This test will allow us to determine if the instructor has any effect on final grades in the course.

Click on *Options* to get the Options **dialog box** shown above. Click *Descriptive*. This will give you means for the **dependent variable** at each **level** of the **independent variable**. Checking this box prevents us from having to run a separate *Means* command. Click *Continue* to return to the main **dialog box**.

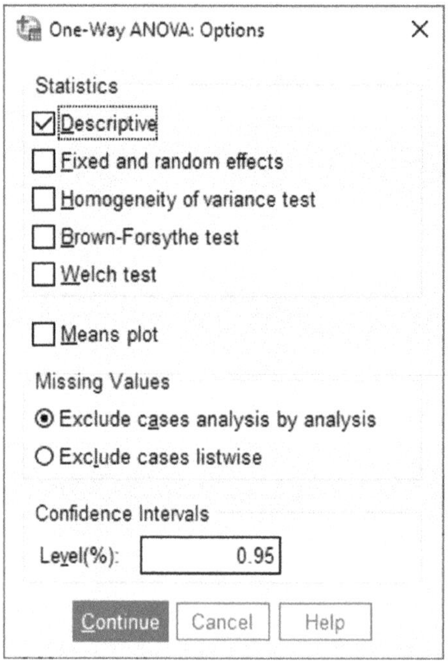

Next, click *Post Hoc* to bring up the Post-hoc Multiple Comparisons **dialog box** shown above. Click *Tukey*, then *Continue*.

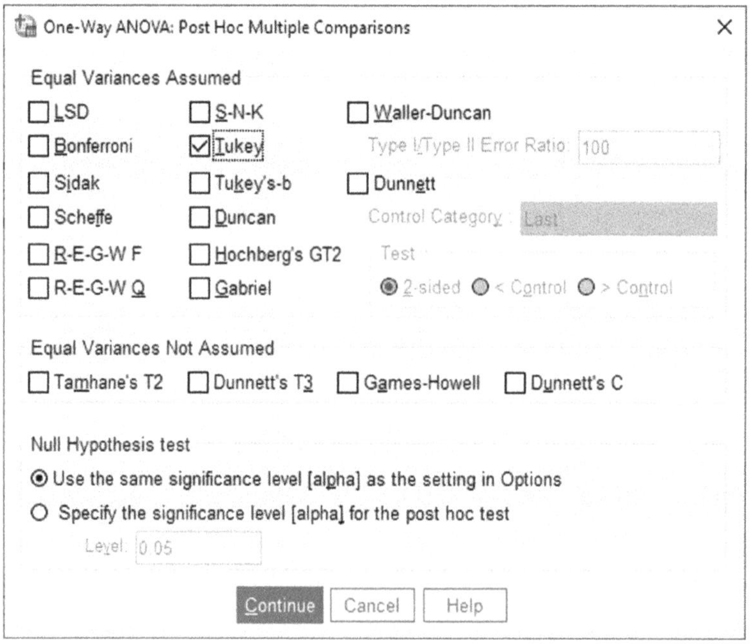

Post-hoc tests are necessary in the event of a significant ANOVA. The ANOVA only indicates if any group is different from any other group. If it is significant, we need to determine which groups are different from which other groups. We could do *t* tests to determine that, but we would have the same problem as mentioned before with inflating the **Type I error** rate.

There are several post-hoc comparisons that correct for multiple comparisons. The most widely used is **Tukey's *HSD***. SPSS will calculate a variety of post-hoc tests for you. Consult an advanced statistics text for a discussion of the differences between these various tests.

Reading the Output

Descriptive statistics will be given for each instructor (i.e., level of the **independent variable**) and the total. For instance, in his or her class, Instructor 1 had an average final exam score of 79.57.

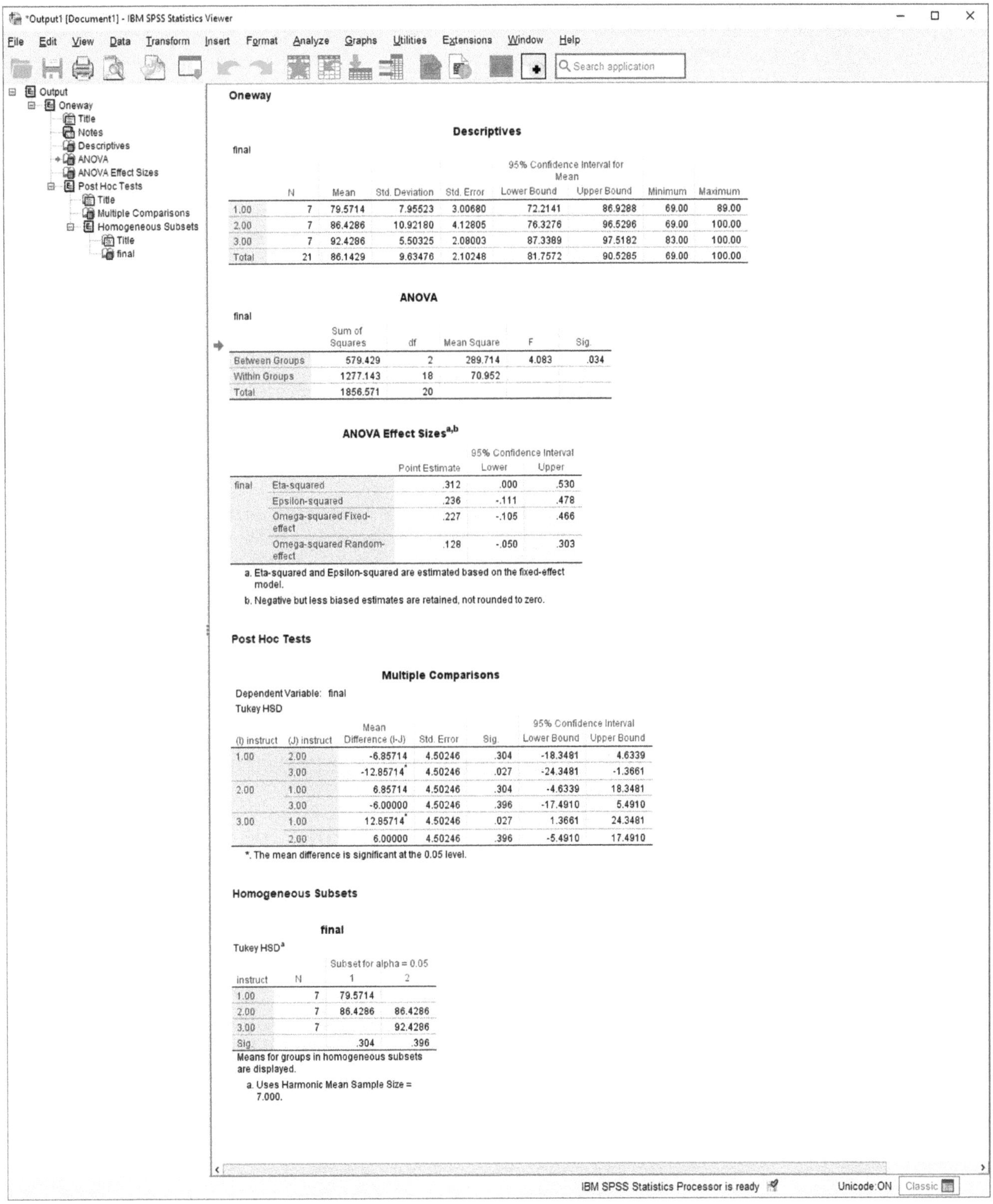

Oneway

Descriptives

final

	N	Mean	Std. Deviation	Std. Error	95% Confidence Interval for Mean Lower Bound	95% Confidence Interval for Mean Upper Bound	Minimum	Maximum
1.00	7	79.5714	7.95523	3.00680	72.2141	86.9288	69.00	89.00
2.00	7	86.4286	10.92180	4.12805	76.3276	96.5296	69.00	100.00
3.00	7	92.4286	5.50325	2.08003	87.3389	97.5182	83.00	100.00
Total	21	86.1429	9.63476	2.10248	81.7572	90.5285	69.00	100.00

ANOVA

final

	Sum of Squares	df	Mean Square	F	Sig.
Between Groups	579.429	2	289.714	4.083	.034
Within Groups	1277.143	18	70.952		
Total	1856.571	20			

ANOVA Effect Sizes[a,b]

		Point Estimate	95% Confidence Interval Lower	95% Confidence Interval Upper
final	Eta-squared	.312	.000	.530
	Epsilon-squared	.236	-.111	.478
	Omega-squared Fixed-effect	.227	-.105	.466
	Omega-squared Random-effect	.128	-.050	.303

a. Eta-squared and Epsilon-squared are estimated based on the fixed-effect model.

b. Negative but less biased estimates are retained, not rounded to zero.

Post Hoc Tests

Multiple Comparisons

Dependent Variable: final

Tukey HSD

(I) instruct	(J) instruct	Mean Difference (I-J)	Std. Error	Sig.	95% Confidence Interval Lower Bound	95% Confidence Interval Upper Bound
1.00	2.00	-6.85714	4.50246	.304	-18.3481	4.6339
	3.00	-12.85714*	4.50246	.027	-24.3481	-1.3661
2.00	1.00	6.85714	4.50246	.304	-4.6339	18.3481
	3.00	-6.00000	4.50246	.396	-17.4910	5.4910
3.00	1.00	12.85714*	4.50246	.027	1.3661	24.3481
	2.00	6.00000	4.50246	.396	-5.4910	17.4910

*. The mean difference is significant at the 0.05 level.

Homogeneous Subsets

final

Tukey HSD[a]

instruct	N	Subset for alpha = 0.05 1	2
1.00	7	79.5714	
2.00	7	86.4286	86.4286
3.00	7		92.4286
Sig.		.304	.396

Means for groups in homogeneous subsets are displayed.

a. Uses Harmonic Mean Sample Size = 7.000.

The next section of the output is the ANOVA **source table**. This is where the various components of the **variance** have been listed, along with their relative sizes. For a one-way ANOVA, there are two components to the variance: Between Groups (which represents the differences due to our **independent variable**) and Within Groups (which represents differences within each level of our **independent variable**). For our example, the Between Groups variance represents differences due to different instructors. The Within Groups variance represents individual differences among students.

The primary answer is *F*. *F* is a ratio of explained **variance** to unexplained **variance**. Consult a statistics text for more information on how it is determined. The *F* has two different degrees of freedom, one for Between Groups (in this case, 2 is the number of **levels** of our **independent variable** <3–1>) and another for Within Groups (18 is the number of participants minus the number of **levels** of our **independent variable**).

If you selected *Estimate effect size for overall tests*, then the next section of output will be the ANOVA Effect Sizes. As discussed in Appendix A, *Eta-Squared* is normally the most appropriate measure of **Effect Size**. In the example here, our effect size is .312, a large effect size given our cut-off of >.25 for a large Eta-Squared. (We generally use <.01 as a small effect size, >.25 as a large, and between them as a moderate effect size.)

The next part of the output consists of the results of our **Tukey's *HSD*** post-hoc comparison. This table presents us with every possible combination of **levels** of our **independent variable**. The first row represents Instructor 1 compared to Instructor 2. Next is Instructor 1 compared to Instructor 3. Next is Instructor 2 compared to Instructor 1. (Note that this is redundant with the first row.) Next is Instructor 2 compared to Instructor 3, and so on.

The column labeled *Sig.* represents the **Type I error** (*p*) rate for the simple (2-level) comparison in that row. In our example above, Instructor 1 is significantly different from Instructor 3, but Instructor 1 is not significantly different from Instructor 2, and Instructor 2 is not significantly different from Instructor 3.

Drawing Conclusions

Drawing conclusions for ANOVA requires that we indicate the value of *F*, the degrees of freedom, and the **significance** level. A significant ANOVA should be followed by the results of a post-hoc analysis and a verbal statement of the results.

Phrasing Results That Are Significant

In the preceding example, we could state the following:

> We computed a one-way ANOVA comparing the final exam scores of participants who took a course from one of three different instructors. A significant difference was found among the instructors (F (2,18) = 4.083, $p < .05$). Tukey's *HSD* was used to determine the nature of the differences between the instructors. This analysis revealed that students who had Instructor 1 scored lower ($M = 79.57$, $sd = 7.96$) than students who had Instructor 3 ($M = 92.43$, $sd = 5.50$). Students who had Instructor 2 ($M = 86.43$, $sd = 10.92$) were not significantly different from either of the other two groups. This difference was a large effect size ($\eta^2 = .312$)

Phrasing Results That Are Not Significant

If we had conducted the analysis using PRETEST as our **dependent variable** instead of FINAL, we would have received the following output:

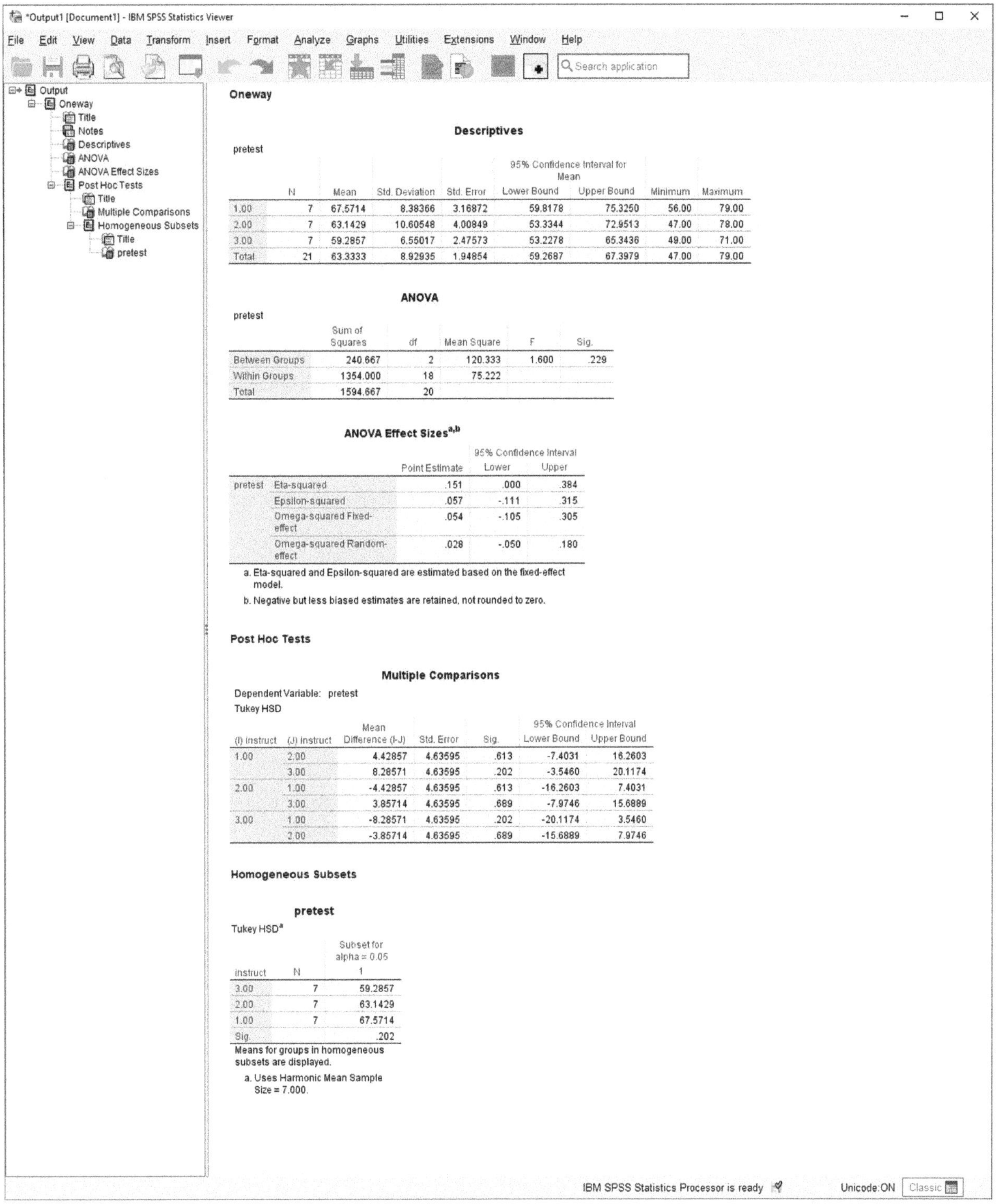

Oneway

Descriptives

pretest

	N	Mean	Std. Deviation	Std. Error	95% Confidence Interval for Mean Lower Bound	95% Confidence Interval for Mean Upper Bound	Minimum	Maximum
1.00	7	67.5714	8.38366	3.16872	59.8178	75.3250	56.00	79.00
2.00	7	63.1429	10.60548	4.00849	53.3344	72.9513	47.00	78.00
3.00	7	59.2857	6.55017	2.47573	53.2278	65.3436	49.00	71.00
Total	21	63.3333	8.92935	1.94854	59.2687	67.3979	47.00	79.00

ANOVA

pretest

	Sum of Squares	df	Mean Square	F	Sig.
Between Groups	240.667	2	120.333	1.600	.229
Within Groups	1354.000	18	75.222		
Total	1594.667	20			

ANOVA Effect Sizes[a,b]

		Point Estimate	95% Confidence Interval Lower	95% Confidence Interval Upper
pretest	Eta-squared	.151	.000	.384
	Epsilon-squared	.057	-.111	.315
	Omega-squared Fixed-effect	.054	-.105	.305
	Omega-squared Random-effect	.028	-.050	.180

a. Eta-squared and Epsilon-squared are estimated based on the fixed-effect model.

b. Negative but less biased estimates are retained, not rounded to zero.

Post Hoc Tests

Multiple Comparisons

Dependent Variable: pretest
Tukey HSD

(I) instruct	(J) instruct	Mean Difference (I-J)	Std. Error	Sig.	95% Confidence Interval Lower Bound	95% Confidence Interval Upper Bound
1.00	2.00	4.42857	4.63595	.613	-7.4031	16.2603
	3.00	8.28571	4.63595	.202	-3.5460	20.1174
2.00	1.00	-4.42857	4.63595	.613	-16.2603	7.4031
	3.00	3.85714	4.63595	.689	-7.9746	15.6889
3.00	1.00	-8.28571	4.63595	.202	-20.1174	3.5460
	2.00	-3.85714	4.63595	.689	-15.6889	7.9746

Homogeneous Subsets

pretest

Tukey HSD[a]

instruct	N	Subset for alpha = 0.05 1
3.00	7	59.2857
2.00	7	63.1429
1.00	7	67.5714
Sig.		.202

Means for groups in homogeneous subsets are displayed.

a. Uses Harmonic Mean Sample Size = 7.000.

The ANOVA was not significant, so there is no need to refer to the Multiple Comparisons table. Given this result, we may state the following:

> The pretest means of students who took a course from three different instructors were compared using a one-way ANOVA. No significant difference was found (F (2,18) = 1.60, $p > .05$). The students from the three different classes did not differ significantly at the start of the term. Students who had Instructor 1 had a mean score of 67.57 (*sd* = 8.38). Students who had Instructor 2 had a mean score of 63.14 (*sd* = 10.61). Students who had Instructor 3 had a mean score of 59.29 (*sd* = 6.55). There was a moderate effect size (η^2=.151).

Practice Exercise

Using Practice Dataset 1 in Appendix B, determine if the average math scores of single, married, and divorced participants are significantly different. Write a statement of results.

Section 7.3 Factorial ANOVA

Description

The factorial ANOVA is one in which there is more than one **independent variable**. A 2 × 2 ANOVA, for instance, has two **independent variables**, each with two **levels**. A 3 × 2 × 2 ANOVA has three **independent variables**. One has three **levels**, and the other two have two **levels**. Factorial ANOVA is very powerful because it allows us to assess the effects of each **independent variable**, plus the effects of the **interaction**.

Interactions occur when the effect of one variable impacts the effect of another variable on the **dependent variable**. For instance, there is an **interaction** between soap and water, influencing their effects on their cleaning ability. In this instance, the effect of soap and water together is greater than their individual effects added together.

Assumptions

Factorial ANOVA requires all of the assumptions of one-way ANOVA (i.e., the **dependent variable** must be at the **interval** or **ratio** level and normally distributed). In addition, the **independent variables** should be independent of each other.

SPSS Data Format

SPSS requires one variable for the **dependent variable** and one variable for each **independent variable**. If we have *any* **independent variable** that is represented as multiple variables (e.g., PRETEST and POSTTEST), we must use the repeated-measures ANOVA.

Running the Command

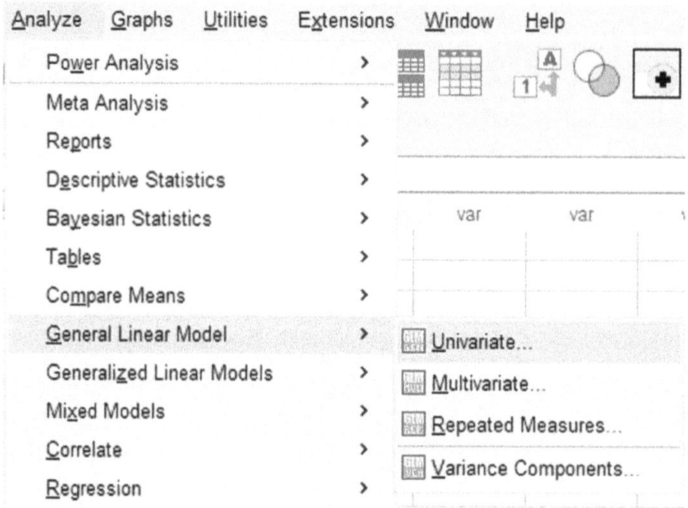

This example uses the GRADES.sav data file from earlier in this chapter. Click *Analyze*, then *General Linear Model*, then *Univariate*.

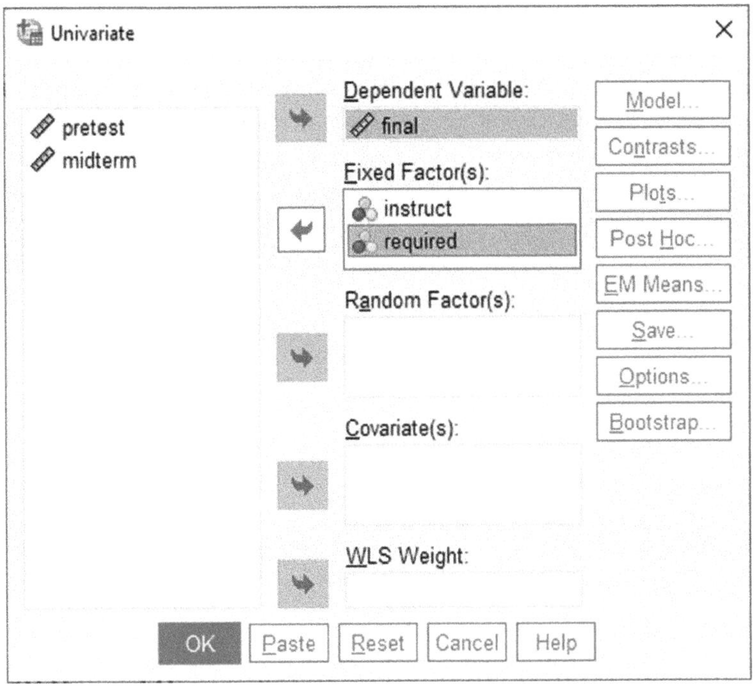

This will bring up the main **dialog box** for Univariate ANOVA. Select the **dependent variable** and place it in the *Dependent Variable* blank (use FINAL for this example). Select one of your **independent variables** (INSTRUCT in this case) and place it in the *Fixed Factor(s)* box. Place the second **independent variable** (REQUIRED) in the *Fixed Factor(s)* box. Having defined the analysis, now click *Options*.

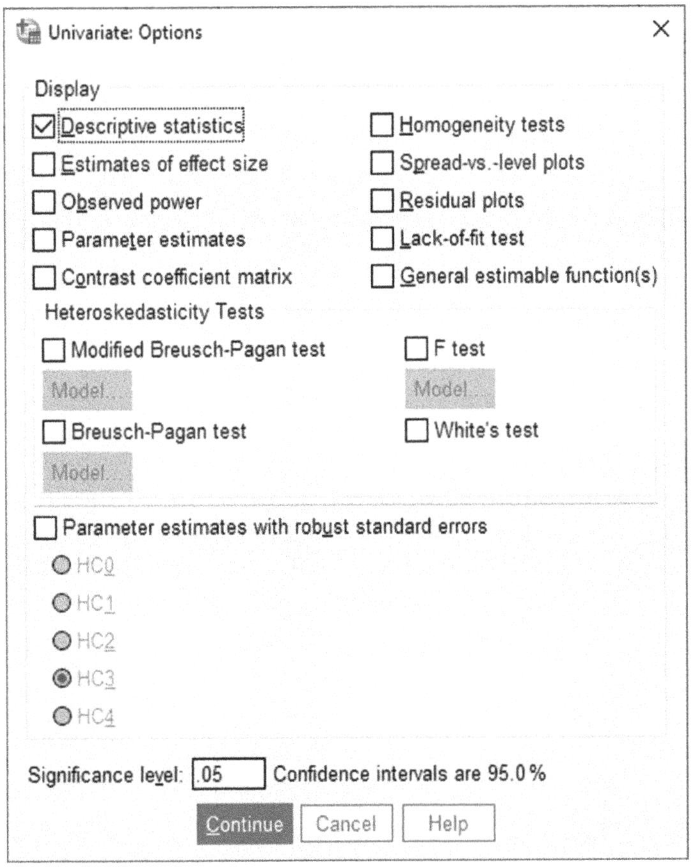

When the Options **dialog box** comes up, click *Descriptive statistics.* This will provide you with **means** for each main effect and **interaction** term. Click *Continue.*

If you were to select *Post Hoc,* SPSS would allow you to elect to run post-hoc analyses for any main effects but not for the **interaction** term. (We will select Tukey's HSD for the INSTRUCT variable here.)

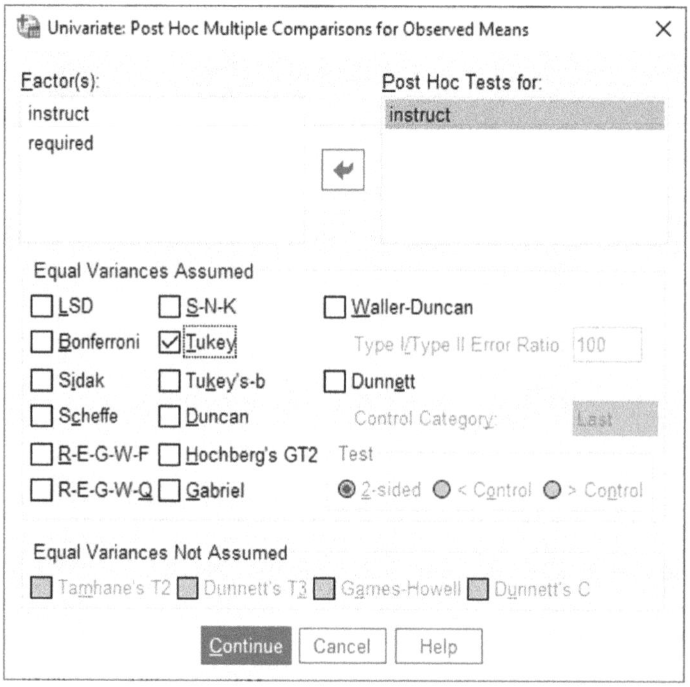

Click *OK* to run the analysis.

Reading the Output

In the middle of the output, you will find the means. We have six **means** representing the **interaction** of the two variables (this was a 3 × 2 design).

Participants who had Instructor 1 (for whom the class was not required) had a **mean** final exam score of 79.67. Students who had Instructor 1 (for whom it was required) had a **mean** final exam score of 79.50, and so on.

The example we just ran is called a two-way ANOVA. This is because we had two **independent variables**. With a two-way ANOVA, we get three answers: A main effect for INSTRUCT, a main effect for REQUIRED, and an **interaction** result for INSTRUCT*REQUIRED.

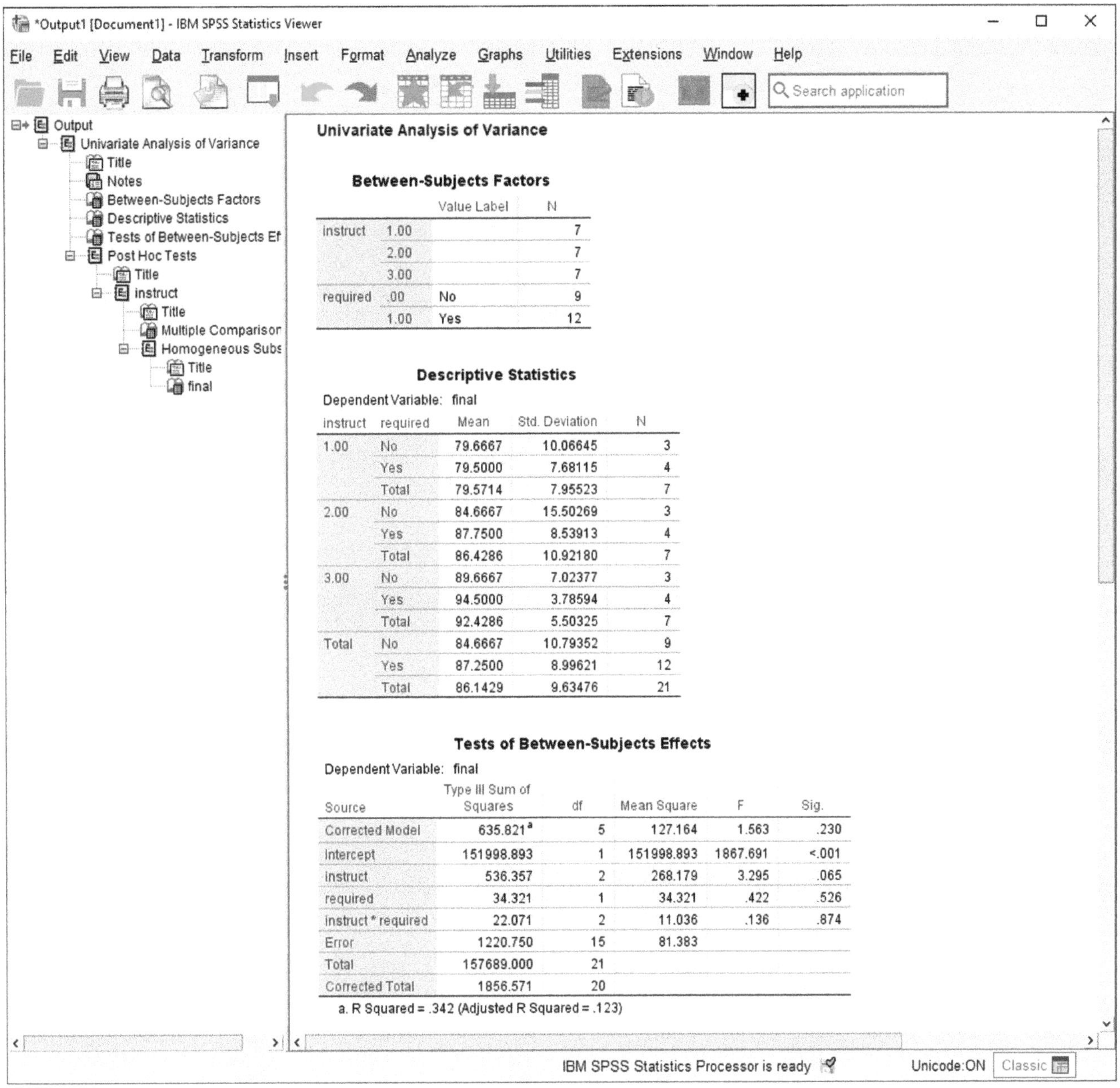

Univariate Analysis of Variance

Between-Subjects Factors

		Value Label	N
instruct	1.00		7
	2.00		7
	3.00		7
required	.00	No	9
	1.00	Yes	12

Descriptive Statistics

Dependent Variable: final

instruct	required	Mean	Std. Deviation	N
1.00	No	79.6667	10.06645	3
	Yes	79.5000	7.68115	4
	Total	79.5714	7.95523	7
2.00	No	84.6667	15.50269	3
	Yes	87.7500	8.53913	4
	Total	86.4286	10.92180	7
3.00	No	89.6667	7.02377	3
	Yes	94.5000	3.78594	4
	Total	92.4286	5.50325	7
Total	No	84.6667	10.79352	9
	Yes	87.2500	8.99621	12
	Total	86.1429	9.63476	21

Tests of Between-Subjects Effects

Dependent Variable: final

Source	Type III Sum of Squares	df	Mean Square	F	Sig.
Corrected Model	635.821[a]	5	127.164	1.563	.230
Intercept	151998.893	1	151998.893	1867.691	<.001
instruct	536.357	2	268.179	3.295	.065
required	34.321	1	34.321	.422	.526
instruct * required	22.071	2	11.036	.136	.874
Error	1220.750	15	81.383		
Total	157689.000	21			
Corrected Total	1856.571	20			

a. R Squared = .342 (Adjusted R Squared = .123)

The source table at the bottom of the output gives us these three answers (in the INSTRUCT, REQUIRED, and INSTRUCT * REQUIRED rows).

In the example here, the main effect for Instructor is significant, but the main effect for whether or not the class is required and the **interaction** are not significant. In the statements of results, you must indicate *F*, two degrees of freedom (effect and residual/error), the **significance** level, and a verbal statement for each of the answers (three in this case). Note that most statistics books give a much simpler version of an ANOVA source table where the Corrected Model, Intercept, and Corrected Total rows are not included. Because one of the main effects was significant, there are also post-hoc results at the very end of the output (as shown on the next page).

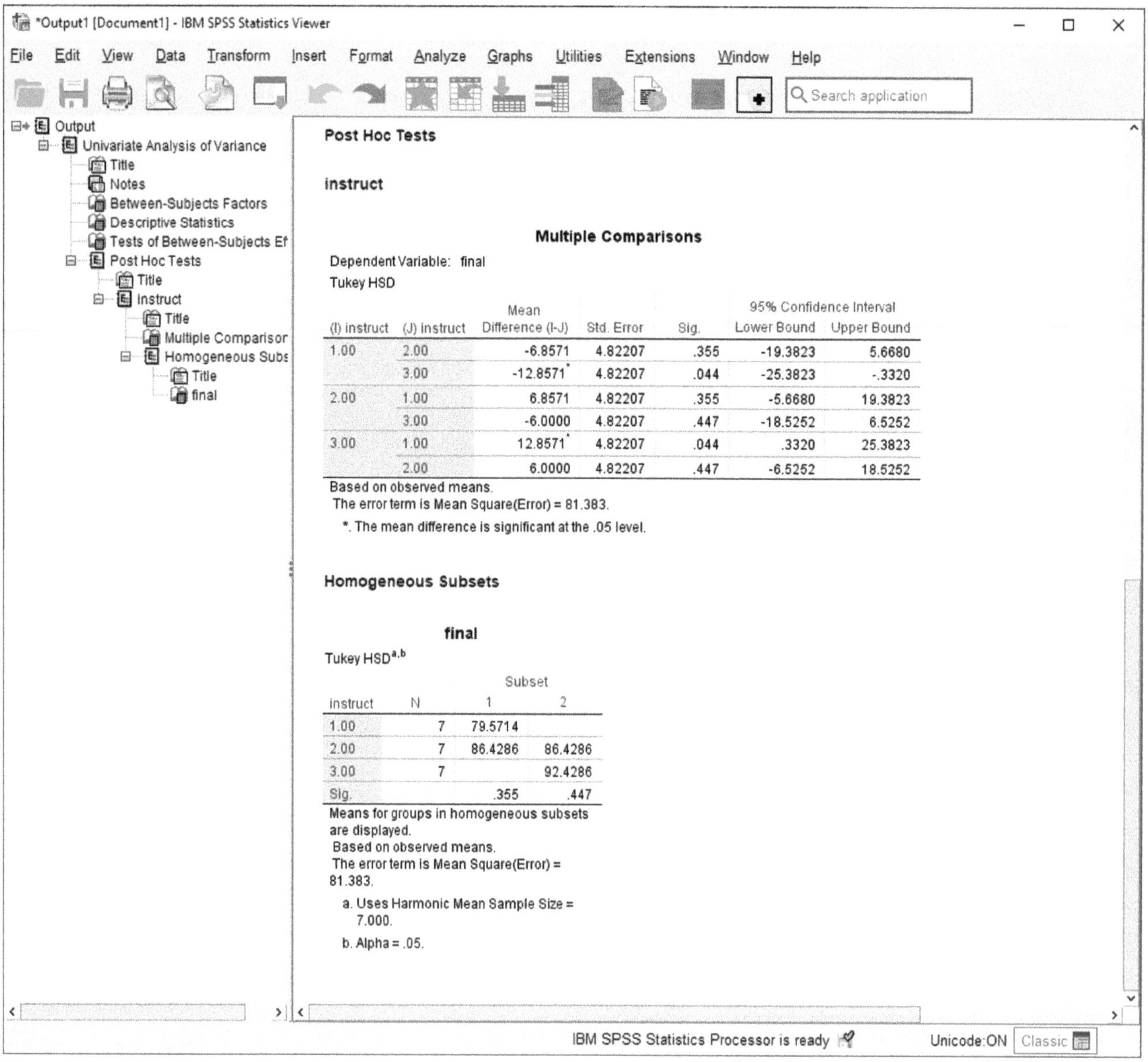
*Output1 [Document1] - IBM SPSS Statistics Viewer

File Edit View Data Transform Insert Format Analyze Graphs Utilities Extensions Window Help

Output
Univariate Analysis of Variance
Title
Notes
Between-Subjects Factors
Descriptive Statistics
Tests of Between-Subjects Ef
Post Hoc Tests
Title
instruct
Title
Multiple Comparisor
Homogeneous Subs
Title
final

Post Hoc Tests

instruct

Multiple Comparisons

Dependent Variable: final
Tukey HSD

(I) instruct	(J) instruct	Mean Difference (I-J)	Std. Error	Sig.	95% Confidence Interval Lower Bound	Upper Bound
1.00	2.00	-6.8571	4.82207	.355	-19.3823	5.6680
	3.00	-12.8571*	4.82207	.044	-25.3823	-.3320
2.00	1.00	6.8571	4.82207	.355	-5.6680	19.3823
	3.00	-6.0000	4.82207	.447	-18.5252	6.5252
3.00	1.00	12.8571*	4.82207	.044	.3320	25.3823
	2.00	6.0000	4.82207	.447	-6.5252	18.5252

Based on observed means.
The error term is Mean Square(Error) = 81.383.
*. The mean difference is significant at the .05 level.

Homogeneous Subsets

final

Tukey HSD[a,b]

instruct	N	Subset 1	2
1.00	7	79.5714	
2.00	7	86.4286	86.4286
3.00	7		92.4286
Sig.		.355	.447

Means for groups in homogeneous subsets are displayed.
Based on observed means.
The error term is Mean Square(Error) = 81.383.
a. Uses Harmonic Mean Sample Size = 7.000.
b. Alpha = .05.

IBM SPSS Statistics Processor is ready   Unicode:ON   Classic

Phrasing Results That Are Significant

For the example we just ran we could say:

> A 3 (instructor) × 2 (required course) between-subjects factorial ANOVA was calculated comparing the final exam scores for participants who had one of three instructors and who took the course either as a required course or as an elective. A significant main effect for instructor was found (F (2,15) = 3.723, p < .05). Students who had Instructor 1 had lower final exam scores (M = 79.14, sd = 7.43) than students who had Instructor 3 (M = 92.43, sd = 5.50). Students who had Instructor 2 (M = 86.43, sd = 10.92) were not significantly different from either of the other two groups. A significant main effect for whether or not the course was required was not found (F (1,15) = .561, p > .05). Students who took the course because it was required did not do significantly better (M = 87.25, sd = 9.61) than students who took the course as an elective (M = 84.33, sd = 10.69). The interaction was not significant (F (2,15) = 0.088, p > .05). The effect of the instructor was not influenced by whether or not the students took the course because it was required.

Phrasing Results That Are Not Significant

If you repeat the analysis using MIDTERM as the **Dependent Variable**, you will receive the following non-significant

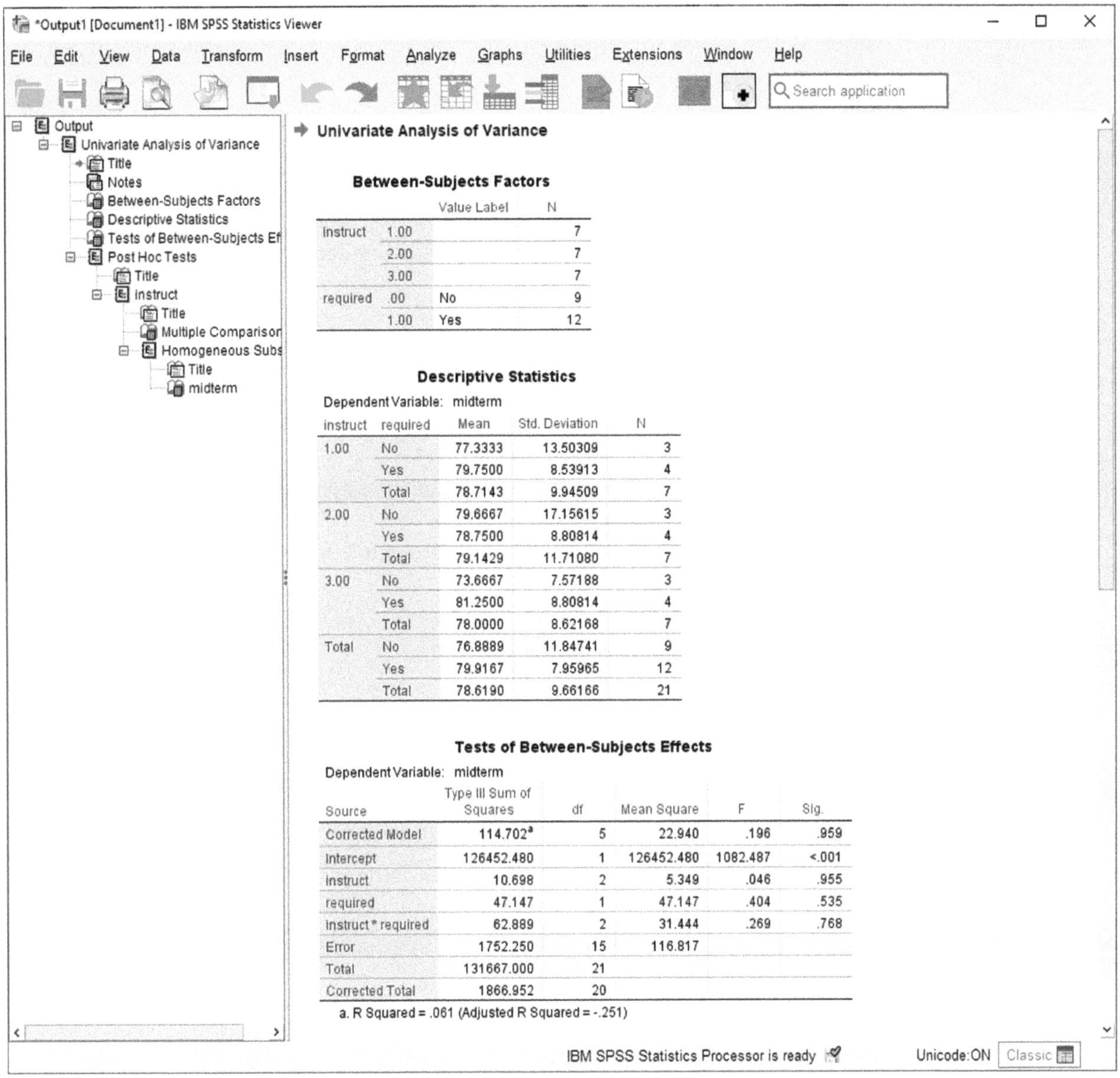

Univariate Analysis of Variance

Between-Subjects Factors

		Value Label	N
instruct	1.00		7
	2.00		7
	3.00		7
required	.00	No	9
	1.00	Yes	12

Descriptive Statistics

Dependent Variable: midterm

instruct	required	Mean	Std. Deviation	N
1.00	No	77.3333	13.50309	3
	Yes	79.7500	8.53913	4
	Total	78.7143	9.94509	7
2.00	No	79.6667	17.15615	3
	Yes	78.7500	8.80814	4
	Total	79.1429	11.71080	7
3.00	No	73.6667	7.57188	3
	Yes	81.2500	8.80814	4
	Total	78.0000	8.62168	7
Total	No	76.8889	11.84741	9
	Yes	79.9167	7.95965	12
	Total	78.6190	9.66166	21

Tests of Between-Subjects Effects

Dependent Variable: midterm

Source	Type III Sum of Squares	df	Mean Square	F	Sig.
Corrected Model	114.702[a]	5	22.940	.196	.959
Intercept	126452.480	1	126452.480	1082.487	<.001
instruct	10.698	2	5.349	.046	.955
required	47.147	1	47.147	.404	.535
instruct * required	62.889	2	31.444	.269	.768
Error	1752.250	15	116.817		
Total	131667.000	21			
Corrected Total	1866.952	20			

a. R Squared = .061 (Adjusted R Squared = -.251)

A 3 (instructor) × 2 (required course) between-subjects factorial ANOVA was calculated comparing the midterm exam scores for participants who had one of three instructors and who took the course as a required course or as an elective. The main effect for instructor was not significant (F (2,15) = .046, p > .05). The main effect for whether or not it was a required course was also not significant (F (1,15) = .404, p > .05). Finally, the interaction was not significant (F (2,15) = .269, p > .05). Thus, it appears that neither the instructor nor whether or not the course is required has any significant effect on midterm exam scores.

Practice Exercise

Using Practice Dataset 2 in Appendix B, determine if salaries are influenced by sex, job classification, or an **interaction** between sex and job classification. Write a statement of results.

Section 7.4 Repeated-Measures ANOVA

Description

Repeated-measures ANOVA extends the basic ANOVA procedure to a within-subjects **independent variable** (when participants provide data for more than one **level** of an **independent variable)**. It functions like a paired-samples *t* test when more than two **levels** are being compared.

Assumptions

The **dependent variable** should be normally distributed and measured on an **interval** or **ratio scale**. Multiple measurements of the **dependent variable** should be from the same (or related) participants.

SPSS Data Format

At least three variables are required. Each variable in the SPSS data file should represent a single **dependent variable** at a single **level** of the **independent variable**. Thus, an analysis of a design with four **levels** of an **independent variable** would require four variables in the SPSS data file.

If any variable represents a between-subjects effect, use the *Repeated Measures ANOVA* command instead.

Running the Command

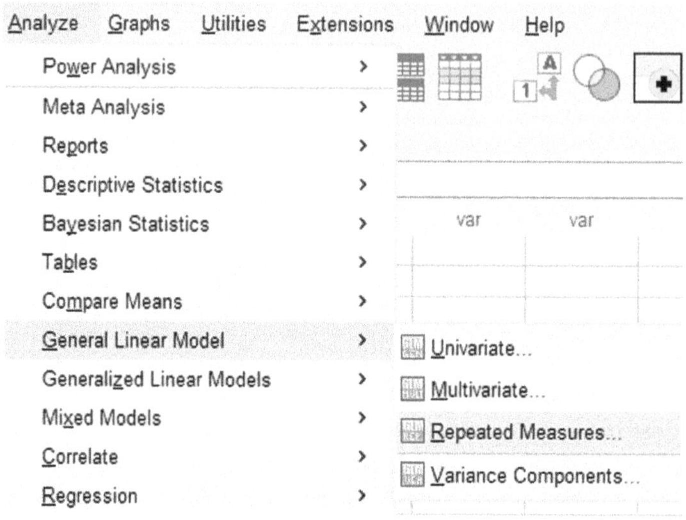

This example uses the GRADES.sav sample dataset. Recall that GRADES.sav includes three sets of grades—PRETEST, MIDTERM, and FINAL—that represent three different times during the semester. This allows us to analyze the effects of time on the test performance of our sample population (hence the within-groups comparison). Click *Analyze*, then *General Linear Model*, then *Repeated Measures*.

Note that this procedure requires an optional module. If you do not have this command, you do not have the necessary module installed. This procedure is NOT included in the SPSS Statistics GradPack Base.

After selecting the command, you will be presented with the Repeated Measures Define Factor(s) **dialog box**. This is where you identify the Within-Subject Factor (we will call it TIME). Enter 3 for the Number of Levels (three exams) and click *Add*.

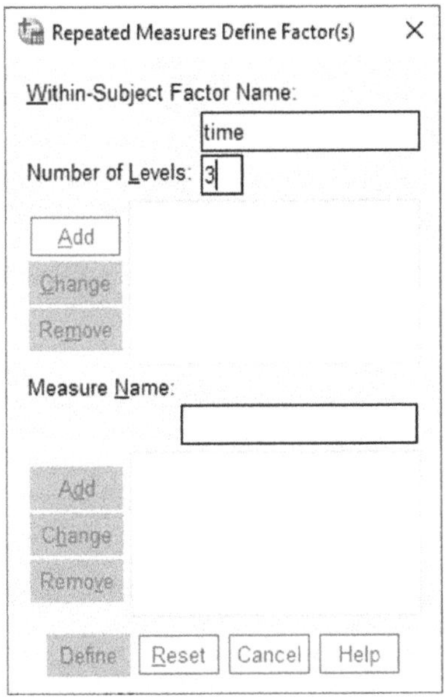

If we had more than one **independent variable** that had repeated measures, we could enter its name and click *Add*. Now click *Define*.

You will be presented with the Repeated Measures **dialog box**. Transfer PRETEST, MIDTERM, and FINAL to the *Within-Subjects Variables* section. The variable names should be ordered according to when they occurred in time (i.e., the values of the **independent variable** that they represent).

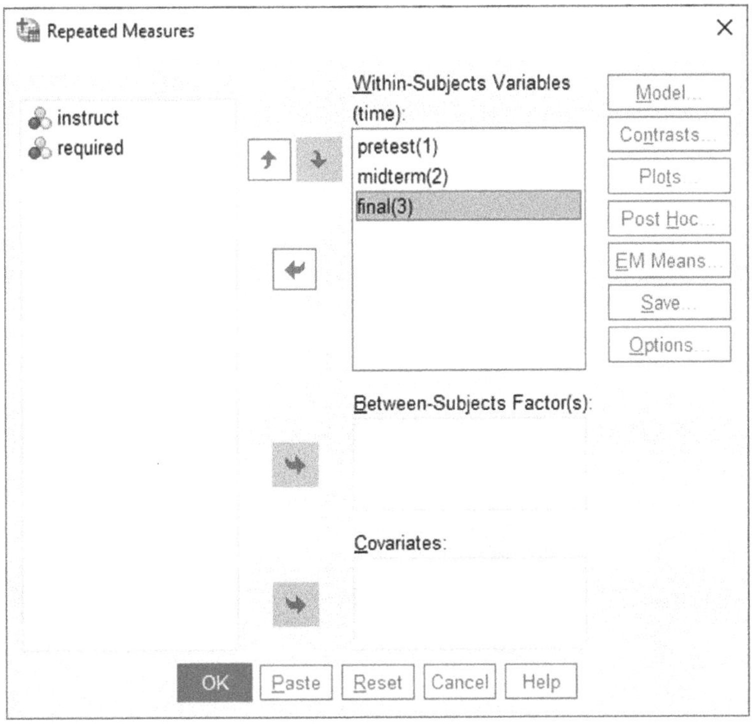

Click *Options*, and the dialog box here will appear. Select *Descriptive statistics*, SPSS will calculate the relevant means. Click *CONTINUE* to close the **dialog box** and *OK* to run the command.

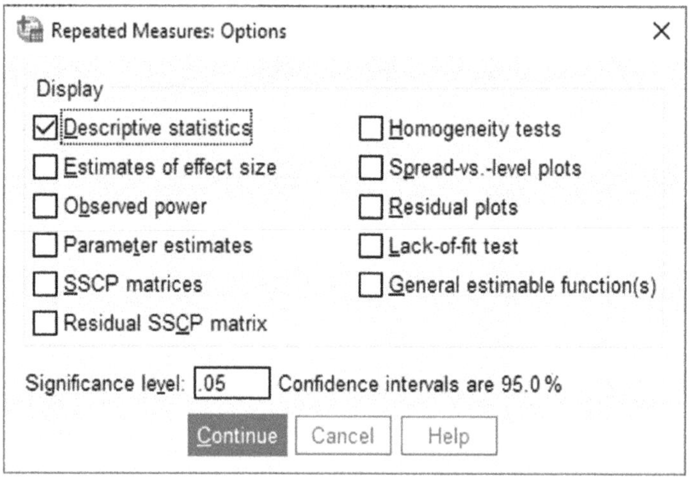

Reading the Output

This procedure uses the *GLM* command. GLM stands for *General Linear Model.* It is a very powerful command, and many sections of output are beyond the scope of this book. However, for the basic repeated-measures ANOVA, we are interested only in the *Tests of Within-Subjects Effects* (near the bottom of the output shown here). Note that the SPSS output will include many other sections of output, which you can ignore at this point.

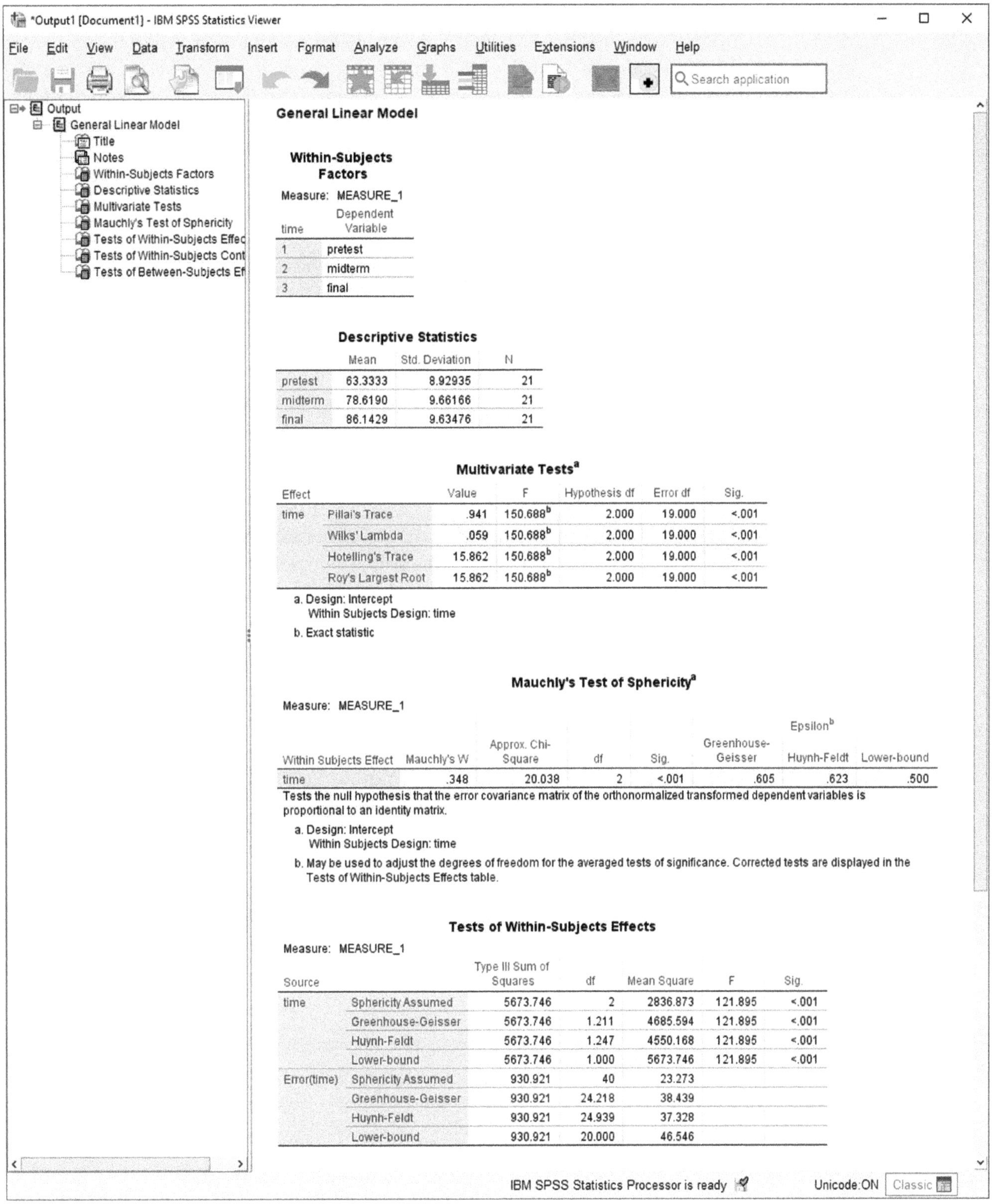

General Linear Model

Within-Subjects Factors

Measure: MEASURE_1

time	Dependent Variable
1	pretest
2	midterm
3	final

Descriptive Statistics

	Mean	Std. Deviation	N
pretest	63.3333	8.92935	21
midterm	78.6190	9.66166	21
final	86.1429	9.63476	21

Multivariate Tests[a]

Effect		Value	F	Hypothesis df	Error df	Sig.
time	Pillai's Trace	.941	150.688[b]	2.000	19.000	<.001
	Wilks' Lambda	.059	150.688[b]	2.000	19.000	<.001
	Hotelling's Trace	15.862	150.688[b]	2.000	19.000	<.001
	Roy's Largest Root	15.862	150.688[b]	2.000	19.000	<.001

a. Design: Intercept
Within Subjects Design: time

b. Exact statistic

Mauchly's Test of Sphericity[a]

Measure: MEASURE_1

Within Subjects Effect	Mauchly's W	Approx. Chi-Square	df	Sig.	Epsilon[b] Greenhouse-Geisser	Huynh-Feldt	Lower-bound
time	.348	20.038	2	<.001	.605	.623	.500

Tests the null hypothesis that the error covariance matrix of the orthonormalized transformed dependent variables is proportional to an identity matrix.

a. Design: Intercept
Within Subjects Design: time

b. May be used to adjust the degrees of freedom for the averaged tests of significance. Corrected tests are displayed in the Tests of Within-Subjects Effects table.

Tests of Within-Subjects Effects

Measure: MEASURE_1

Source		Type III Sum of Squares	df	Mean Square	F	Sig.
time	Sphericity Assumed	5673.746	2	2836.873	121.895	<.001
	Greenhouse-Geisser	5673.746	1.211	4685.594	121.895	<.001
	Huynh-Feldt	5673.746	1.247	4550.168	121.895	<.001
	Lower-bound	5673.746	1.000	5673.746	121.895	<.001
Error(time)	Sphericity Assumed	930.921	40	23.273		
	Greenhouse-Geisser	930.921	24.218	38.439		
	Huynh-Feldt	930.921	24.939	37.328		
	Lower-bound	930.921	20.000	46.546		

The *Tests of Within-Subjects Effects* output should look very similar to the output from the other *ANOVA* commands. In the example above, the effect of TIME has an *F* value of 121.895 with 2 and 40 degrees of freedom (we use the line for *Sphericity Assumed*). It is significant at less than the .001 level. When describing these results, we should indicate the type of test, *F* value, degrees of freedom, and **significance** level.

Phrasing Results That Are Significant

Because the ANOVA results were significant, we need to conduct some sort of post-hoc analysis. One of the main limitations of SPSS is the difficulty in performing post-hoc analyses for within-subjects factors. With SPSS, the easiest solution to this problem is to conduct **protected dependent *t* tests** with repeated-measures ANOVA. There are more powerful (and more appropriate) post-hoc analyses, but SPSS will not compute them for us. For more information, consult your instructor or a more advanced statistics text.

To conduct the protected *t* tests, we will compare PRETEST to MIDTERM, MIDTERM to FINAL, and PRETEST to FINAL, using paired-samples *t* tests. Because we are conducting three tests and, therefore, inflating our **Type I error** rate, we will use a **significance** level of .017 (.05/3) instead of .05. Note that to conduct this analysis, we will need to use the paired-samples *t* test command and enter all three pairs.

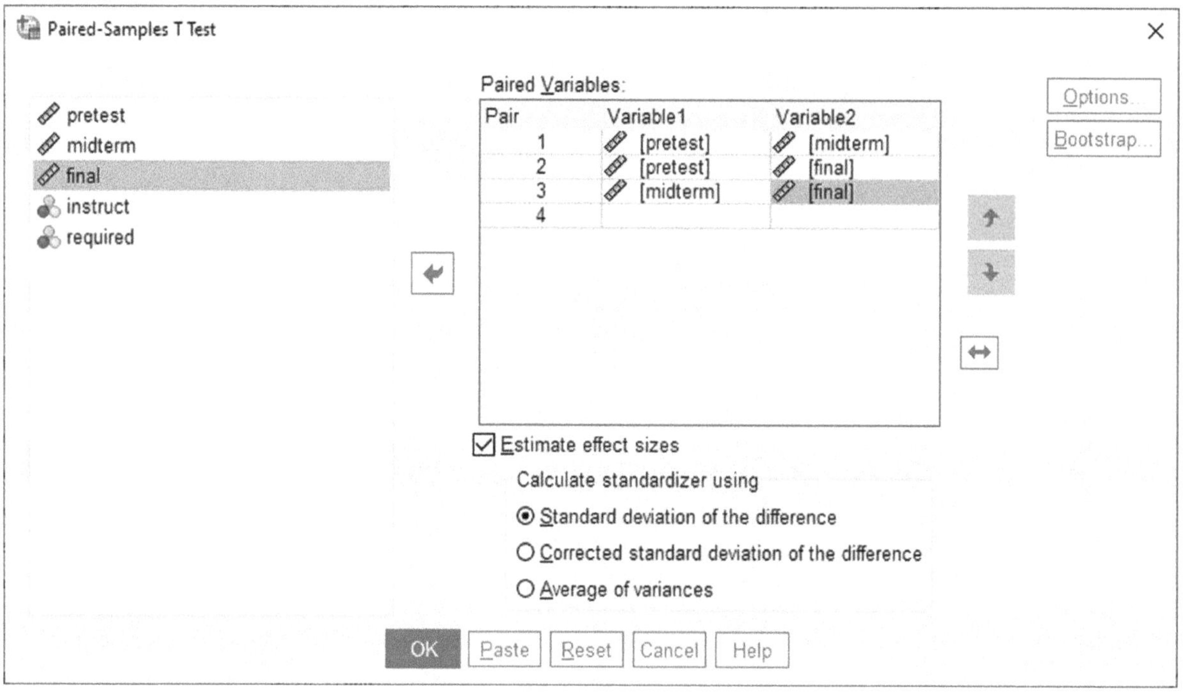

The three comparisons each had a **significance** level of less than .017, so we can conclude that the scores improved from pretest to midterm and again from midterm to final. To generate the **descriptive statistics**, we have to run the *Descriptives* command for each variable.

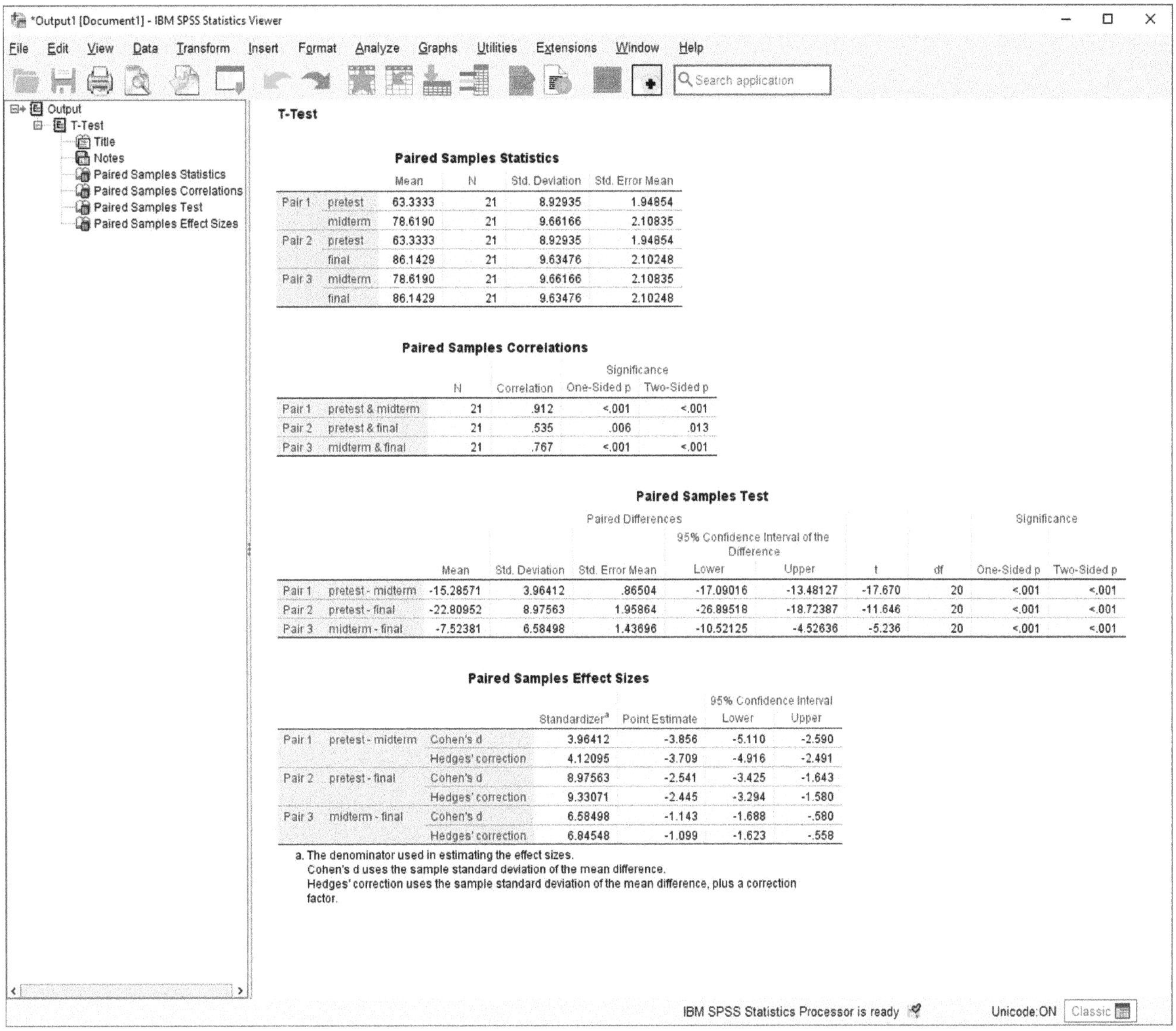

T-Test

Paired Samples Statistics

		Mean	N	Std. Deviation	Std. Error Mean
Pair 1	pretest	63.3333	21	8.92935	1.94854
	midterm	78.6190	21	9.66166	2.10835
Pair 2	pretest	63.3333	21	8.92935	1.94854
	final	86.1429	21	9.63476	2.10248
Pair 3	midterm	78.6190	21	9.66166	2.10835
	final	86.1429	21	9.63476	2.10248

Paired Samples Correlations

		N	Correlation	Significance One-Sided p	Two-Sided p
Pair 1	pretest & midterm	21	.912	<.001	<.001
Pair 2	pretest & final	21	.535	.006	.013
Pair 3	midterm & final	21	.767	<.001	<.001

Paired Samples Test

		Paired Differences							Significance	
					95% Confidence Interval of the Difference					
		Mean	Std. Deviation	Std. Error Mean	Lower	Upper	t	df	One-Sided p	Two-Sided p
Pair 1	pretest - midterm	-15.28571	3.96412	.86504	-17.09016	-13.48127	-17.670	20	<.001	<.001
Pair 2	pretest - final	-22.80952	8.97563	1.95864	-26.89518	-18.72387	-11.646	20	<.001	<.001
Pair 3	midterm - final	-7.52381	6.58498	1.43696	-10.52125	-4.52636	-5.236	20	<.001	<.001

Paired Samples Effect Sizes

			Standardizer[a]	Point Estimate	95% Confidence Interval Lower	Upper
Pair 1	pretest - midterm	Cohen's d	3.96412	-3.856	-5.110	-2.590
		Hedges' correction	4.12095	-3.709	-4.916	-2.491
Pair 2	pretest - final	Cohen's d	8.97563	-2.541	-3.425	-1.643
		Hedges' correction	9.33071	-2.445	-3.294	-1.580
Pair 3	midterm - final	Cohen's d	6.58498	-1.143	-1.688	-.580
		Hedges' correction	6.84548	-1.099	-1.623	-.558

a. The denominator used in estimating the effect sizes.
Cohen's d uses the sample standard deviation of the mean difference.
Hedges' correction uses the sample standard deviation of the mean difference, plus a correction factor.

Because the results of our example were significant, we could state the following:

> A one-way repeated-measures ANOVA was calculated comparing the exam scores of participants at three different times: pretest, midterm, and final. A significant effect was found (F (2,40) = 121.895, $p < .001$). Follow-up protected t tests revealed that scores increased significantly from pretest (M = 63.33, sd = 8.93) to midterm (M = 78.62, sd = 9.66), and again from midterm to final (M = 86.14, sd = 9.63).

Phrasing Results That Are Not Significant

With results that are not significant, we could state the following (the output here comes from the TIME.sav data file in Appendix C):

> A one-way repeated-measures ANOVA was calculated comparing the scores of participants at three different times: before, during, and after. No significant effect was found (F (2,10) = .044, $p > .05$). No significant difference exists among before (M = 3.50, sd = 1.05, during (M = 3.67, sd = 1.03), and after (M = 3.67, sd = 1.03).

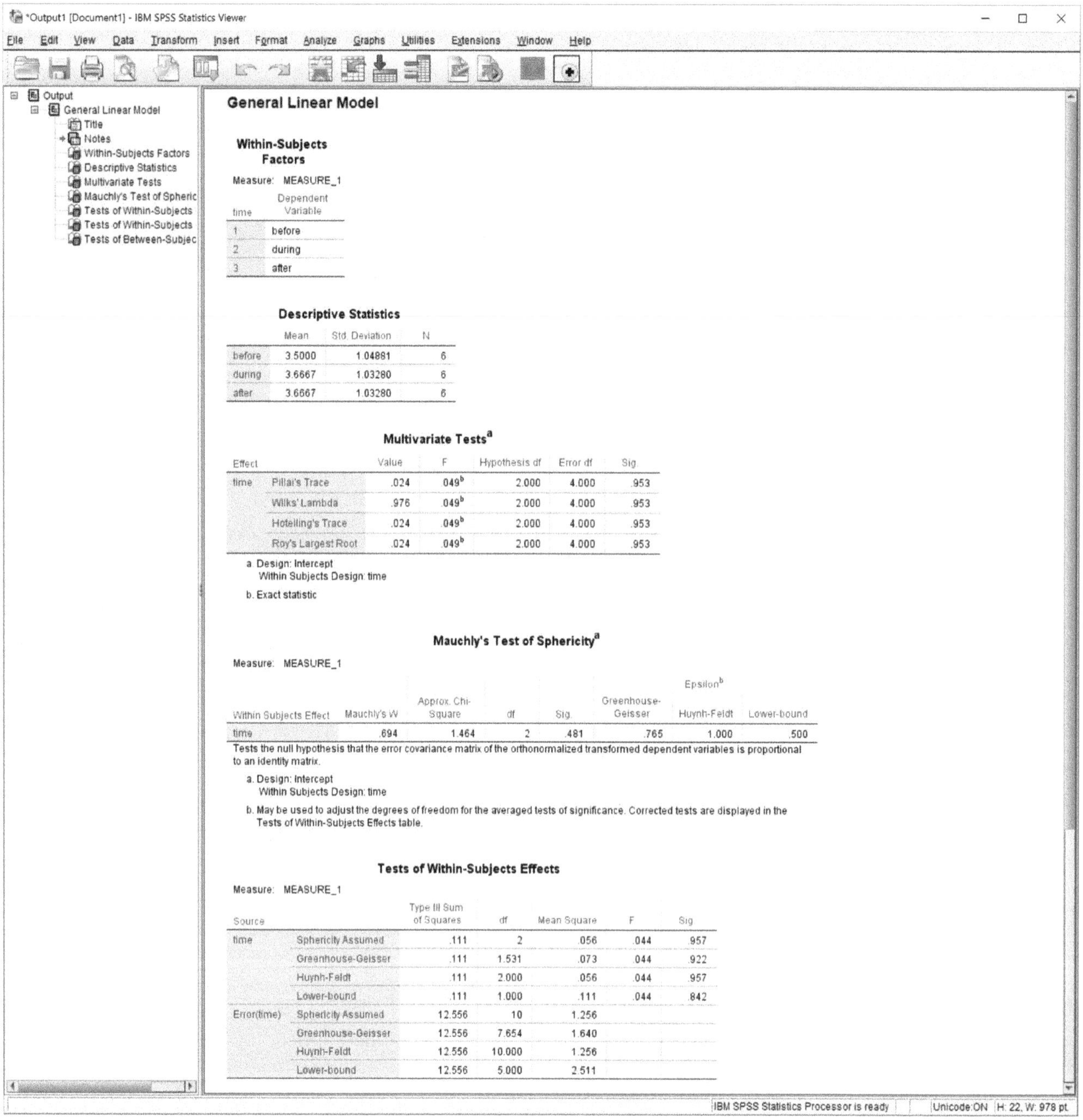

General Linear Model

Within-Subjects Factors

Measure: MEASURE_1

time	Dependent Variable
1	before
2	during
3	after

Descriptive Statistics

	Mean	Std. Deviation	N
before	3.5000	1.04881	6
during	3.6667	1.03280	6
after	3.6667	1.03280	6

Multivariate Tests[a]

Effect		Value	F	Hypothesis df	Error df	Sig.
time	Pillai's Trace	.024	.049[b]	2.000	4.000	.953
	Wilks' Lambda	.976	.049[b]	2.000	4.000	.953
	Hotelling's Trace	.024	.049[b]	2.000	4.000	.953
	Roy's Largest Root	.024	.049[b]	2.000	4.000	.953

a. Design: Intercept
Within Subjects Design: time

b. Exact statistic

Mauchly's Test of Sphericity[a]

Measure: MEASURE_1

					Epsilon[b]		
Within Subjects Effect	Mauchly's W	Approx. Chi-Square	df	Sig.	Greenhouse-Geisser	Huynh-Feldt	Lower-bound
time	.694	1.464	2	.481	.765	1.000	.500

Tests the null hypothesis that the error covariance matrix of the orthonormalized transformed dependent variables is proportional to an identity matrix.

a. Design: Intercept
Within Subjects Design: time

b. May be used to adjust the degrees of freedom for the averaged tests of significance. Corrected tests are displayed in the Tests of Within-Subjects Effects table.

Tests of Within-Subjects Effects

Measure: MEASURE_1

Source		Type III Sum of Squares	df	Mean Square	F	Sig.
time	Sphericity Assumed	.111	2	.056	.044	.957
	Greenhouse-Geisser	.111	1.531	.073	.044	.922
	Huynh-Feldt	.111	2.000	.056	.044	.957
	Lower-bound	.111	1.000	.111	.044	.842
Error(time)	Sphericity Assumed	12.556	10	1.256		
	Greenhouse-Geisser	12.556	7.654	1.640		
	Huynh-Feldt	12.556	10.000	1.256		
	Lower-bound	12.556	5.000	2.511		

Practice Exercise

Use Practice Dataset 3 in Appendix B. Determine if the anxiety level of participants changed over time (regardless of which treatment they received) using a one-way repeated-measures ANOVA and protected dependent *t* tests. Write a statement of results.

Section 7.5 Mixed-Design ANOVA

Description

The mixed-design ANOVA (sometimes called a split-plot design) tests the effects of more than one **independent variable**. At least one of the **independent variables** must be within-subjects (repeated measures). At least one of the **independent variables** must be between-subjects.

Assumptions

The **dependent variable** should be normally distributed and measured on an **interval** or **ratio scale**.

SPSS Data Format

The **dependent variable** should be represented as one variable (one column) for each level of the within-subjects **independent variables**. Another variable (column) should be present in the data file for each between-subjects variable. Thus, a 2 × 2 mixed-design ANOVA would require three columns: Two representing the **dependent variable** (one at each level) and one representing the between-subjects **independent variable**.

Running the Command

The *General Linear Model* command runs the *Repeated Measures ANOVA* command. Click *Analyze*, then *General Linear Model*, then *Repeated Measures*.

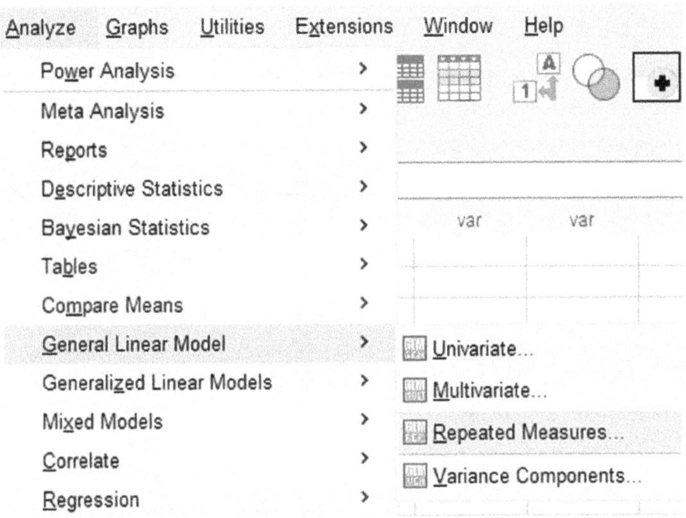

Note that this procedure requires an optional module. If you do not have this command, you do not have the necessary module installed. This procedure is NOT included in the SPSS Statistics GradPack Base.

The *Repeated Measures* command should be used if any of the **independent variables** are repeated measures (within-subjects).

This example also uses the GRADES.sav data file. Enter PRE-TEST, MIDTERM, and FINAL in the *Within-Subjects Variables* block. (See the *Repeated Measures ANOVA* command in Section 7.4 for an explanation.) This example is a 3 × 3 mixed-design. There are two **independent variables** (TIME and INSTRUCT), each with three **levels**. We previously entered the information for TIME in the Repeated Measures Define Factors **dialog box**.

We need to transfer INSTRUCT into the *Between-Subjects Factor(s)* block.

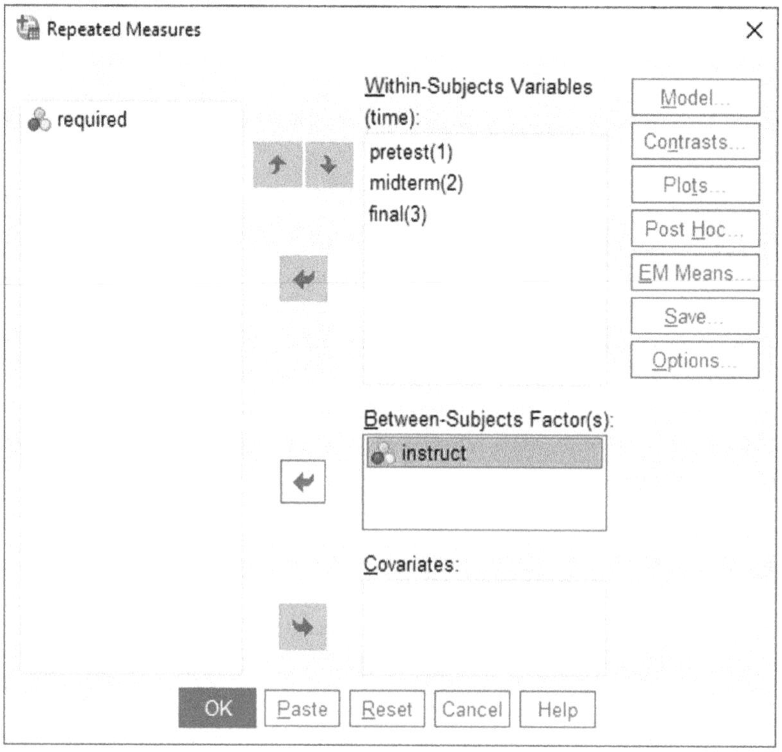

The *Descriptive statistics* option we used for one-way repeated-measures ANOVA will not work here because it only provides descriptive statistics for repeated-measures variables. Instead, we will select *EM Means* and move each of our main effects and the interaction over.

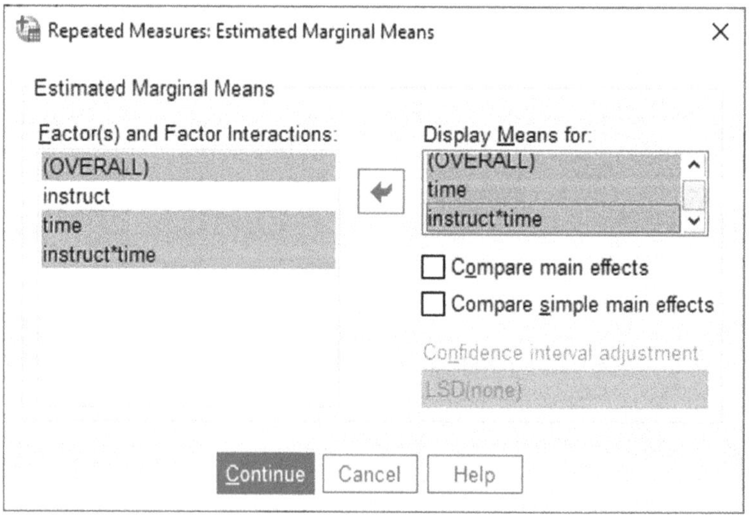

Click *OK* to run the command.

Reading the Output

As with the standard *Repeated Measures* command, the *GLM* procedure provides a significant amount of output we will not use. For a mixed-design ANOVA, we are interested in two sections. The first is *Tests of Within-Subjects Effects*, and the second is *Tests of Between-Subjects Effects*. Both of those are shown here.

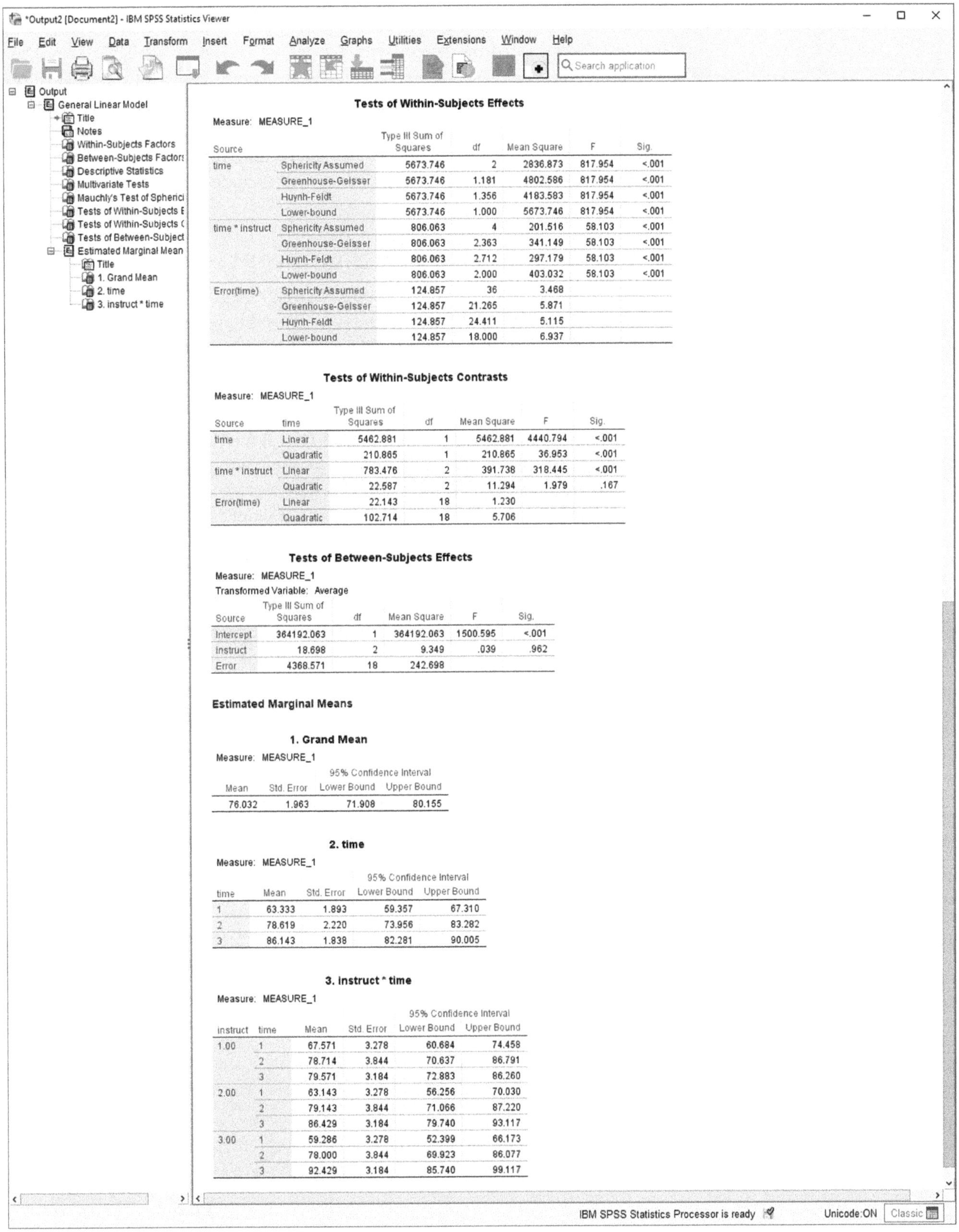

Tests of Within-Subjects Effects

Measure: MEASURE_1

Source		Type III Sum of Squares	df	Mean Square	F	Sig.
time	Sphericity Assumed	5673.746	2	2836.873	817.954	<.001
	Greenhouse-Geisser	5673.746	1.181	4802.586	817.954	<.001
	Huynh-Feldt	5673.746	1.356	4183.583	817.954	<.001
	Lower-bound	5673.746	1.000	5673.746	817.954	<.001
time * instruct	Sphericity Assumed	806.063	4	201.516	58.103	<.001
	Greenhouse-Geisser	806.063	2.363	341.149	58.103	<.001
	Huynh-Feldt	806.063	2.712	297.179	58.103	<.001
	Lower-bound	806.063	2.000	403.032	58.103	<.001
Error(time)	Sphericity Assumed	124.857	36	3.468		
	Greenhouse-Geisser	124.857	21.265	5.871		
	Huynh-Feldt	124.857	24.411	5.115		
	Lower-bound	124.857	18.000	6.937		

Tests of Within-Subjects Contrasts

Measure: MEASURE_1

Source	time	Type III Sum of Squares	df	Mean Square	F	Sig.
time	Linear	5462.881	1	5462.881	4440.794	<.001
	Quadratic	210.865	1	210.865	36.953	<.001
time * instruct	Linear	783.476	2	391.738	318.445	<.001
	Quadratic	22.587	2	11.294	1.979	.167
Error(time)	Linear	22.143	18	1.230		
	Quadratic	102.714	18	5.706		

Tests of Between-Subjects Effects

Measure: MEASURE_1
Transformed Variable: Average

Source	Type III Sum of Squares	df	Mean Square	F	Sig.
Intercept	364192.063	1	364192.063	1500.595	<.001
instruct	18.698	2	9.349	.039	.962
Error	4368.571	18	242.698		

Estimated Marginal Means

1. Grand Mean

Measure: MEASURE_1

Mean	Std. Error	95% Confidence Interval Lower Bound	Upper Bound
76.032	1.963	71.908	80.155

2. time

Measure: MEASURE_1

time	Mean	Std. Error	95% Confidence Interval Lower Bound	Upper Bound
1	63.333	1.893	59.357	67.310
2	78.619	2.220	73.956	83.282
3	86.143	1.838	82.281	90.005

3. instruct * time

Measure: MEASURE_1

instruct	time	Mean	Std. Error	95% Confidence Interval Lower Bound	Upper Bound
1.00	1	67.571	3.278	60.684	74.458
	2	78.714	3.844	70.637	86.791
	3	79.571	3.184	72.883	86.260
2.00	1	63.143	3.278	56.256	70.030
	2	79.143	3.844	71.066	87.220
	3	86.429	3.184	79.740	93.117
3.00	1	59.286	3.278	52.399	66.173
	2	78.000	3.844	69.923	86.077
	3	92.429	3.184	85.740	99.117

This section gives two of the three answers we need (the main effect for TIME and the **interaction** result for TIME × INSTRUCTOR). The second section of output is *Tests of Between-Subjects Effects* (sample output is above). Here, we get the answers that do not contain any within-subjects effects. For our example, we get the main effect for INSTRUCT. Both of these sections must be combined to produce the full answer for our analysis.

If we obtain significant effects, we must perform some sort of post-hoc analysis. Again, this is one of the limitations of SPSS. No easy way to perform the appropriate post-hoc test for repeated-measures (within-subjects) factors is available (You can have it conduct post-hoc tests on between-subjects factors). Ask your instructor for assistance with this.

When describing the results, you should include *F*, the degrees of freedom, and the **significance** level for each main effect and **interaction**. In addition, some **descriptive statistics** must be included (either give means or include a figure).

Phrasing Results That Are Significant

There are (at least) three answers for all mixed-design ANOVAs. See Section 7.3 on factorial ANOVA for more details about how to interpret and phrase the results. For the above example, we could state the following in the results section (note that this assumes that appropriate post-hoc tests have been conducted):

> A 3 × 3 mixed-design ANOVA was calculated to examine the effects of the instructor (Instructors 1, 2, and 3) and time (pretest, midterm, and final) on scores. A significant time × instructor interaction was present ($F\ (4{,}36) = 58.103$, $p < .001$). In addition, the main effect for time was significant ($F\ (2{,}36) = 817.954$, $p < .001$). The main effect for instructor was not significant ($F\ (2{,}18) = .039$, $p > .05$).

With significant **interactions**, it is often helpful to provide a graph with the **descriptive statistics**. By selecting the *Plots* option in the main **dialog box**, you can make graphs of the **interaction** like the one on the next page. **Interactions** add considerable complexity to the interpretation of statistical results. Consult a research methods text or ask your instructor for more help with **interactions**.

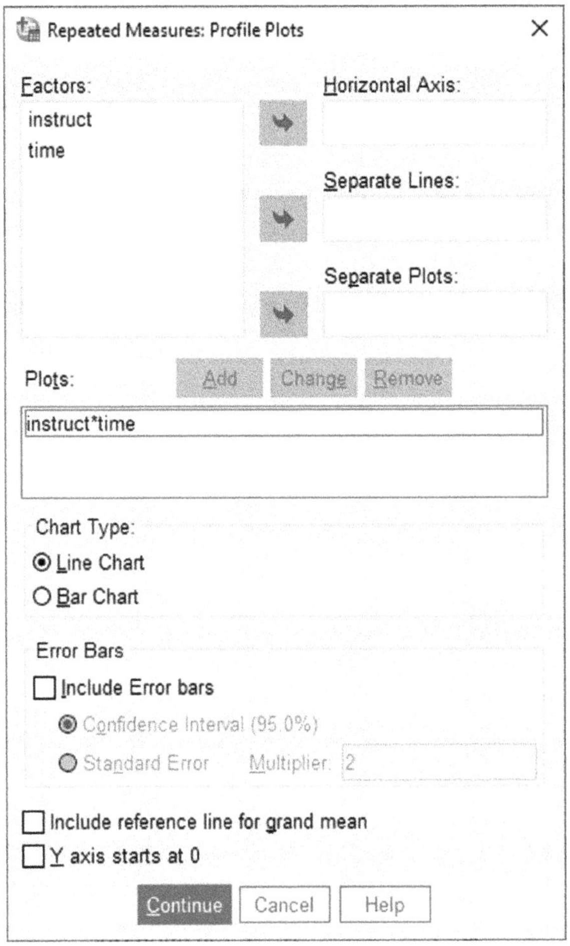
Repeated Measures: Profile Plots
Factors:
instruct
time
Horizontal Axis:
Separate Lines:
Separate Plots:
Plots:
Add
Change
Remove
instruct*time
Chart Type:
Line Chart
Bar Chart
Error Bars
Include Error bars
Confidence Interval (95.0%)
Standard Error
Multiplier: 2
Include reference line for grand mean
Y axis starts at 0
Continue
Cancel
Help

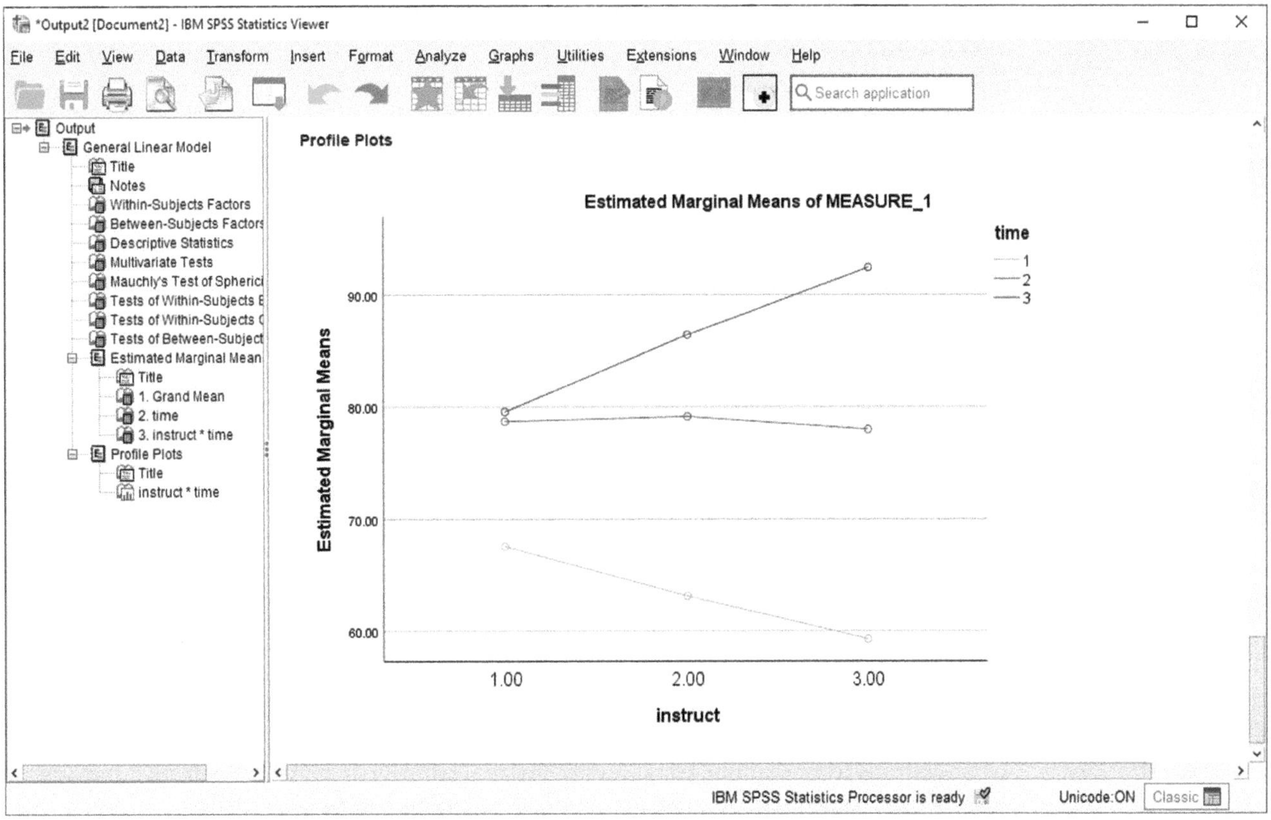
*Output2 [Document2] - IBM SPSS Statistics Viewer
File Edit View Data Transform Insert Format Analyze Graphs Utilities Extensions Window Help
Search application
Output
General Linear Model
Title
Notes
Within-Subjects Factors
Between-Subjects Factors
Descriptive Statistics
Multivariate Tests
Estimated Marginal Means
Title
1. Grand Mean
2. time
3. instruct * time
Profile Plots
Title
instruct * time
Profile Plots
Estimated Marginal Means of MEASURE_1
time
1
2
3
Estimated Marginal Means
90.00
80.00
70.00
60.00
1.00
2.00
3.00
instruct
IBM SPSS Statistics Processor is ready
Unicode:ON
Classic

Phrasing Results That Are Not Significant

If our results had not been significant (this output is from TIME.sav in Appendix C), we could have stated the following:

> A 3 × 2 mixed-design ANOVA was calculated to examine the effects of the group (1 or 2) and time (before, during, after) on scores. No significant main effects or interactions were found. The time × group interaction (F (2,8) = .809, $p > .05$), the main effect for time (F (2,8) = .043, $p > .05$), and the main effect for group (F (1,4) = .2.50, $p > .05$) were not significant. Scores were not influenced by either time or group.

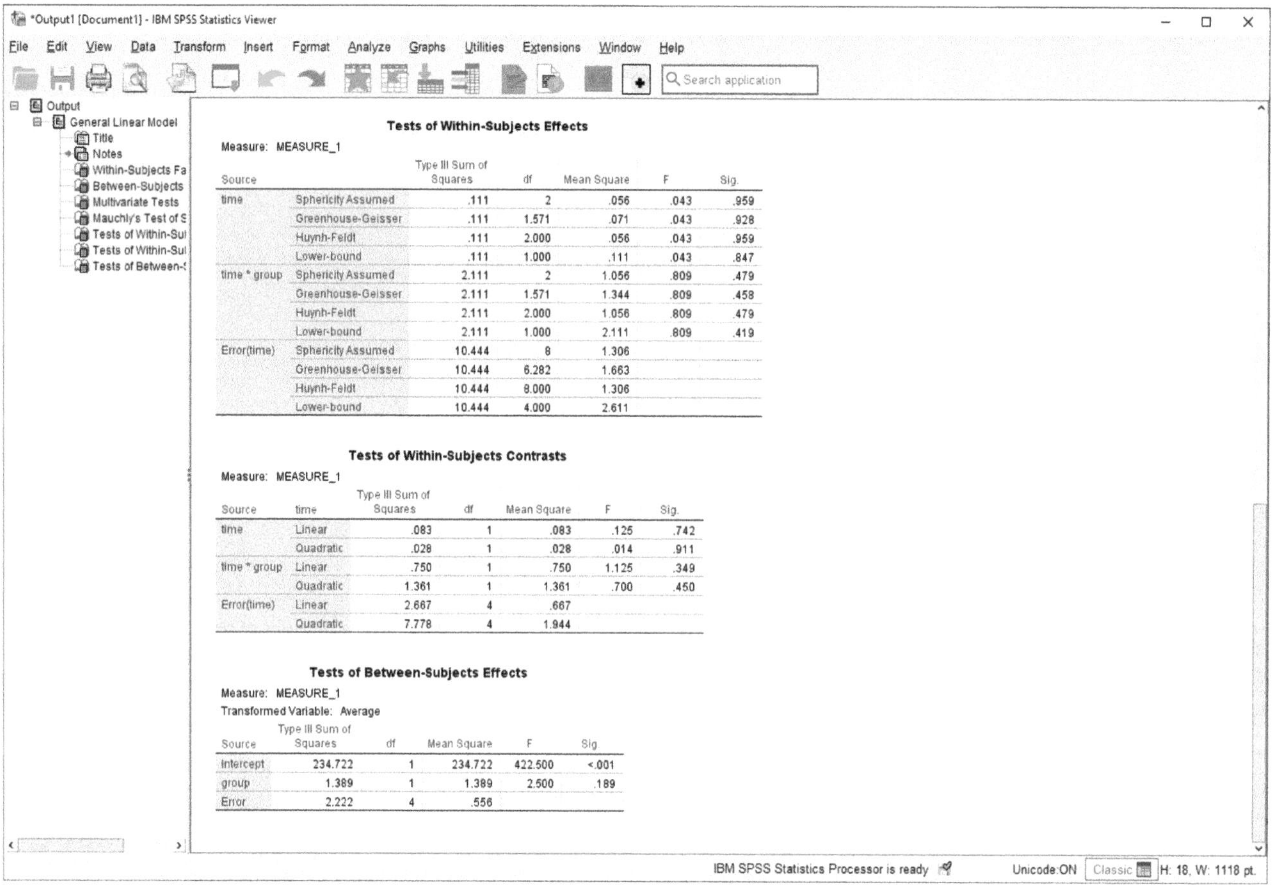

Tests of Within-Subjects Effects

Measure: MEASURE_1

Source		Type III Sum of Squares	df	Mean Square	F	Sig.
time	Sphericity Assumed	.111	2	.056	.043	.959
	Greenhouse-Geisser	.111	1.571	.071	.043	.928
	Huynh-Feldt	.111	2.000	.056	.043	.959
	Lower-bound	.111	1.000	.111	.043	.847
time * group	Sphericity Assumed	2.111	2	1.056	.809	.479
	Greenhouse-Geisser	2.111	1.571	1.344	.809	.458
	Huynh-Feldt	2.111	2.000	1.056	.809	.479
	Lower-bound	2.111	1.000	2.111	.809	.419
Error(time)	Sphericity Assumed	10.444	8	1.306		
	Greenhouse-Geisser	10.444	6.282	1.663		
	Huynh-Feldt	10.444	8.000	1.306		
	Lower-bound	10.444	4.000	2.611		

Tests of Within-Subjects Contrasts

Measure: MEASURE_1

Source	time	Type III Sum of Squares	df	Mean Square	F	Sig.
time	Linear	.083	1	.083	.125	.742
	Quadratic	.028	1	.028	.014	.911
time * group	Linear	.750	1	.750	1.125	.349
	Quadratic	1.361	1	1.361	.700	.450
Error(time)	Linear	2.667	4	.667		
	Quadratic	7.778	4	1.944		

Tests of Between-Subjects Effects

Measure: MEASURE_1
Transformed Variable: Average

Source	Type III Sum of Squares	df	Mean Square	F	Sig.
Intercept	234.722	1	234.722	422.500	<.001
group	1.389	1	1.389	2.500	.189
Error	2.222	4	.556		

Practice Exercise

Use Practice Dataset 3 in Appendix B. Determine if anxiety **levels** changed over time for each of the treatment (CONDITION) types. How did time change anxiety **levels** for each treatment? Write a statement of results.

Section 7.6 Analysis of Covariance

Description

Analysis of covariance (ANCOVA) allows you to remove the effect of a known **covariate**. In this way, it becomes a statistical method of control. With methodological controls (e.g., **random assignment**), internal **validity** is gained. When such methodological controls are not possible, statistical controls can be used.

ANCOVA can be performed by using the *GLM* command if you have repeated-measures factors. Because the *GLM* command is not included in the Base Statistics module, it is not included here.

Assumptions

ANCOVA requires that the **covariate** be significantly correlated with the **dependent variable**. The **dependent variable** and the **covariate** should be at the **interval** or **ratio levels**. In addition, both should be normally distributed.

SPSS Data Format

The SPSS data file must contain one variable for each **independent variable**, one variable representing the **dependent variable**, and at least one **covariate**.

Running the Command

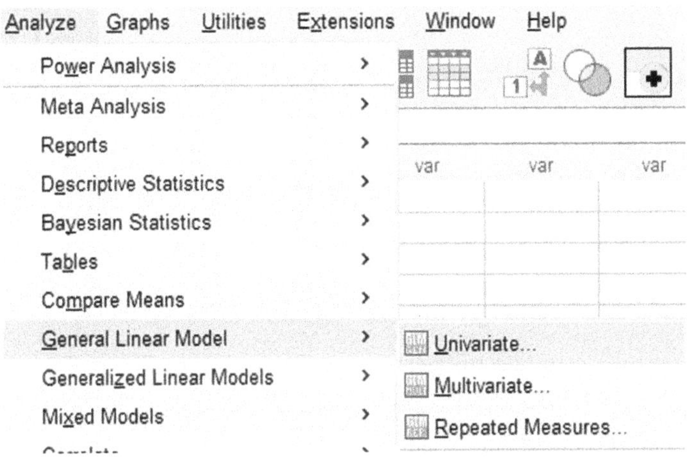

The *Factorial ANOVA* command is used to run ANCOVA. To run it, click *Analyze*, then *General Linear Model*, then *Univariate*. Follow the directions discussed for factorial ANOVA, using the HEIGHT.sav sample data file. Place the variable HEIGHT as your *Dependent Variable*. Enter SEX as your *Fixed Factor*, then WEIGHT as the *Covariate*. This last step determines the difference between regular factorial ANOVA and ANCOVA. To obtain descriptive statistics check *Descriptive statistics* under *Options*. Click *OK* to run the ANCOVA.

Reading the Output

The output consists of one main source table (shown on the next page). This table gives you the main effects and **interactions** you would have received with a normal factorial ANOVA. In addition, there is a row for each **covariate**. In our example, we have one main effect (SEX) and one **covariate** (WEIGHT). Normally, we examine the **covariate** line only to confirm that the **covariate** is significantly related to the **dependent variable**.

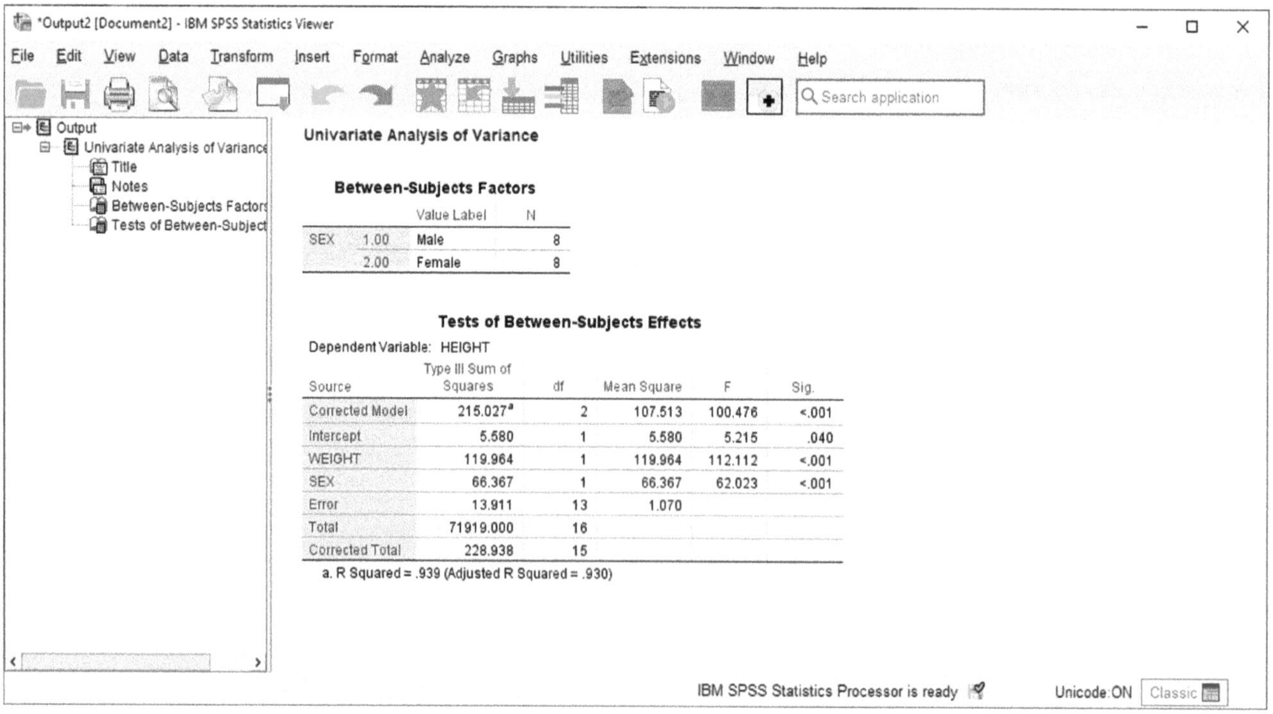

Univariate Analysis of Variance

Between-Subjects Factors

		Value Label	N
SEX	1.00	Male	8
	2.00	Female	8

Tests of Between-Subjects Effects

Dependent Variable: HEIGHT

Source	Type III Sum of Squares	df	Mean Square	F	Sig.
Corrected Model	215.027[a]	2	107.513	100.476	<.001
Intercept	5.580	1	5.580	5.215	.040
WEIGHT	119.964	1	119.964	112.112	<.001
SEX	66.367	1	66.367	62.023	<.001
Error	13.911	13	1.070		
Total	71919.000	16			
Corrected Total	228.938	15			

a. R Squared = .939 (Adjusted R Squared = .930)

Drawing Conclusions

This sample analysis was performed to determine if males and females differ in height after weight is accounted for. We know that weight is related to height. Rather than match participants or use methodological controls, we can statistically remove the effect of weight.

When giving the results of ANCOVA, we must give *F*, degrees of freedom, and **significance** levels for all main effects, **interactions**, and **covariates**. If main effects or **interactions** are significant, post-hoc tests must be conducted. **Descriptive statistics** (**mean** and **standard deviation**) for each **level** of the **independent variable** should also be given.

Phrasing Results That Are Significant

The previous example obtained a significant result, so we could state the following:

> A one-way between-subjects ANCOVA was calculated to examine the effect of sex on height, covarying out the effect of weight. Weight was significantly related to height (F (1,13) = 112.11, $p < .001$). The main effect for sex was significant (F (1,13) = 62.02, $p < .001$), with males significantly taller ($M = 69.38$, $sd = 3.70$) than females ($M = 64.50$, $sd = 2.33$).

Phrasing Results That Are Not Significant

If the **covariate** is not significant, we need to repeat the analysis without including the **covariate** (i.e., run a normal ANOVA).

For ANCOVA results that are not significant (but the covariate is significant), you could state the following (this output is obtained from the GRADES.sav data file running an ANCOVA on FINAL grades, covarying out PRETEST to see if REQUIRED has an effect):

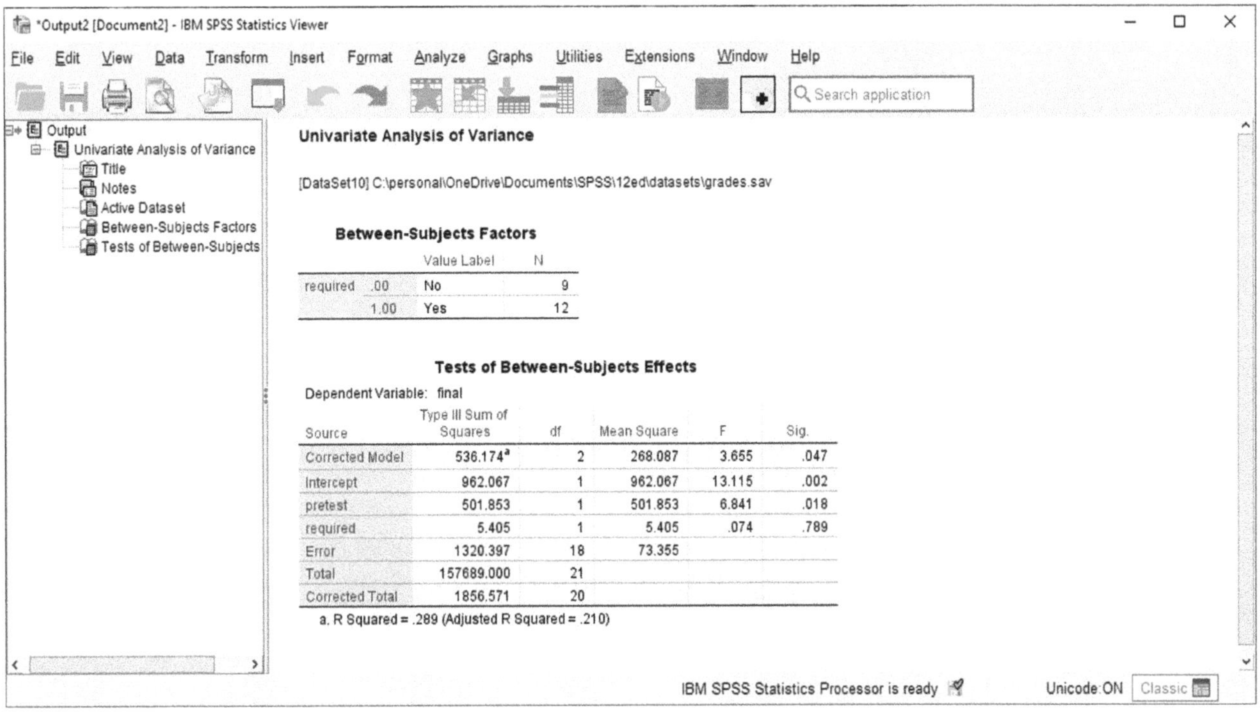

Univariate Analysis of Variance

[DataSet10] C:\personal\OneDrive\Documents\SPSS\12ed\datasets\grades.sav

Between-Subjects Factors

		Value Label	N
required	.00	No	9
	1.00	Yes	12

Tests of Between-Subjects Effects

Dependent Variable: final

Source	Type III Sum of Squares	df	Mean Square	F	Sig.
Corrected Model	536.174[a]	2	268.087	3.655	.047
Intercept	962.067	1	962.067	13.115	.002
pretest	501.853	1	501.853	6.841	.018
required	5.405	1	5.405	.074	.789
Error	1320.397	18	73.355		
Total	157689.000	21			
Corrected Total	1856.571	20			

a. R Squared = .289 (Adjusted R Squared = .210)

A one-way between-subjects ANCOVA was calculated to examine the effect of being a required course on final exam scores, covarying out the pretest scores. Pretest scores were significantly related to final exam scores (F (1,18) = 6.841, $p < .05$). The main effect for being a required course was not significant (F (1,18) = .074, $p > .05$).

Practice Exercise

Using Practice Dataset 2 in Appendix B, determine if salaries are different for males and females. Repeat the analysis, statistically controlling for years of service. Write a statement of results for each. Compare and contrast your two answers.

Section 7.7 Multivariate Analysis of Variance (MANOVA)

Description

Multivariate tests are those that involve more than one **dependent variable**. While it is possible to conduct several univariate tests (one for each **dependent variable**), this causes **Type I error** inflation. Multivariate tests look at all **dependent variables** at once, in much the same way that ANOVA looks at all **levels** of an **independent variable** at once.

Assumptions

MANOVA assumes that you have multiple **dependent variables** that are related to each other. Each **dependent variable** should be normally distributed and measured on an **interval** or **ratio scale**.

SPSS Data Format

The SPSS data file should have a variable for each **dependent variable**. One additional variable is required for each between-subjects **independent variable**. It is also possible to do a MANCOVA, a repeated-measures MANOVA, and a repeated-measures MANCOVA. These extensions require additional variables in the data file.

Running the Command

Note that this procedure requires an optional module. If you do not have this command, you do not have the necessary module installed. This procedure is NOT included in the SPSS Statistics GradPack Base.

The data below represent SAT and GRE scores for 18 participants. Six participants received no special training, six received short-term training before taking the tests, and six received long-term training. GROUP is coded 0 = no training, 1 = short-term, 2 = long-term. Enter the data and save them as SAT.sav.

SAT	GRE	GROUP
580	600	0
520	520	0
500	510	0
410	400	0
650	630	0
480	480	0
500	490	1
640	650	1
500	480	1
500	510	1
580	570	1
490	500	1
520	520	2
620	630	2
550	560	2
500	510	2
540	560	2
600	600	2

Locate the *Multivariate* command by clicking *Analyze,* then *General Linear Model,* and then *Multivariate.*

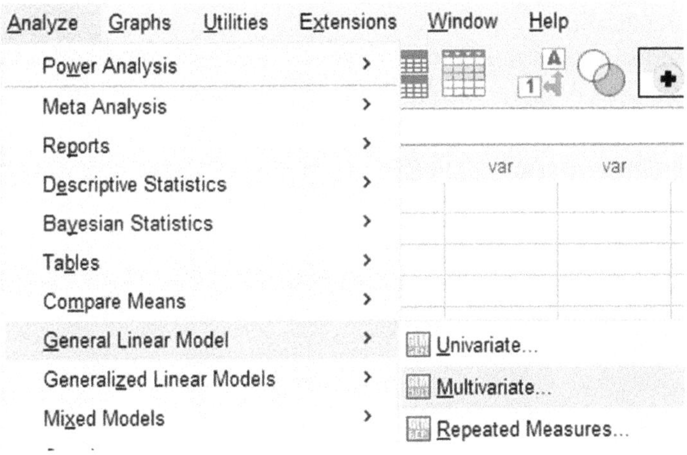

This will bring up the main **dialog box**. Enter the **dependent variables** (SAT and GRE, in this case) in the *Dependent Variables* box. Enter the **independent variable** (GROUP, in this case) in the *Fixed Factor(s)* box. Click *OK* to run the command.

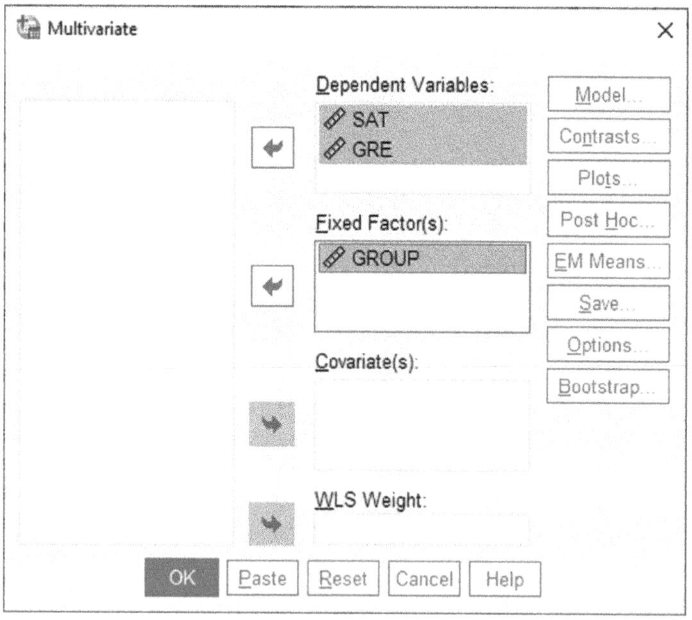

Reading the Output

We are interested in two primary sections of output. The first one gives the results of the multivariate tests. The section labeled GROUP is the one we want. This tells us whether GROUP had an effect on any of our **dependent variables**. Four different types of multivariate test results are given. The most widely used is *Wilks' Lambda*. Thus, the answer for the MANOVA is a *Lambda* of .828, with 4 and 28 degrees of freedom. That value is not significant.

The second section of output we want gives the results of the Between-Subjects univariate tests (ANOVAs) for each **dependent variable**.

*Output2 [Document2] - IBM SPSS Statistics Viewer

General Linear Model

Between-Subjects Factors

		Value Label	N
GROUP	.00	No Training	6
	1.00	Short Term	6
	2.00	Long Term	6

Multivariate Tests[a]

Effect		Value	F	Hypothesis df	Error df	Sig.
Intercept	Pillai's Trace	.988	569.187[b]	2.000	14.000	<.001
	Wilks' Lambda	.012	569.187[b]	2.000	14.000	<.001
	Hotelling's Trace	81.312	569.187[b]	2.000	14.000	<.001
	Roy's Largest Root	81.312	569.187[b]	2.000	14.000	<.001
GROUP	Pillai's Trace	.174	.713	4.000	30.000	.590
	Wilks' Lambda	.828	.693[b]	4.000	28.000	.603
	Hotelling's Trace	.206	.669	4.000	26.000	.619
	Roy's Largest Root	.196	1.469[c]	2.000	15.000	.261

a. Design: Intercept + GROUP
b. Exact statistic
c. The statistic is an upper bound on F that yields a lower bound on the significance level.

Tests of Between-Subjects Effects

Source	Dependent Variable	Type III Sum of Squares	df	Mean Square	F	Sig.
Corrected Model	SAT	3077.778[a]	2	1538.889	.360	.703
	GRE	5200.000[b]	2	2600.000	.587	.568
Intercept	SAT	5205688.889	1	5205688.889	1219.448	<.001
	GRE	5248800.000	1	5248800.000	1185.723	<.001
GROUP	SAT	3077.778	2	1538.889	.360	.703
	GRE	5200.000	2	2600.000	.587	.568
Error	SAT	64033.333	15	4268.889		
	GRE	66400.000	15	4426.667		
Total	SAT	5272800.000	18			
	GRE	5320400.000	18			
Corrected Total	SAT	67111.111	17			
	GRE	71600.000	17			

a. R Squared = .046 (Adjusted R Squared = -.081)
b. R Squared = .073 (Adjusted R Squared = -.051)

IBM SPSS Statistics Processor is ready Unicode:ON Classic H: 18, W: 840 pt.

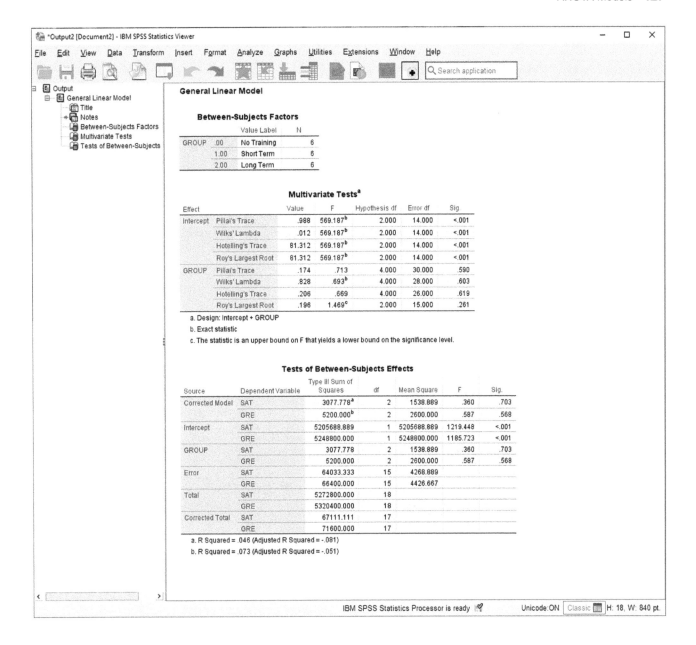

Drawing Conclusions

We interpret the results of the univariate tests only if the group *Wilks' Lambda* is significant. Our results are not significant, but we will first consider how to interpret results that are significant.

Phrasing Results That Are Significant

If we had received the output on the next page (which can be obtained by changing the group 2 values for SAT to 580, 620, 590, 630, 620, and 680), we would have had a significant MANOVA. Note that we would also select *Post Hoc* and had SPSS compute Tukey's HSD for the variable GROUP.

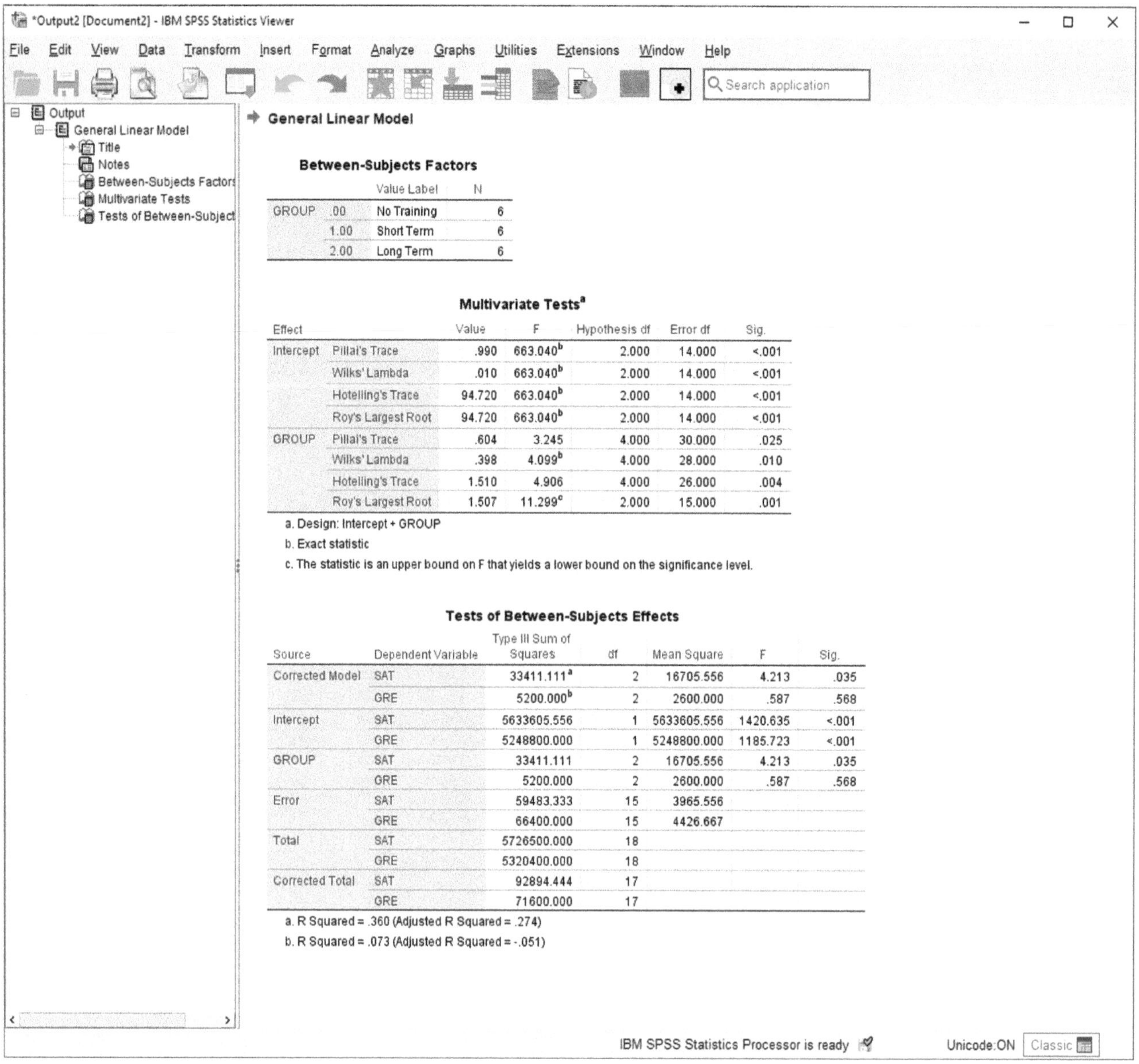

General Linear Model

Between-Subjects Factors

		Value Label	N
GROUP	.00	No Training	6
	1.00	Short Term	6
	2.00	Long Term	6

Multivariate Tests[a]

Effect		Value	F	Hypothesis df	Error df	Sig.
Intercept	Pillai's Trace	.990	663.040[b]	2.000	14.000	<.001
	Wilks' Lambda	.010	663.040[b]	2.000	14.000	<.001
	Hotelling's Trace	94.720	663.040[b]	2.000	14.000	<.001
	Roy's Largest Root	94.720	663.040[b]	2.000	14.000	<.001
GROUP	Pillai's Trace	.604	3.245	4.000	30.000	.025
	Wilks' Lambda	.398	4.099[b]	4.000	28.000	.010
	Hotelling's Trace	1.510	4.906	4.000	26.000	.004
	Roy's Largest Root	1.507	11.299[c]	2.000	15.000	.001

a. Design: Intercept + GROUP
b. Exact statistic
c. The statistic is an upper bound on F that yields a lower bound on the significance level.

Tests of Between-Subjects Effects

Source	Dependent Variable	Type III Sum of Squares	df	Mean Square	F	Sig.
Corrected Model	SAT	33411.111[a]	2	16705.556	4.213	.035
	GRE	5200.000[b]	2	2600.000	.587	.568
Intercept	SAT	5633605.556	1	5633605.556	1420.635	<.001
	GRE	5248800.000	1	5248800.000	1185.723	<.001
GROUP	SAT	33411.111	2	16705.556	4.213	.035
	GRE	5200.000	2	2600.000	.587	.568
Error	SAT	59483.333	15	3965.556		
	GRE	66400.000	15	4426.667		
Total	SAT	5726500.000	18			
	GRE	5320400.000	18			
Corrected Total	SAT	92894.444	17			
	GRE	71600.000	17			

a. R Squared = .360 (Adjusted R Squared = .274)
b. R Squared = .073 (Adjusted R Squared = -.051)

We could state the following:

> A one-way MANOVA was calculated examining the effect of training (none, short-term, or long-term) on *SAT* and *GRE* scores. A significant effect was found (*Lambda* (4,28) = .398, $p = .010$). Follow-up univariate ANOVAs indicated that *SAT* scores were significantly improved by training (F (2,15) = 4.213, $p = .035$). *GRE* scores were not significantly improved by training (F (2,15) = .587, $p > .05$).

Phrasing Results That Are Not Significant

The actual example presented (based upon the SAT.sav data file) was not significant. Therefore, we could state the following in the results section:

> A one-way MANOVA was calculated examining the effect of training (none, short-term, or long-term) on *SAT* and *GRE* scores. No significant effect was found (*Lambda* (4,28) = .828, $p > .05$). Neither *SAT* nor *GRE* scores were significantly influenced by training.

Chapter 8

Nonparametric Inferential Statistics

Nonparametric tests are used when the corresponding parametric procedure is inappropriate. Normally, this is because the **dependent variable** is not **interval-** or **ratio-scaled**. It can also be because the **dependent variable** is not normally distributed. If the data of interest are frequency counts, nonparametric statistics may also be appropriate.

SPSS has a powerful *Nonparametric Tests* command that includes enhanced reports, which help users choose the appropriate test and interpret the result correctly.

Section 8.1 Chi-Square Goodness of Fit

Description

The chi-square goodness of fit test determines whether or not sample proportions match the theoretical values. For instance, it could be used to determine if a die is loaded or fair. It could also be used to compare the proportion of children born with birth defects to the population value (e.g., to determine if a certain neighborhood has a statistically higher-than-normal rate of birth defects).

Assumptions

We need to make very few assumptions. There are no assumptions about the shape of the distribution. The expected frequencies for each category should be at least 1, and no more than 20 percent of the categories should have expected frequencies of less than 5.

SPSS Data Format

SPSS requires only a single variable. It can be measured on any scale, although SPSS requires it to be defined as nominal.

Running the Command

We will create the following dataset and call it COINS.sav. The following data represent the flipping of each of two coins 20 times (H is coded as heads, T as tails).

Coin 1: H T H H T H H T H H H T T T H T H T T H
Coin 2: T T H H T H T H T T H H T H H T H T H H

DOI: 10.4324/9781003450467-8

Name the two variables COIN1 and COIN2, and code H as 1 and T as 2. The data file that you create will have 20 rows of data and two columns, called COIN1 and COIN2. Define the variables as type: Nominal.

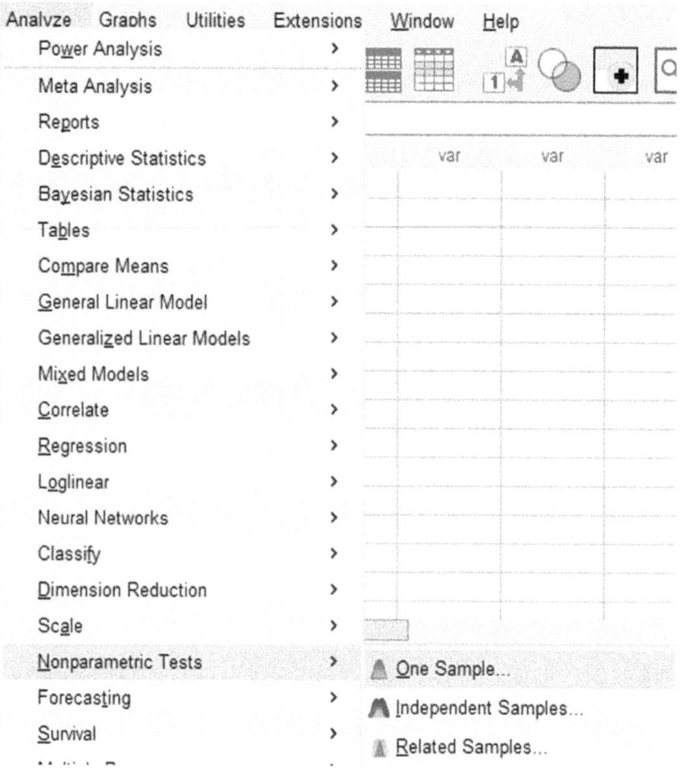

To run the *Chi-Square* command, click *Analyze,* then *Nonparametric Tests,* then *One Sample.* This will bring up the **dialog box** for the *One-Sample Nonparametric Tests* command. You will notice that this new procedure has a **dialog box** that is quite different from the interfaces for other SPSS commands. It was completely rewritten using SPSS's new modeling procedures starting with Version 18.

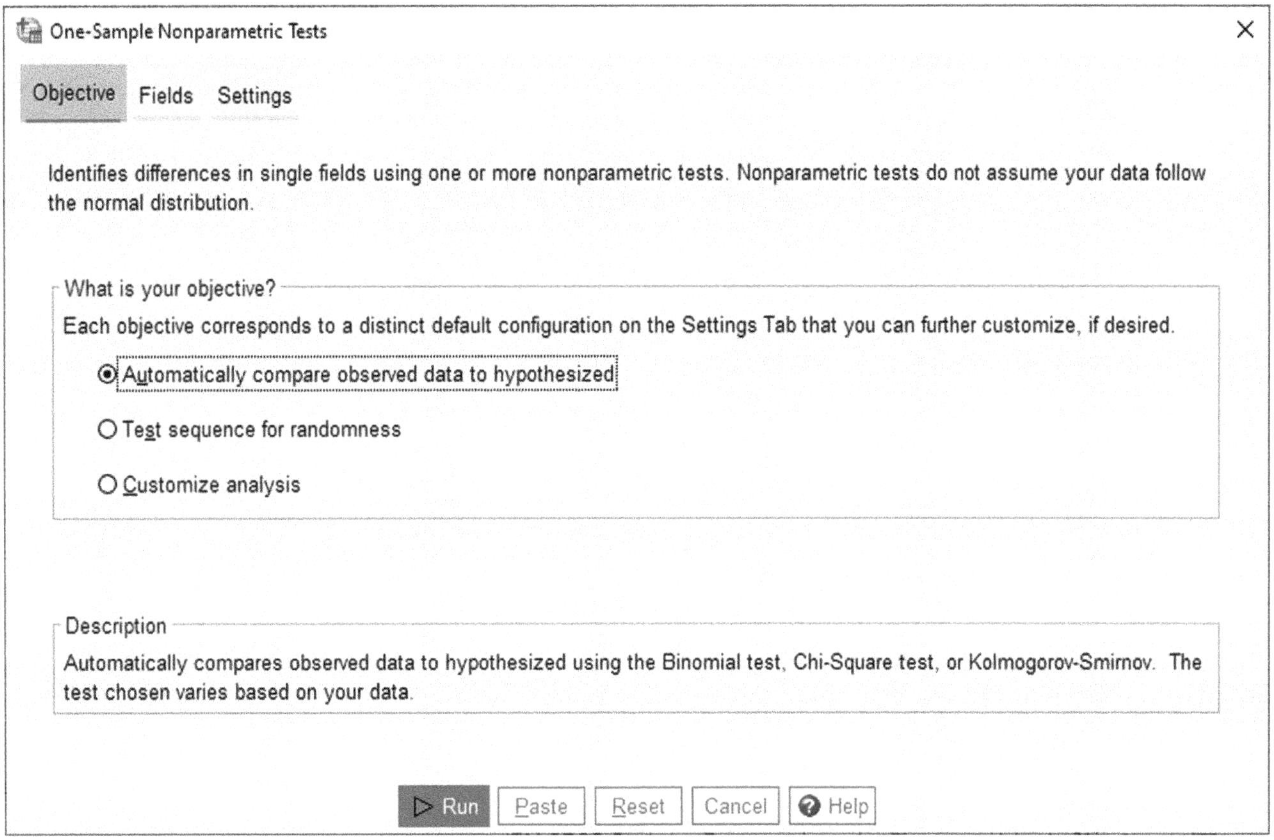

Our objective for a chi-square goodness of fit test is to determine if the distribution of the data is equal to the hypothesized distribution. Therefore, just leave the *Objective* tab the same and click on the *Fields* tab at the top of the window.

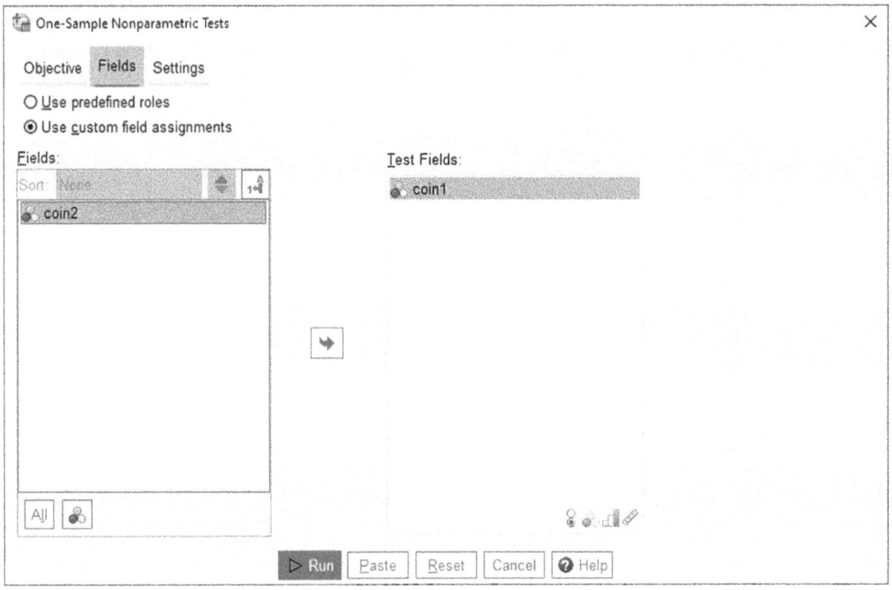

You will need to click *Use custom field assignments* and then move the variable COIN1 to the *Test Fields* area on the right.

This new procedure is actually quite smart and will examine your data and choose the most appropriate test for you. In order to demonstrate the chi-square test of independence, however, we will force it to run only that test by selecting the *Settings* tab across the top, then clicking *Customize tests* under *Choose Tests,* and then checking the *Compare observed probabilities to hypothesized (Chi-Square test)* option.

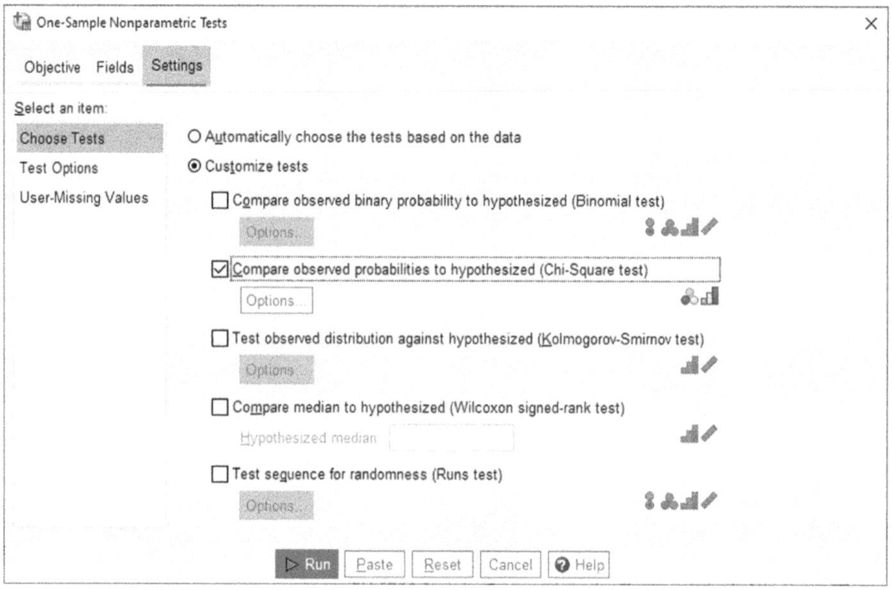

A fair coin has an equal chance of coming up heads or tails. Therefore, there is nothing further we need to do. If, however, different values had different expected probabilities, we would click *Options,* where we would have an opportunity to *Customize expected probability.* Close the Chi-Square Test Options window by clicking *Cancel* and then click *Run.*

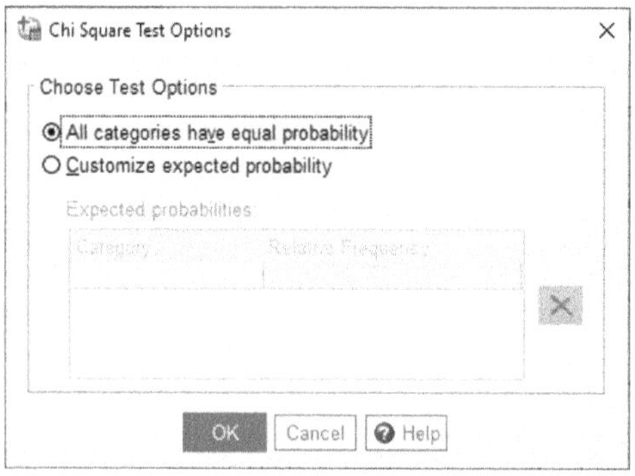

Reading the Output

This new procedure completely changes the way output is presented when compared with other procedures in SPSS. The new output states the **null hypothesis** in standard English, tells you what test was run, gives a **significance** level, and tells you whether or not to reject the **null hypothesis**. In the example shown here, we are clearly told that the one-sample chi-square test (chi-square goodness of fit test) tested whether the values of COIN1 had equal probabilities (50% chance of heads and 50% chance of tails). We were told to "Retain the **null hypothesis**" (which most standard statistics texts would call "fail to reject the **null hypothesis**").

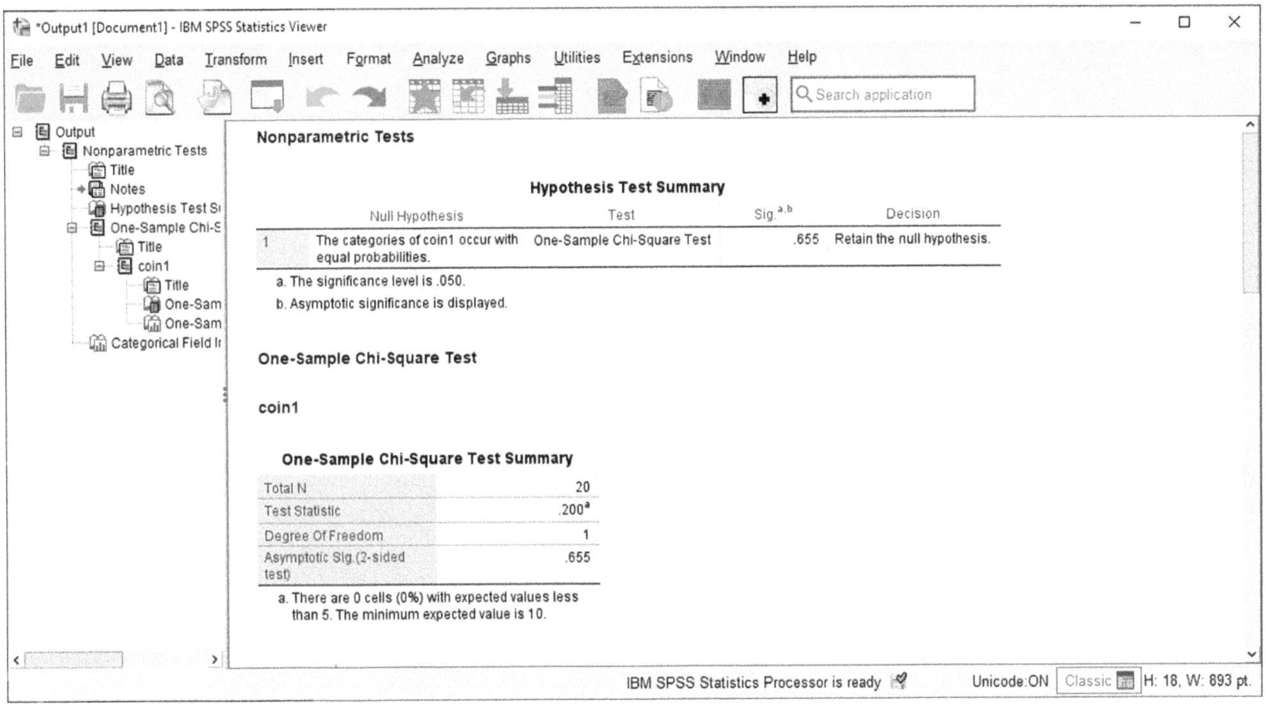

Under the text output is a graphical representation of the analysis.

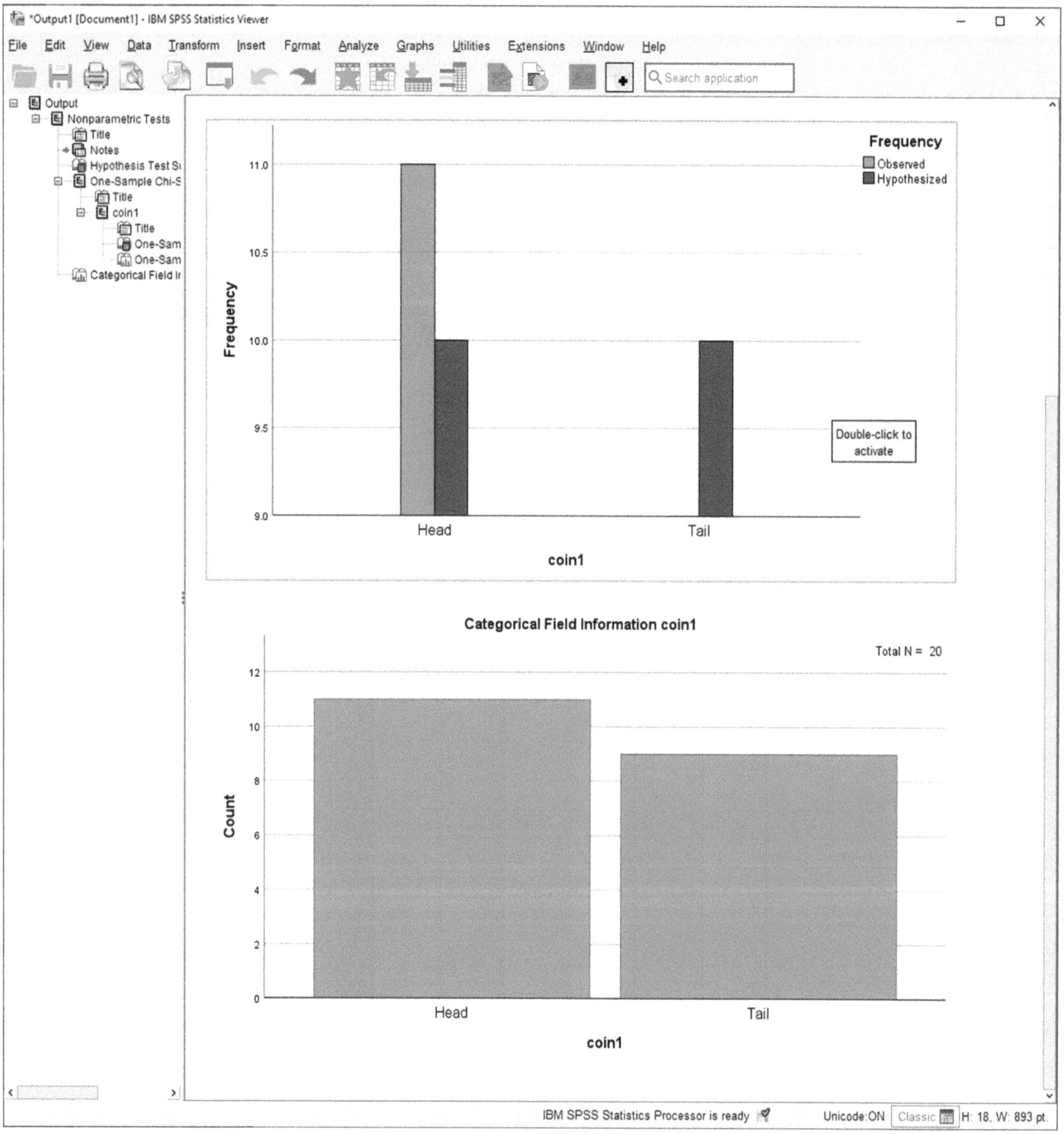

Drawing Conclusions

A significant chi-square test indicates that the data vary from the expected values. A test that is not significant indicates that the data are consistent with the expected values.

Phrasing Results That Are Significant

In describing the results, you should state the value of chi-square (whose symbol is χ^2), the degrees of freedom, the **significance** level, and a description of the results. For instance, with a significant chi-square (this output compares COIN2 to hypothesized proportions of 80 and 20), we could state the following:

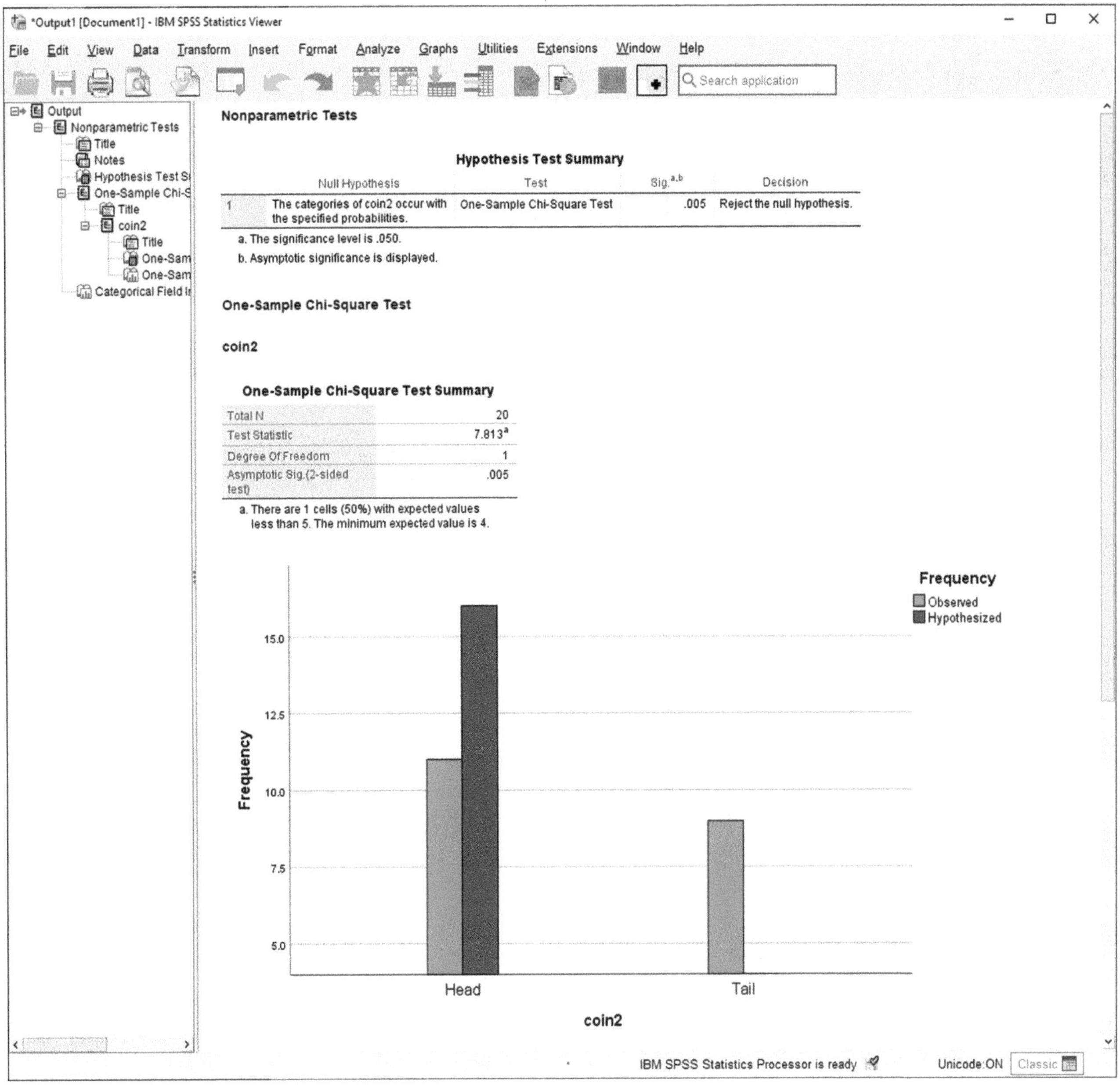

A chi-square goodness of fit test was calculated comparing the frequency of occurrence of each Heads and Tails on a coin. It was hypothesized that Heads would occur 80% of the time. Significant deviation from the hypothesized values was found (χ^2 (1) = 7.813, $p < .05$). The coin does not appear to be biased to come up Head 80% of the time.

Phrasing Results That Are Not Significant

If the analysis produces no significant difference, as in the original example, we could state the following:

> A chi-square goodness of fit test was calculated comparing the frequency of occurrence of heads and tails on a coin. It was hypothesized that each value would occur an equal number of times. No significant deviation from the hypothesized values was found (χ^2 (1) = .20, $p > .05$). The coin appears to be fair.

Practice Exercise

Use Practice Dataset 2 in Appendix B. In the population from which the sample was drawn, 20 percent of employees are clerical, 50 percent are technical, and 30 percent are professional. Determine whether or not the sample drawn conforms to these values. Hint: You will need to *customize expected probabilities* and enter the category values (1, 2, 3) and relative percentages (20, 50, 30). Also, be sure you have entered the variable CLASSIFY as nominal.

Section 8.2 Chi-Square Test of Independence

Description

The chi-square test of independence tests whether or not two variables are independent of each other. For instance, flips of a coin should be **independent events**, so knowing the outcome of one coin toss should not tell us anything about the second coin toss. The chi-square test of independence is essentially a nonparametric version of the **interaction** term in ANOVA.

Assumptions

Very few assumptions are needed. For instance, we make no assumptions about the shape of the distribution. The expected frequencies for each category should be at least 1, and no more than 20 percent of the categories should have expected frequencies of less than 5. (Note: To save space, these examples violate that assumption.)

SPSS Data Format

At least two variables are required.

Running the Command

The chi-square test of independence is a component of the *Crosstabs* command rather than the *Nonparametric Statistics* command. For more details, see Section 3.2 in Chapter 3 on frequency distributions for multiple variables.

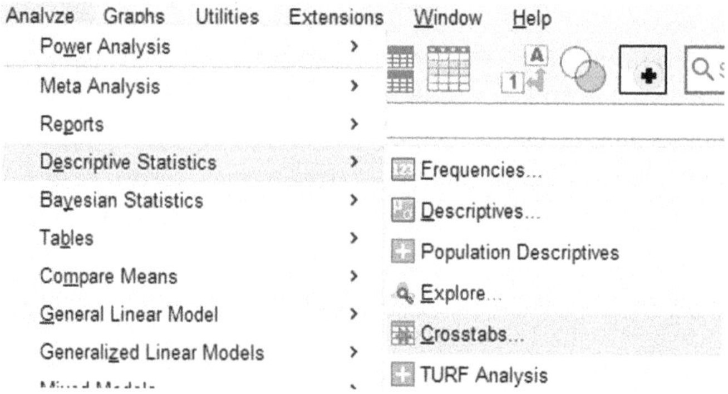

This example uses the COINS.sav file. COIN1 is placed in the *Row(s)* blank, and COIN2 is placed in the *Column(s)* blank.

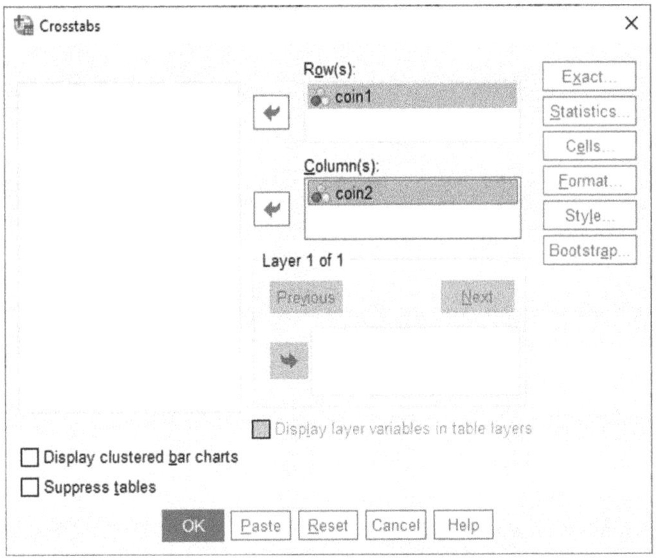

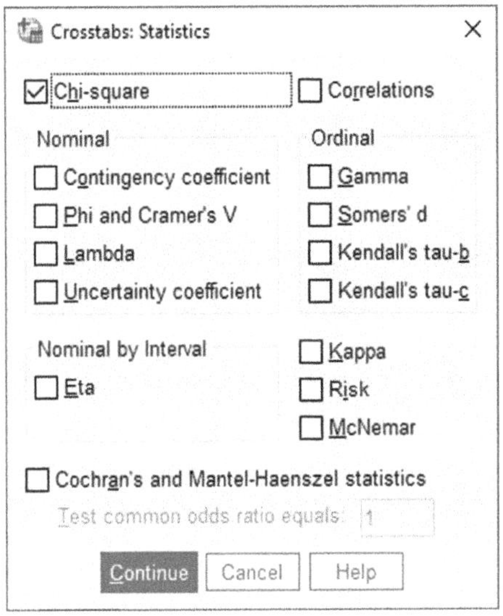

Click *Statistics*, then check the *Chi-square* box. Click *Continue*. You may also want to click *Cells* to select expected frequencies in addition to observed frequencies.

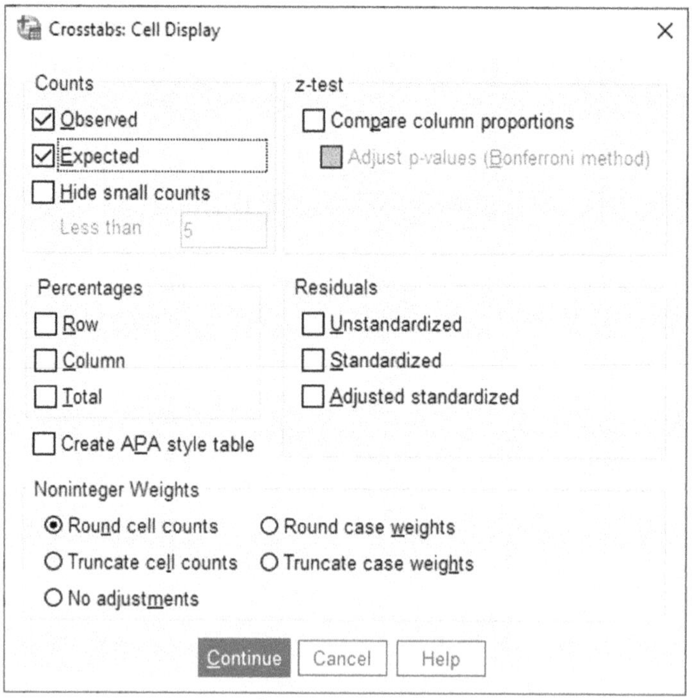

Note that you can also use the *Cells* option to display the percentages of each variable for each value. This is especially useful when your groups are different sizes, but it is unnecessary in this example.

Click *OK* to run the analysis.

Reading the Output

The output consists of two parts. The first part gives you the counts. In this example, the actual and expected frequencies are shown because they were selected with the *Cells* option.

The second part of the output gives the results of the chi-square test. The most commonly used value is the Pearson chi-square, shown in the first row (value of .737).

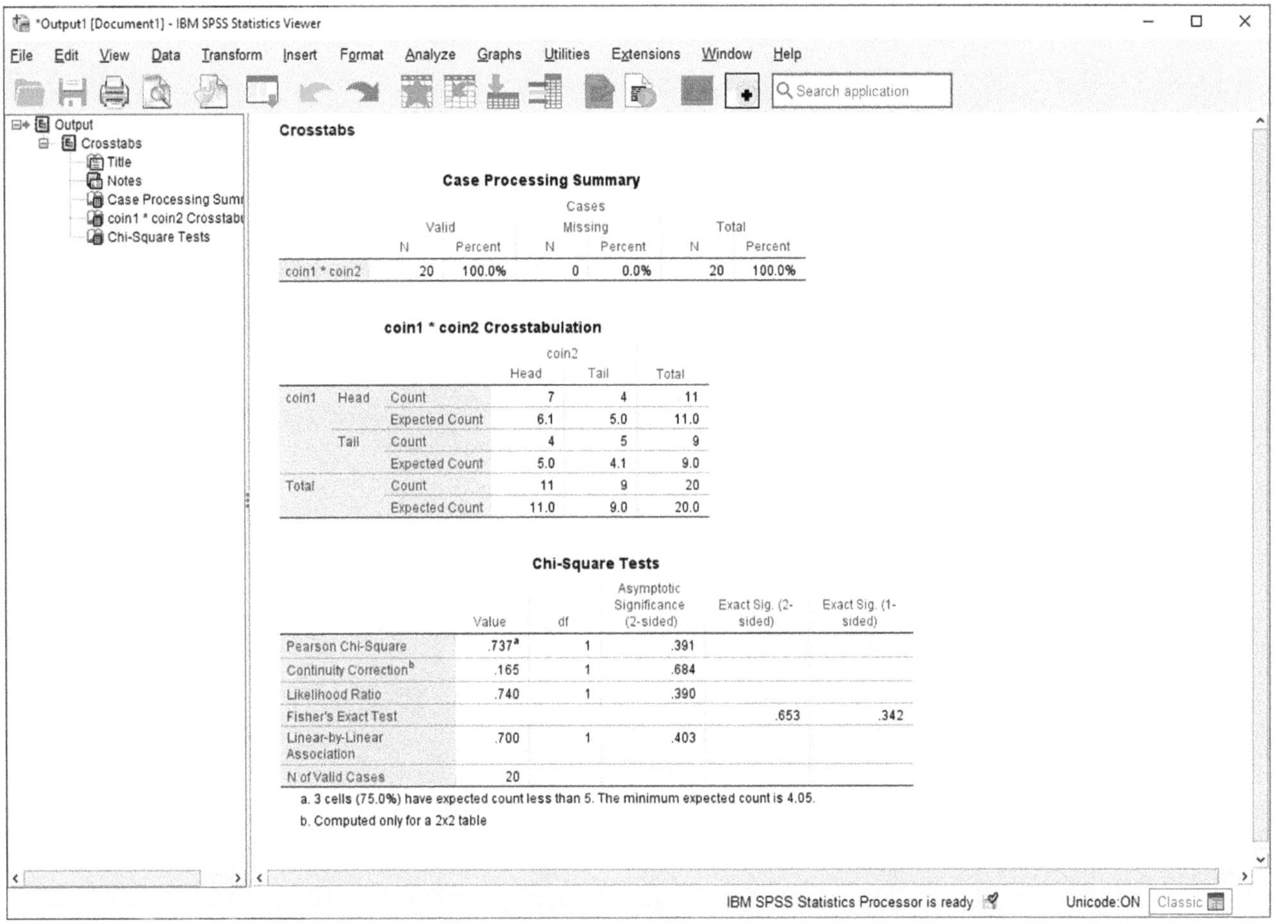

Crosstabs

Case Processing Summary

	Cases					
	Valid		Missing		Total	
	N	Percent	N	Percent	N	Percent
coin1 * coin2	20	100.0%	0	0.0%	20	100.0%

coin1 * coin2 Crosstabulation

			coin2		
			Head	Tail	Total
coin1	Head	Count	7	4	11
		Expected Count	6.1	5.0	11.0
	Tail	Count	4	5	9
		Expected Count	5.0	4.1	9.0
Total		Count	11	9	20
		Expected Count	11.0	9.0	20.0

Chi-Square Tests

	Value	df	Asymptotic Significance (2-sided)	Exact Sig. (2-sided)	Exact Sig. (1-sided)
Pearson Chi-Square	.737[a]	1	.391		
Continuity Correction[b]	.165	1	.684		
Likelihood Ratio	.740	1	.390		
Fisher's Exact Test				.653	.342
Linear-by-Linear Association	.700	1	.403		
N of Valid Cases	20				

a. 3 cells (75.0%) have expected count less than 5. The minimum expected count is 4.05.

b. Computed only for a 2x2 table

Drawing Conclusions

A significant chi-square test result indicates that the two variables are not independent. A value that is not significant indicates that the variables do not vary significantly from independence.

Phrasing Results That Are Significant

In describing the results, you should give the value of chi-square, the degrees of freedom, the **significance** level, and a description of the results. For instance, with a significant chi-square (for a dataset different from the one discussed above), we could state the following: These results can be obtained with the following data, testing COIN1 and COIN3.

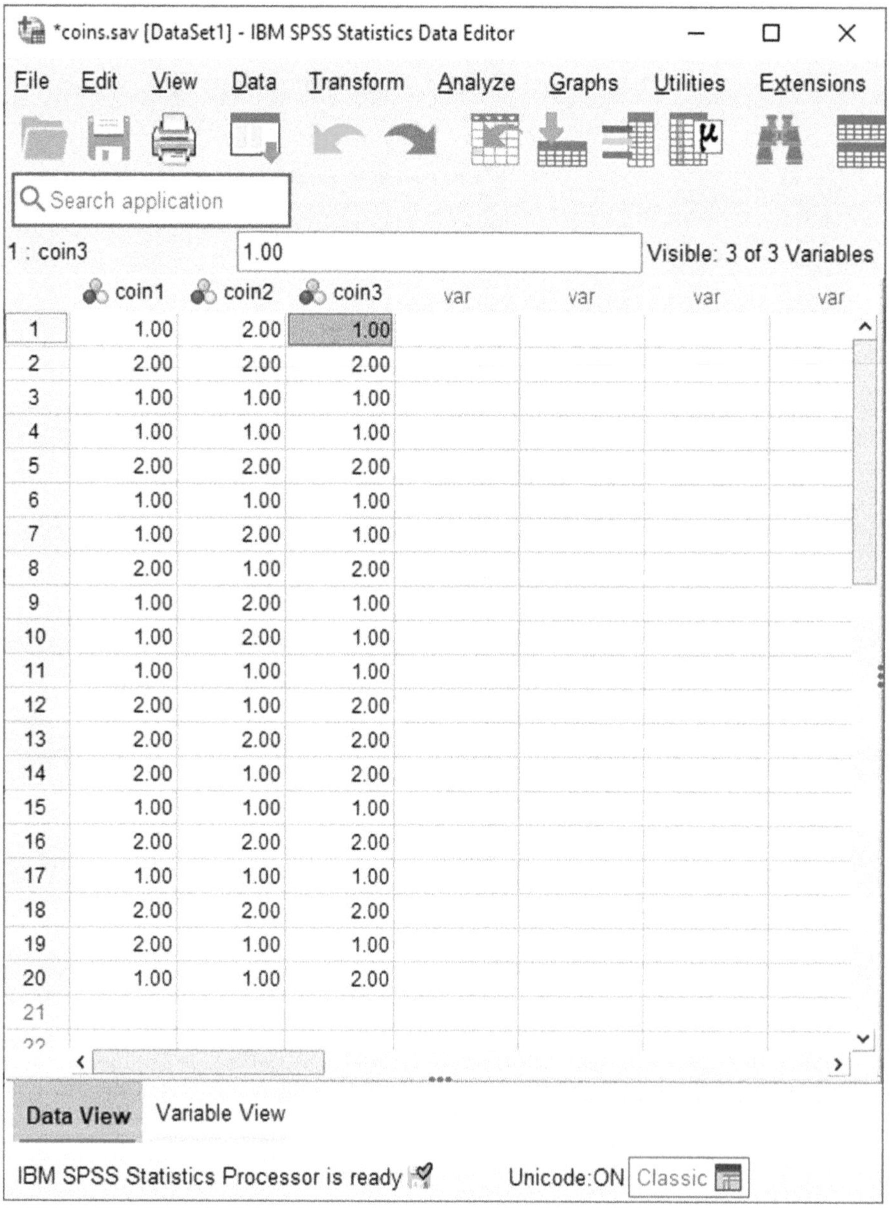

	coin1	coin2	coin3	var	var	var	var
1	1.00	2.00	1.00				
2	2.00	2.00	2.00				
3	1.00	1.00	1.00				
4	1.00	1.00	1.00				
5	2.00	2.00	2.00				
6	1.00	1.00	1.00				
7	1.00	2.00	1.00				
8	2.00	1.00	2.00				
9	1.00	2.00	1.00				
10	1.00	2.00	1.00				
11	1.00	1.00	1.00				
12	2.00	1.00	2.00				
13	2.00	2.00	2.00				
14	2.00	1.00	2.00				
15	1.00	1.00	1.00				
16	2.00	2.00	2.00				
17	1.00	1.00	1.00				
18	2.00	2.00	2.00				
19	2.00	1.00	1.00				
20	1.00	1.00	2.00				
21							

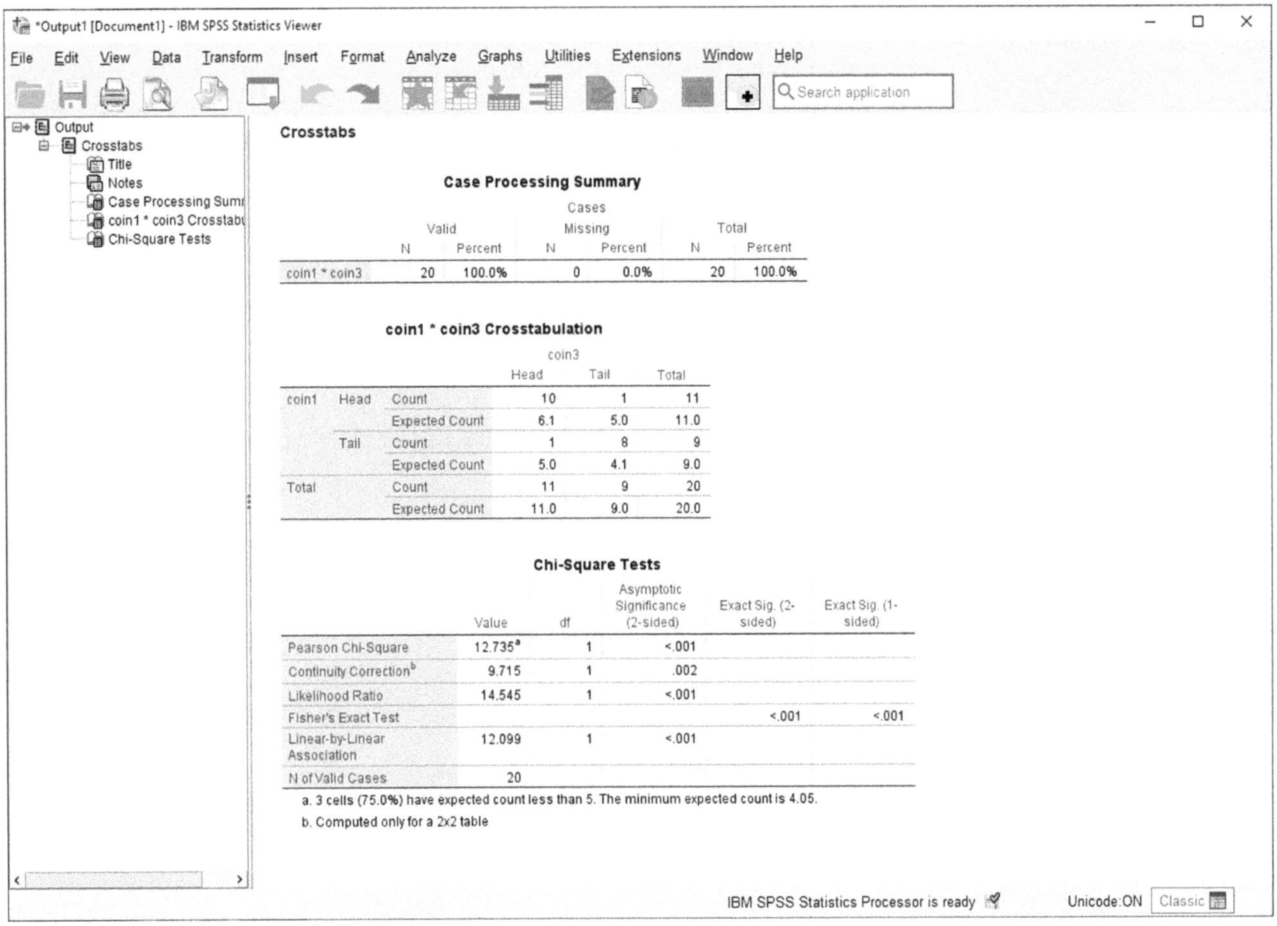

Crosstabs

Case Processing Summary

	Cases					
	Valid		Missing		Total	
	N	Percent	N	Percent	N	Percent
coin1 * coin3	20	100.0%	0	0.0%	20	100.0%

coin1 * coin3 Crosstabulation

			coin3		
			Head	Tail	Total
coin1	Head	Count	10	1	11
		Expected Count	6.1	5.0	11.0
	Tail	Count	1	8	9
		Expected Count	5.0	4.1	9.0
Total		Count	11	9	20
		Expected Count	11.0	9.0	20.0

Chi-Square Tests

	Value	df	Asymptotic Significance (2-sided)	Exact Sig. (2-sided)	Exact Sig. (1-sided)
Pearson Chi-Square	12.735[a]	1	<.001		
Continuity Correction[b]	9.715	1	.002		
Likelihood Ratio	14.545	1	<.001		
Fisher's Exact Test				<.001	<.001
Linear-by-Linear Association	12.099	1	<.001		
N of Valid Cases	20				

a. 3 cells (75.0%) have expected count less than 5. The minimum expected count is 4.05.

b. Computed only for a 2x2 table

> A chi-square test of independence was calculated comparing whether flips of a coin were independent. A significant interaction was found (χ^2 (1) = 12.735, $p < .05$). The flips of a coin were dependent on each other.

Phrasing Results That Are Not Significant

A chi-square test that is not significant indicates that there is no significant dependence of one variable on the other. The coin original (COIN1 and COIN2) example was not significant. Therefore, we could state the following:

> A chi-square test of independence was calculated comparing the results of flipping two coins. No significant relationship was found (χ^2 (1) = .737, $p > .05$). Flips of a coin appear to be independent events.

Practice Exercise

A researcher wants to know if individuals are more likely to help in an emergency when they are indoors or when they are outdoors. Of 28 participants who were outdoors, 19 helped and 9 did not. Of 23 participants who were indoors, 8 helped and 15 did not. Enter these data, and find out if helping behavior is affected by the environment. The key to this problem is in the data entry. Hint: How many participants were there, and how many pieces of information do we know about each? Having SPSS give you cell percentages will help you with the interpretation of this problem.

Section 8.3 Mann-Whitney *U* Test

Description

The Mann-Whitney *U* test is the nonparametric equivalent of the independent *t* test. It tests whether or not two independent samples are from the same distribution. The Mann-Whitney *U* test is weaker than the independent *t* test, and the *t* test should be used if you can meet its assumptions.

Assumptions

The Mann-Whitney *U* test uses the rankings of the data. Therefore, the data for the two samples must at least be **ordinal**. There are no assumptions about the shape of the distribution.

SPSS Data Format

This command requires a single variable representing the **dependent variable** and a second variable indicating group membership. SPSS requires the **dependent variable** to be a scale-type variable even though the variable itself only needs to be **ordinal**.

Running the Command

This example will use a new data file. It represents eight participants in a series of races. There were long races, medium races, and short races. Participants were either experienced (2) or had medium experience (1).

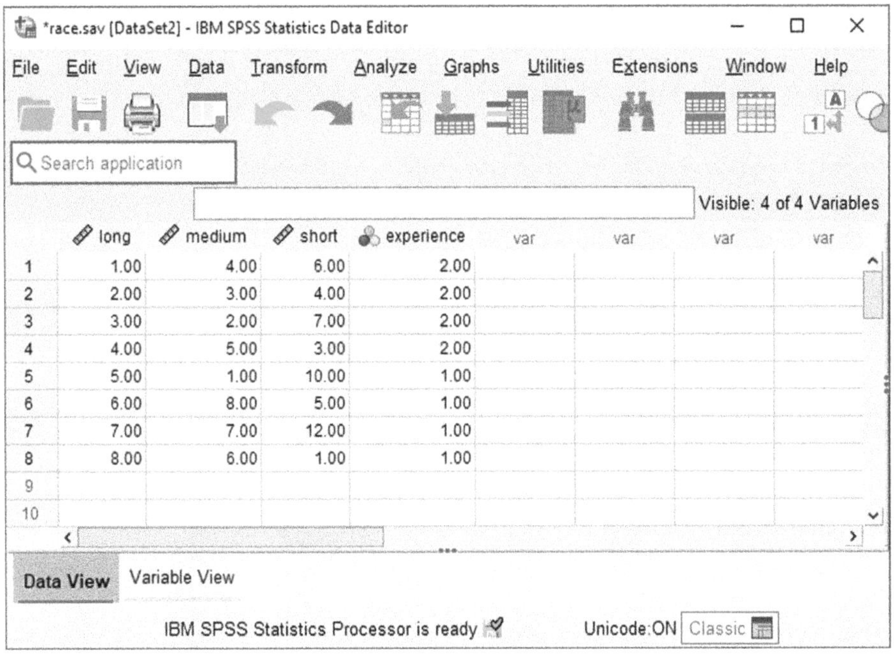

Enter the data from the figure shown above in a new file and save the data file as RACE.sav. The values for LONG, MEDIUM, and SHORT represent the results of the race, with 1 being first place and 8 being last.

Make sure that LONG, MEDIUM, and SHORT are scale-type variables and that EXPERIENCE is a nominal-type variable.

To run the command, click *Analyze*, then *Nonparametric Tests*, then *Independent Samples*. This will bring up the main **dialog box**.

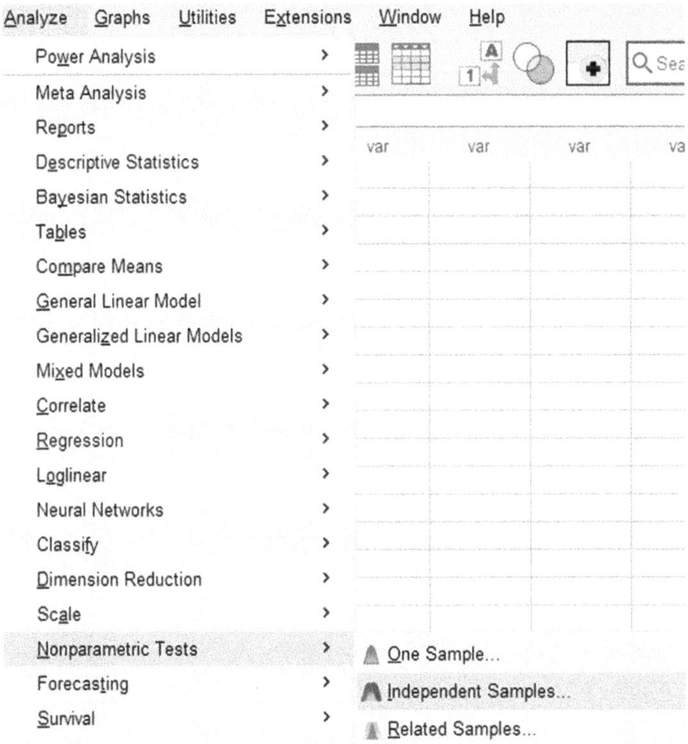

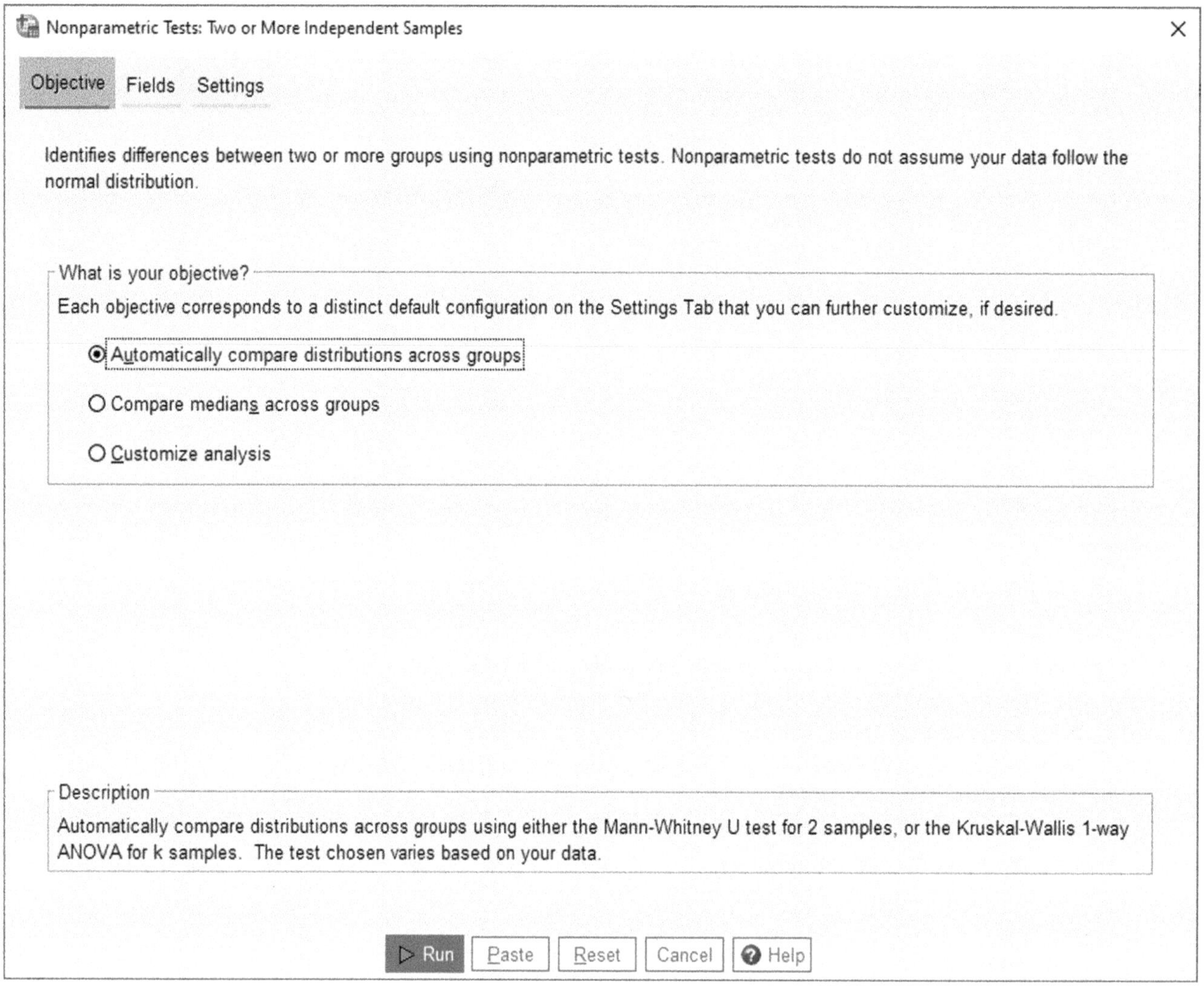
Nonparametric Tests: Two or More Independent Samples
Objective
Fields
Settings
Identifies differences between two or more groups using nonparametric tests. Nonparametric tests do not assume your data follow the normal distribution.
What is your objective?
Each objective corresponds to a distinct default configuration on the Settings Tab that you can further customize, if desired.
Automatically compare distributions across groups
Compare medians across groups
Customize analysis
Description
Automatically compare distributions across groups using either the Mann-Whitney U test for 2 samples, or the Kruskal-Wallis 1-way ANOVA for k samples. The test chosen varies based on your data.
Run
Paste
Reset
Cancel
Help

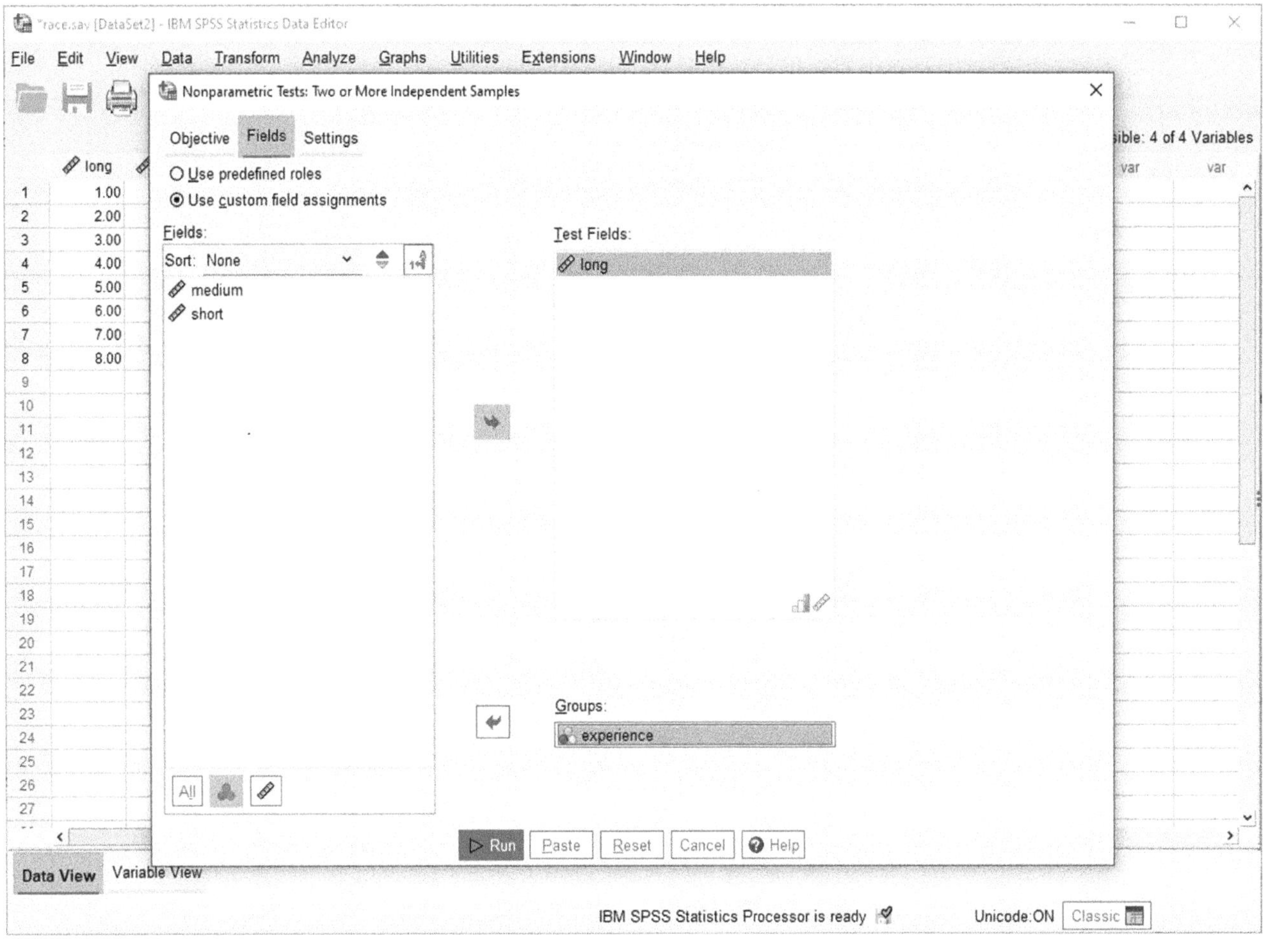

Leave the *Objective* to *Automatically compare distributions across groups.* Then click on the *Fields* tab. Next, enter EXPERIENCE in the *Groups* area. Click on *Run* to perform the analysis.

Reading the Output

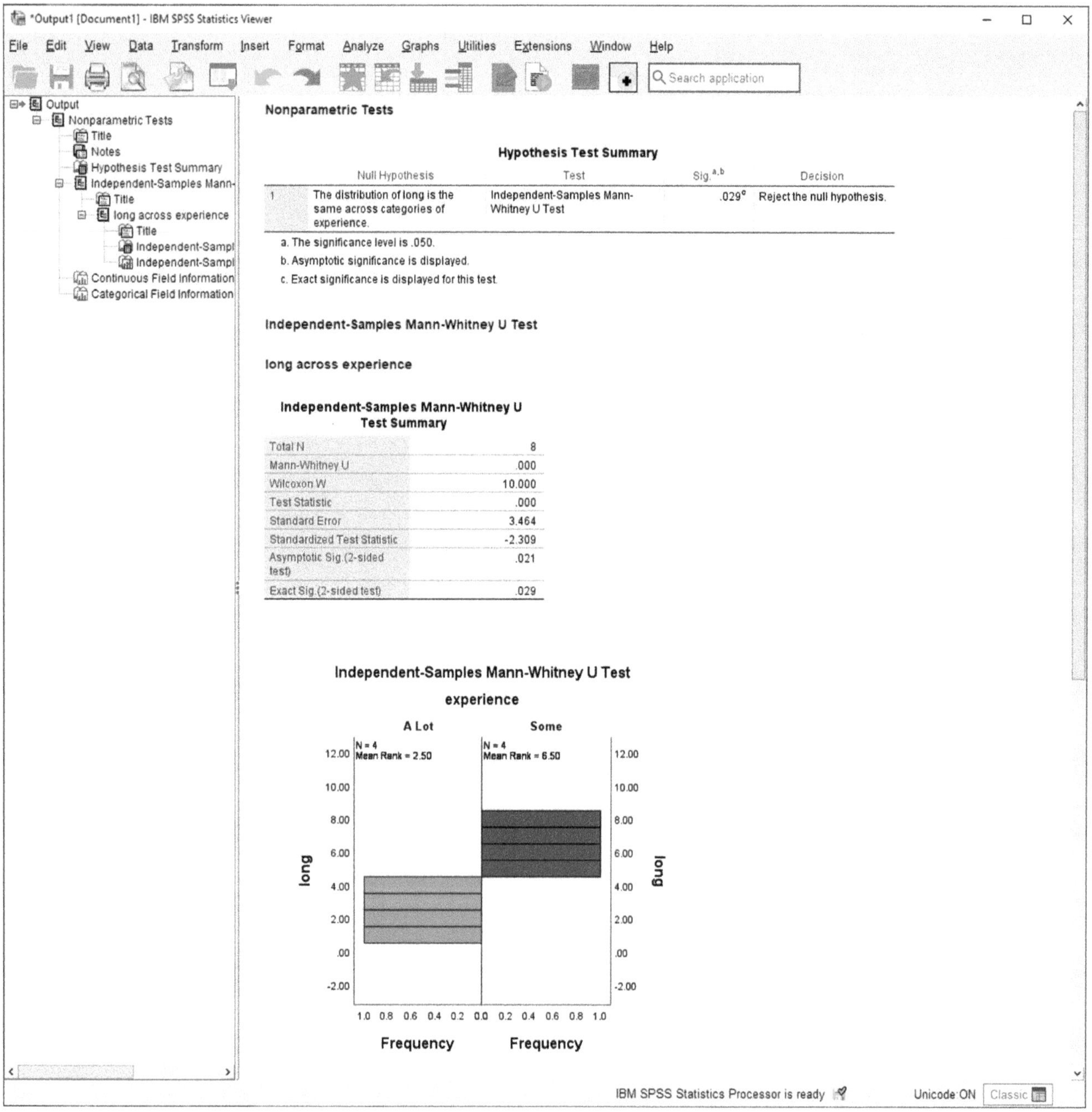

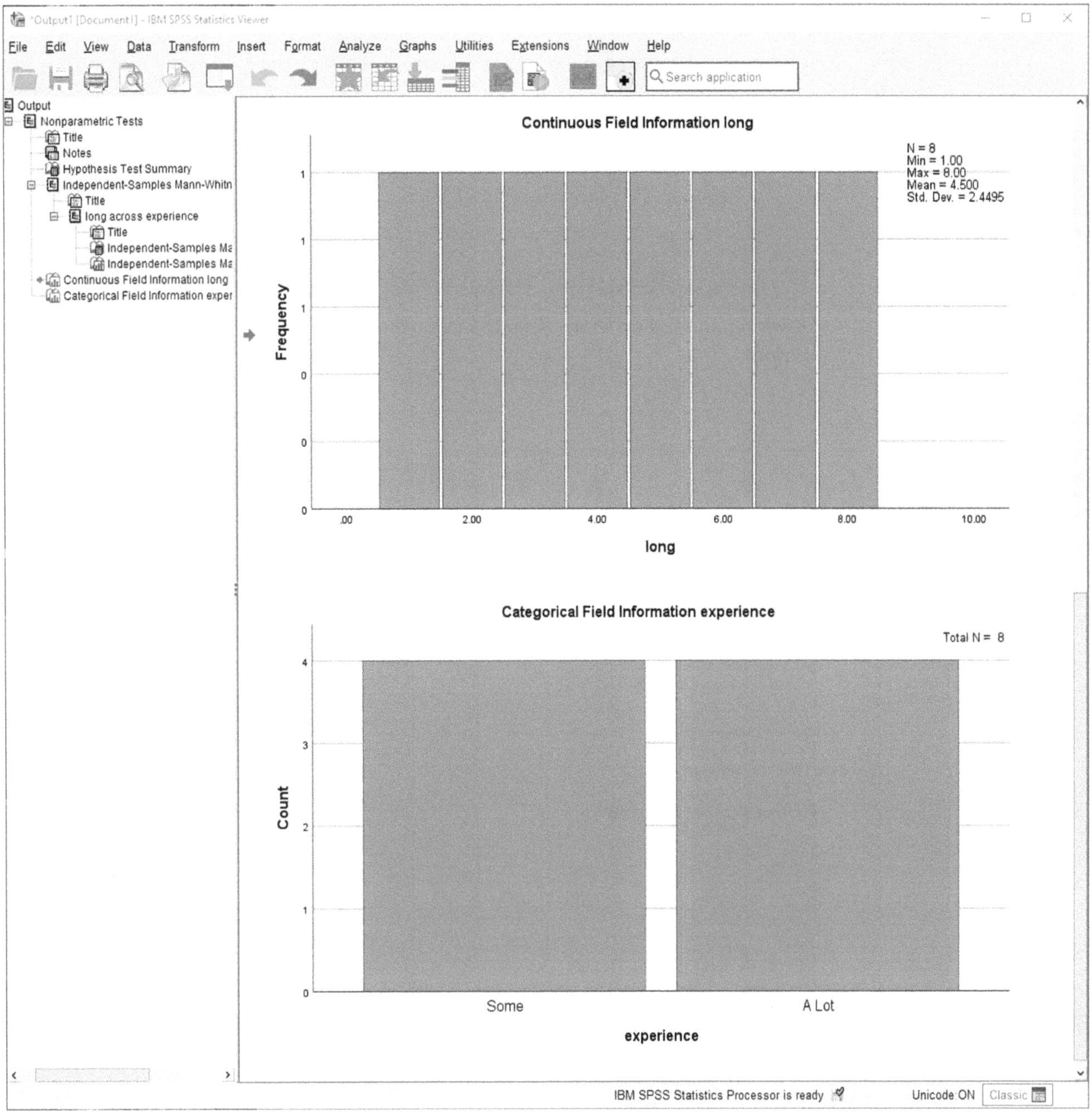

Here you will get a plain-English result. In this case, we reject the **null hypothesis** that the distribution was the same for both levels of experience.

The output consists of several sections. One section gives **descriptive statistics** for the two samples. Because the data are required only to be at least **ordinal**, summaries relating to their ranks are used. Those participants who had medium experience averaged 6.5 as their place in the race. The experienced participants averaged 2.5 as their place in the race.

Another section of the output is the result of the Mann-Whitney *U* test itself. The value obtained was 0.0, with a **significance** level of .029.

Drawing Conclusions

A significant Mann-Whitney *U* result indicates that the two samples are different in terms of their average ranks.

Phrasing Results That Are Significant

Our preceding example is significant, so we could state the following:

> A Mann-Whitney U test was calculated examining the place that runners with varying levels of experience took in a long-distance race. Runners with medium experience did significantly worse (M place = 6.50) than experienced runners (M place = 2.50; $U = 0.00$, $p < .05$).

Phrasing Results That Are Not Significant

If we ran the test on short-distance races instead of long, we would get the following output:

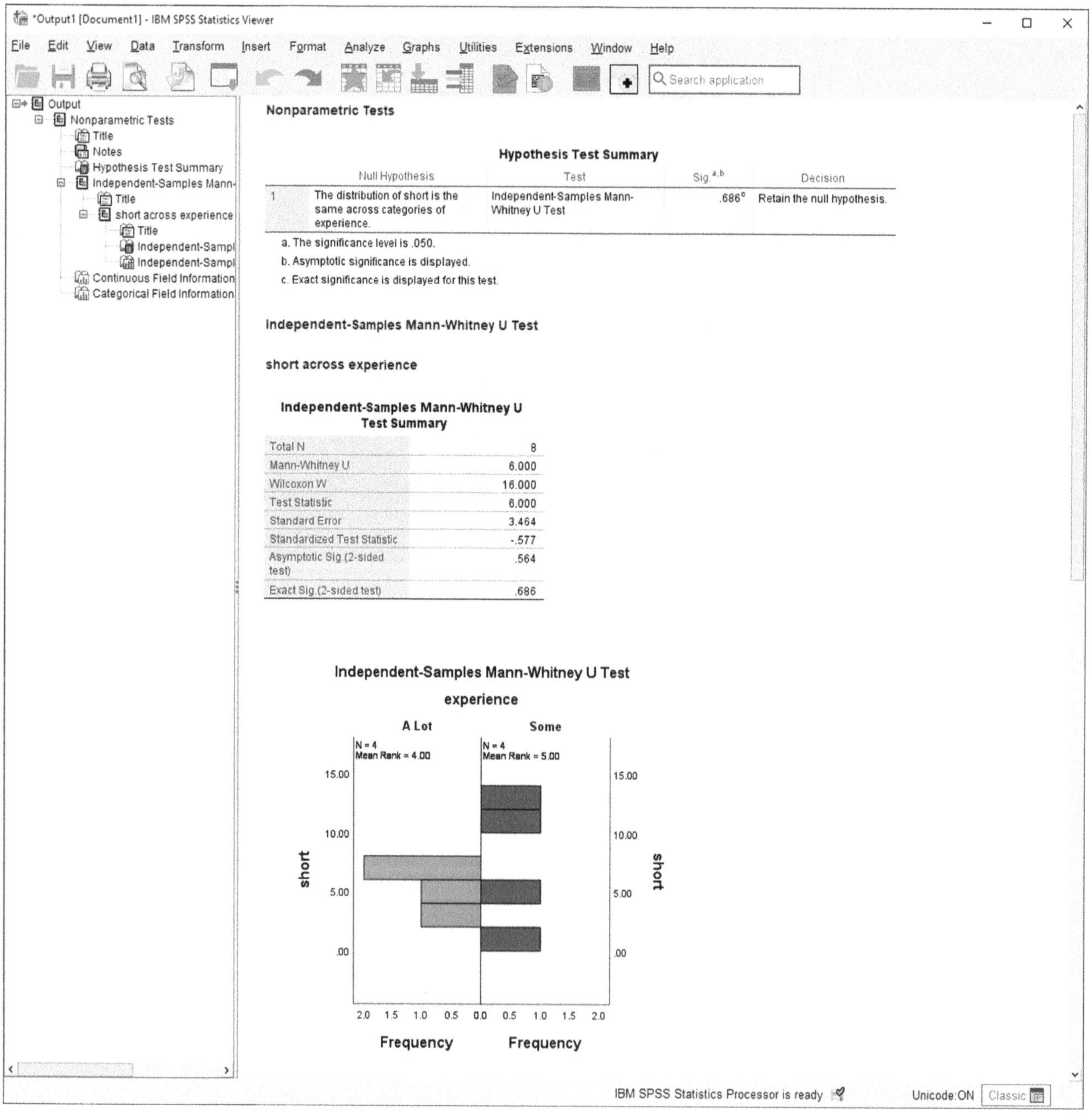

A Mann-Whitney U test was used to examine the difference in the race performance of runners with medium experience and runners with experience in a short-distance race. No significant difference in the results of the race was found ($U = 6.00$, $p > .05$). Runners with medium experience averaged a place of 5. Experienced runners averaged 4.

Practice Exercise

Using Practice Dataset 1 in Appendix B, determine if younger participants (<26) have significantly lower mathematics scores than older participants. (Note: You will need to create a new ordinal variable representing each age group. See Section 2.2 if you need a refresher on how to do that.)

Section 8.4 Wilcoxon Test

Description

The Wilcoxon test is the nonparametric equivalent of the paired-samples (dependent) *t* test. It tests whether or not two related samples are from the same distribution. The Wilcoxon test is weaker than the independent *t* test, so the *t* test should be used if you can meet its assumptions.

Assumptions

The Wilcoxon test is based on the difference in rankings. The data for the two samples must at least be **ordinal**. There are no assumptions about the shape of the distribution.

SPSS Data Format

The test requires two variables. One variable represents the **dependent variable** at the first level of the **independent variable**. The other variable represents the **dependent variable** at the second level of the **independent variable**.

Running the Command

Locate the command by clicking *Analyze*, then *Nonparametric Tests*, then *Related Samples*. This example uses the RACE.sav dataset.

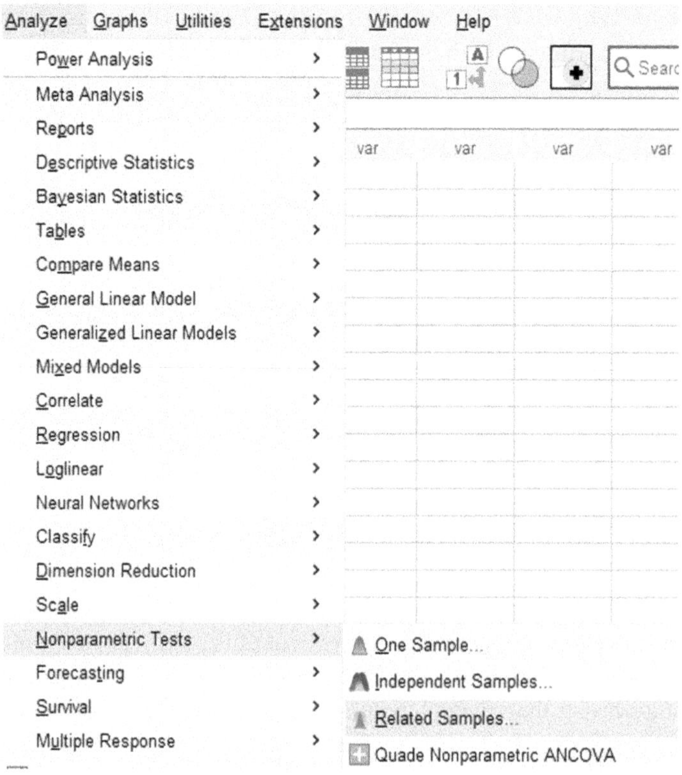

First, you will be presented with the Nonparametric Tests **dialog box**.

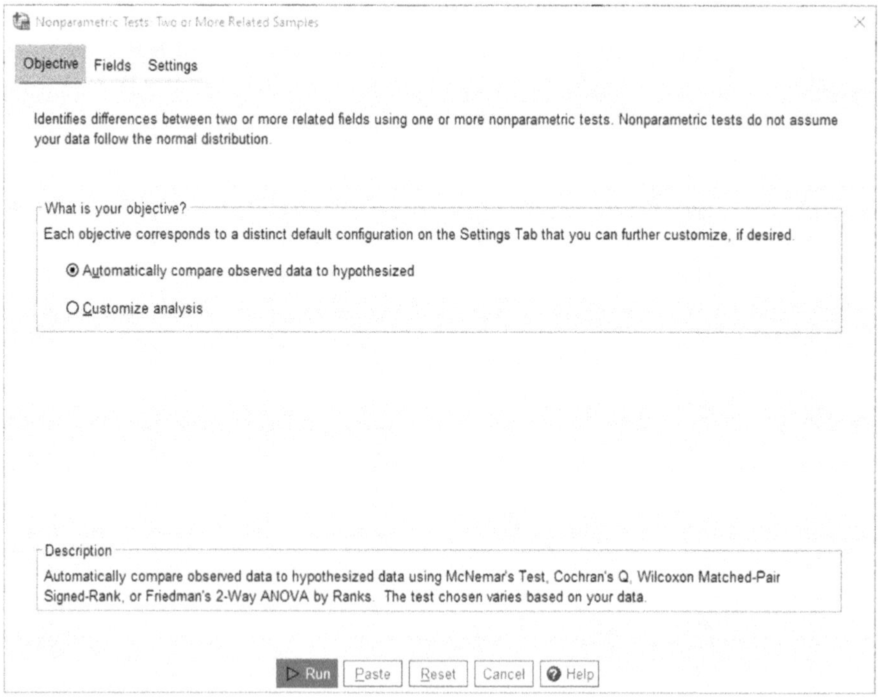

Select the *Fields* tab. Move the variables LONG and MEDIUM over to the *Test Fields* area as shown below.

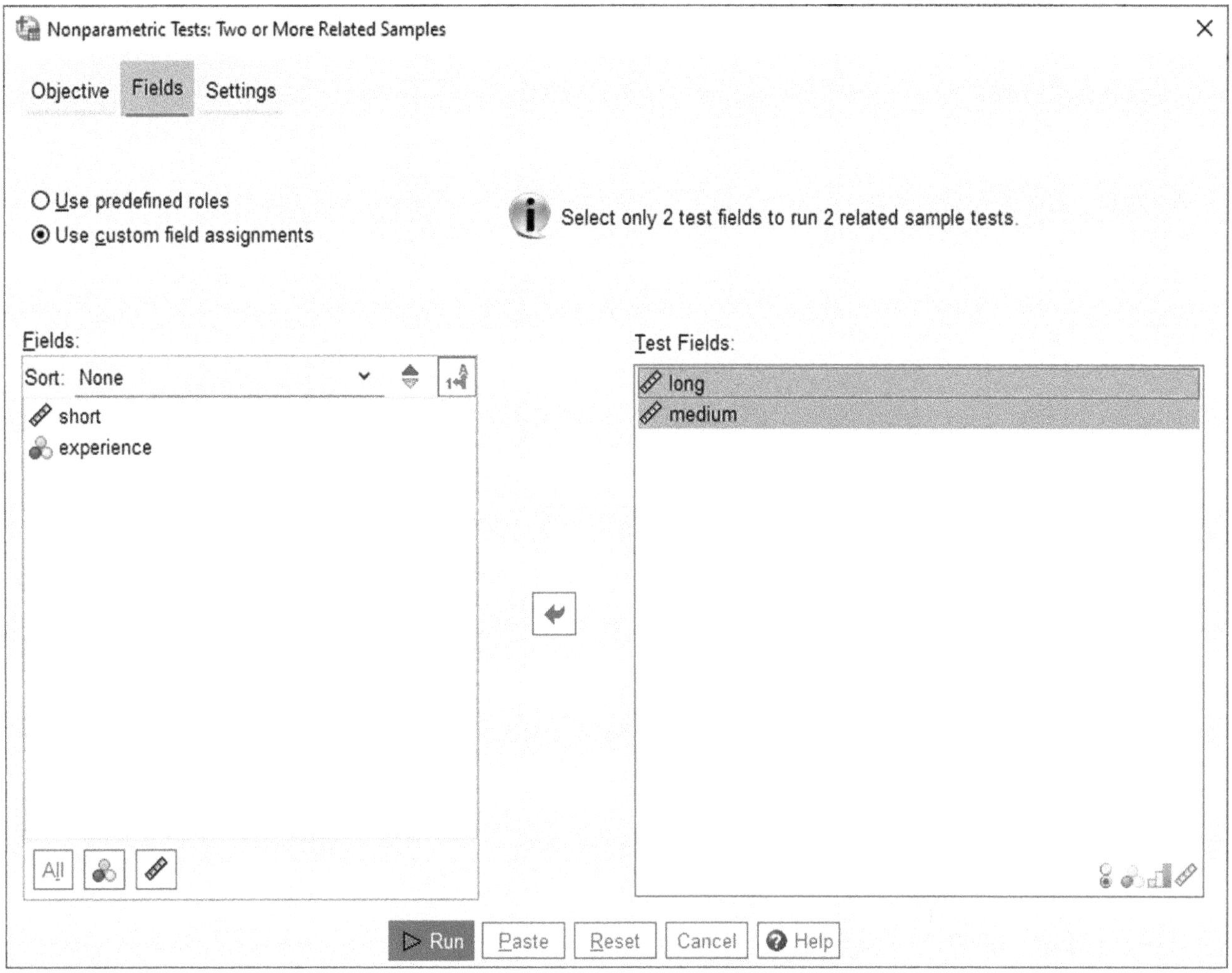

Finally, select the *Settings* tab and *Customize tests* to select *Wilcoxon matched-pair signed-rank*. Click *Run*.

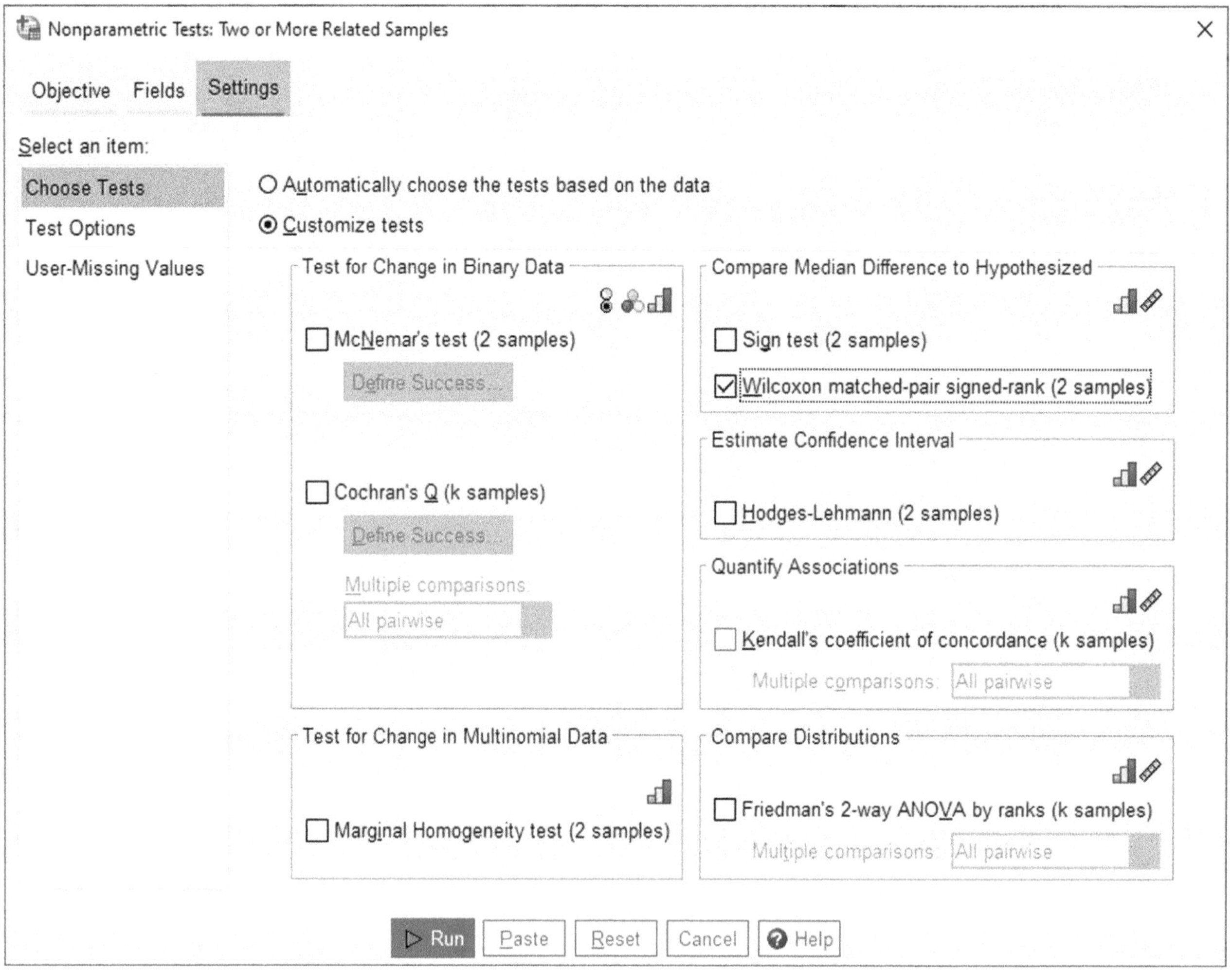

Reading the Output

As part of the new nonparametric procedure, the output will be presented in plain English with graphs.

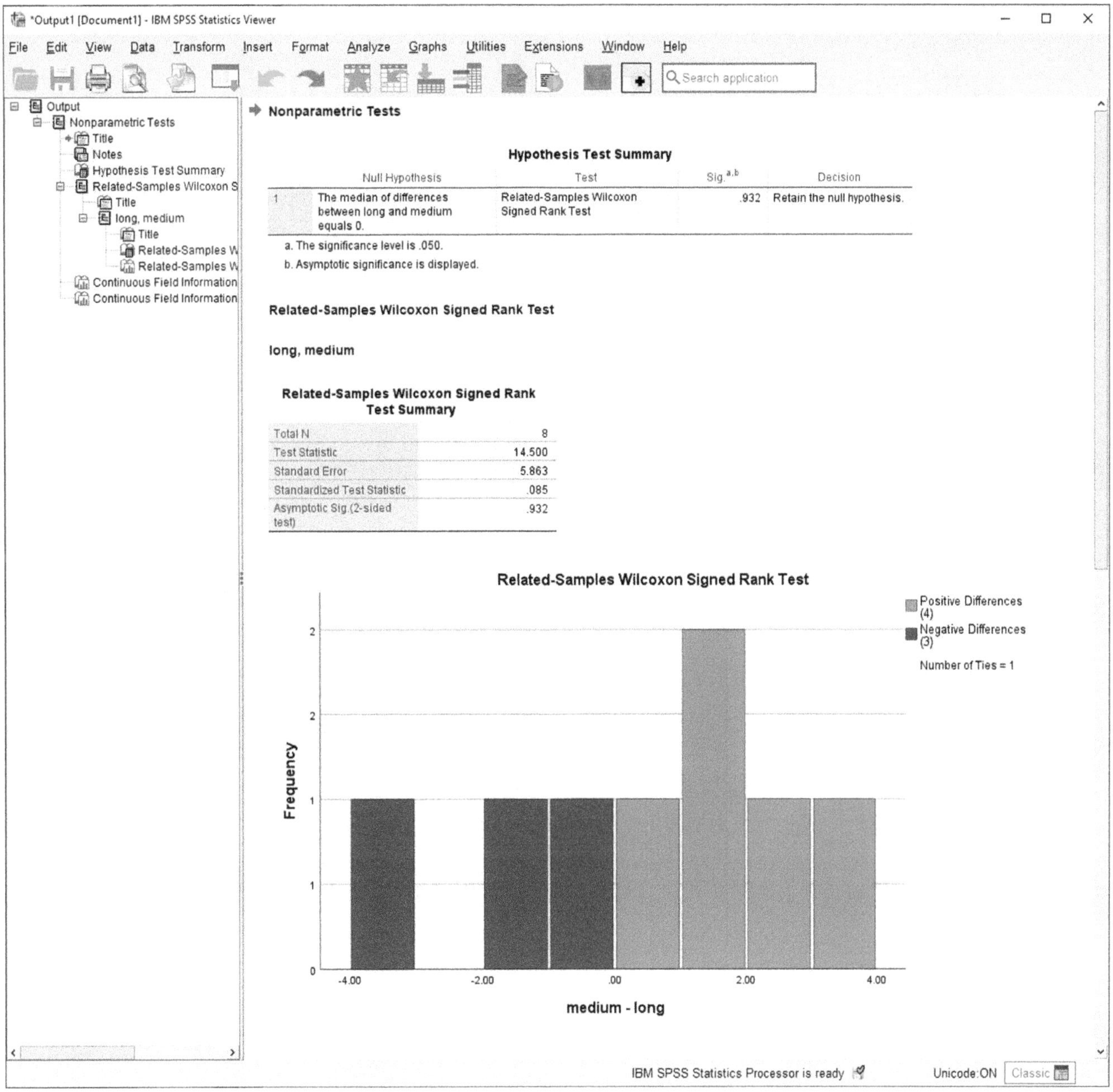

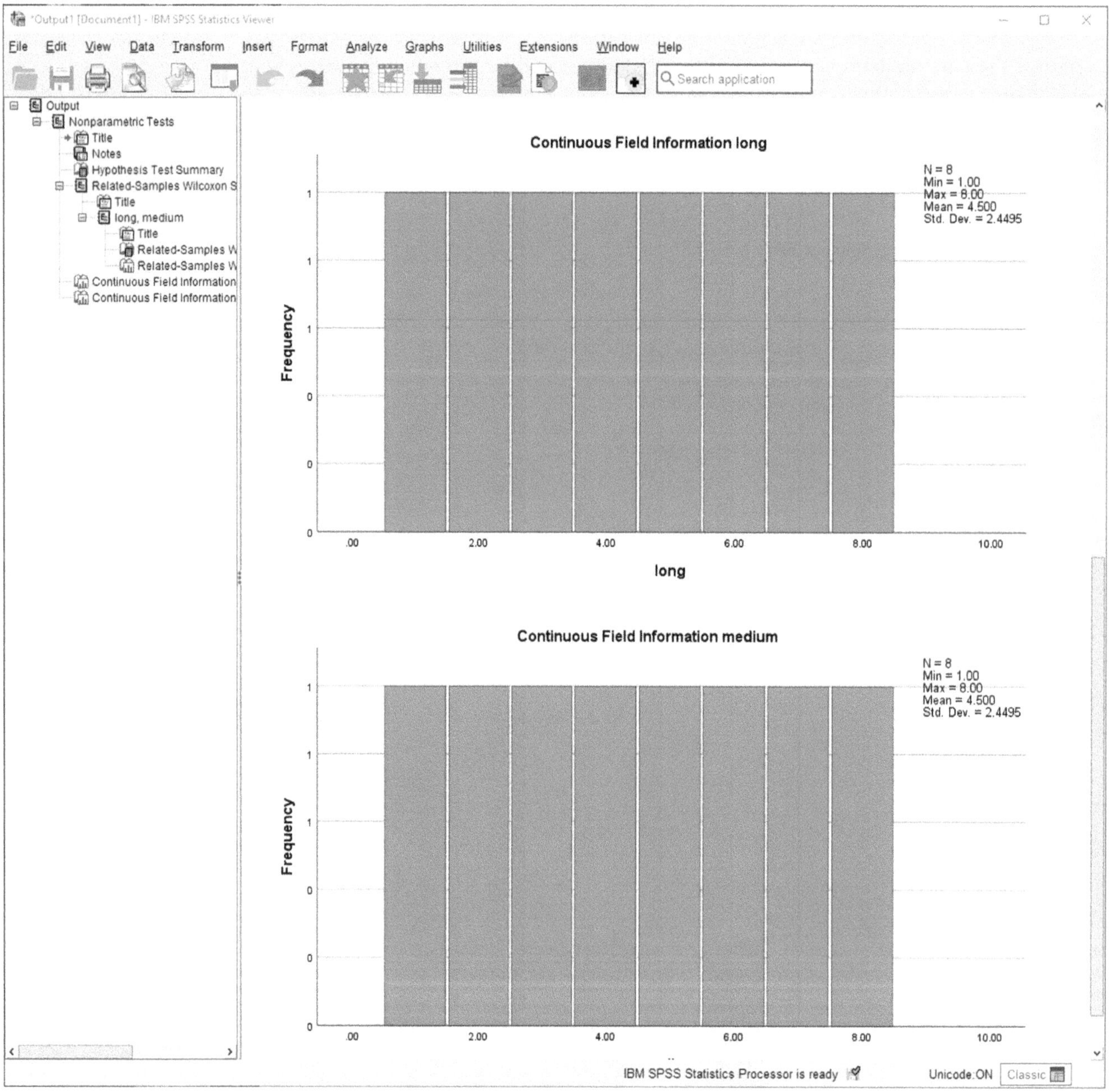

The example above shows that no significant difference was found between the results of the long-distance and medium-distance races.

Phrasing Results That Are Significant

A significant result means that a change has occurred between the two measurements. If we compared PRETEST to FINAL scores using the GRADES.sav dataset, we would obtain:

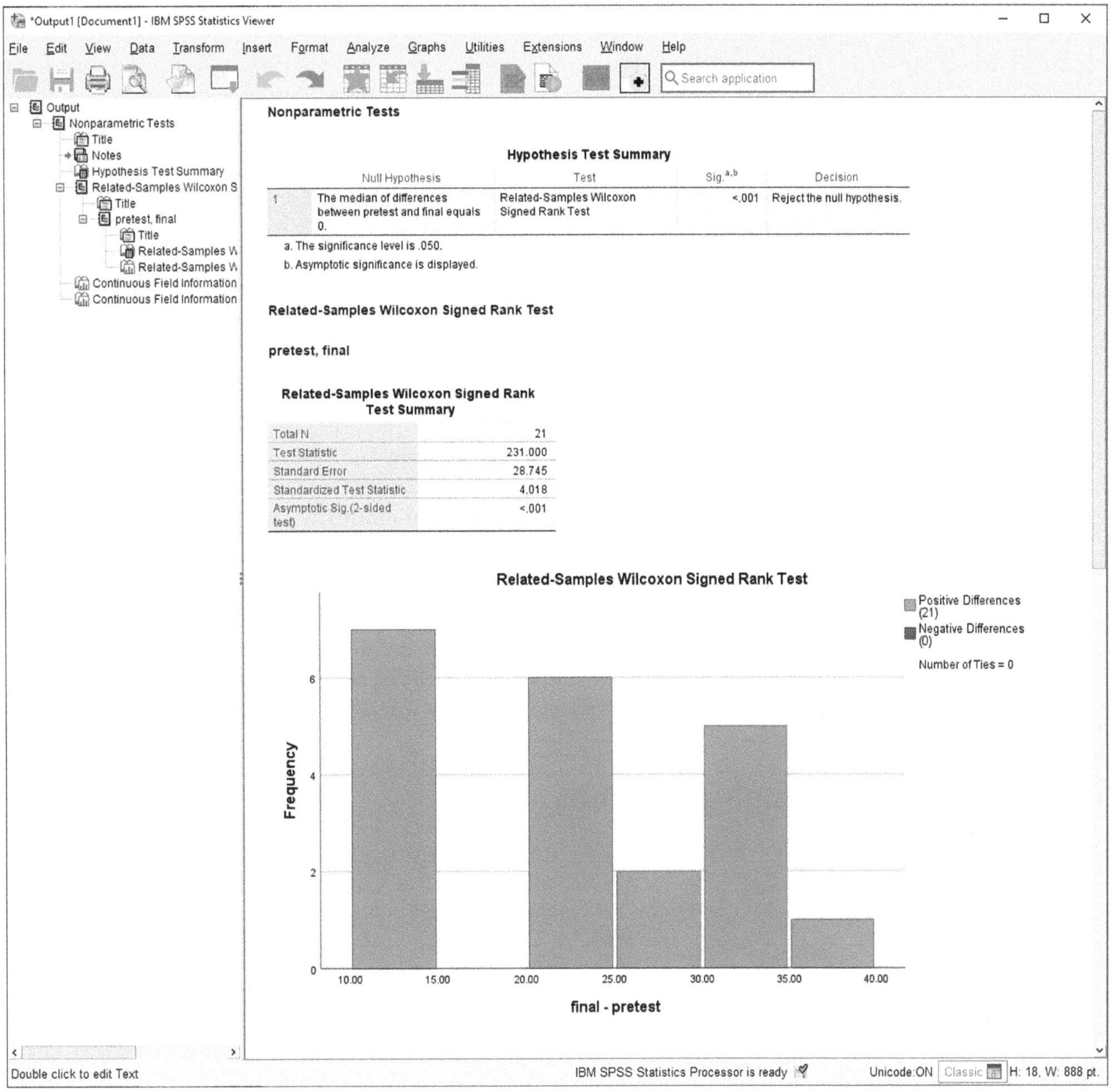

Nonparametric Tests

Hypothesis Test Summary

	Null Hypothesis	Test	Sig.[a,b]	Decision
1	The median of differences between pretest and final equals 0.	Related-Samples Wilcoxon Signed Rank Test	<.001	Reject the null hypothesis.

a. The significance level is .050.

b. Asymptotic significance is displayed.

Related-Samples Wilcoxon Signed Rank Test

pretest, final

Related-Samples Wilcoxon Signed Rank Test Summary

Total N	21
Test Statistic	231.000
Standard Error	28.745
Standardized Test Statistic	4.018
Asymptotic Sig.(2-sided test)	<.001

A Wilcoxon test examined the results of the pretest and final exam scores. A significant difference was found in the results ($Z = 4.018$, $p < .05$). Final exam scores were better than pretest scores.

Phrasing Results That Are Not Significant

In fact, the results in the original example were not significant, so we could state the following:

> A Wilcoxon test examined the results of the medium-distance and long-distance races. No significant difference was found in the results ($Z = .085$, $p > .05$). Medium-distance results were not significantly different from long-distance results.

Practice Exercise

Use the RACE.sav data file to determine whether or not the outcome of short-distance races is different from that of medium-distance races. Phrase your results.

Section 8.5 Kruskal-Wallis *H* Test

Description

The Kruskal-Wallis *H* test is the nonparametric equivalent of the one-way ANOVA. It tests whether or not several independent samples come from the same population.

Assumptions

Because the test is nonparametric, there are very few assumptions. However, the test does assume an **ordinal** level of measurement for the **dependent variable**. The **independent variable** should be **nominal** or **ordinal**.

SPSS Data Format

SPSS requires one variable to represent the **dependent variable** and another to represent the **levels** of the **independent variable**. The **dependent variable** must be entered as a scale-type variable even if it is actually **ordinal**.

Running the Command

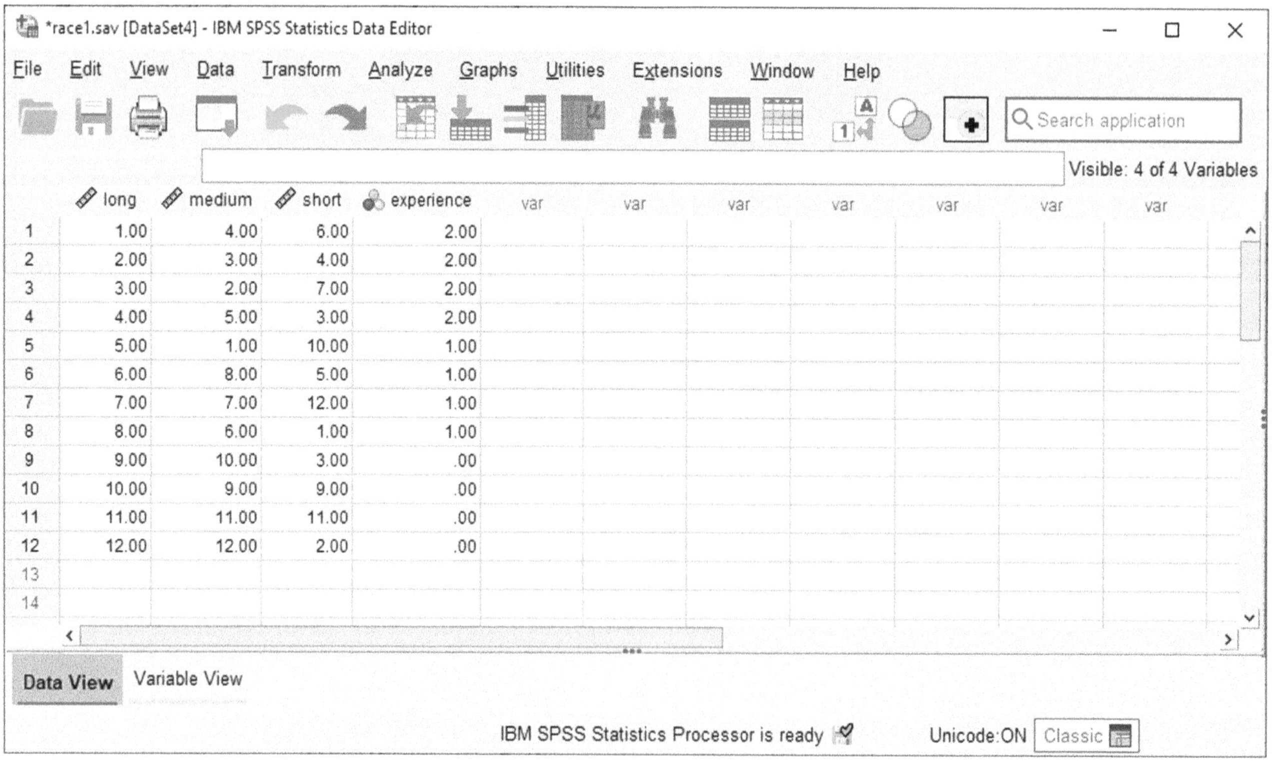

This example modifies the RACE.sav data file to include a third group of four additional participants with no experience (EXPERIENCE = 0). Make sure your data look like the screenshot shown above.

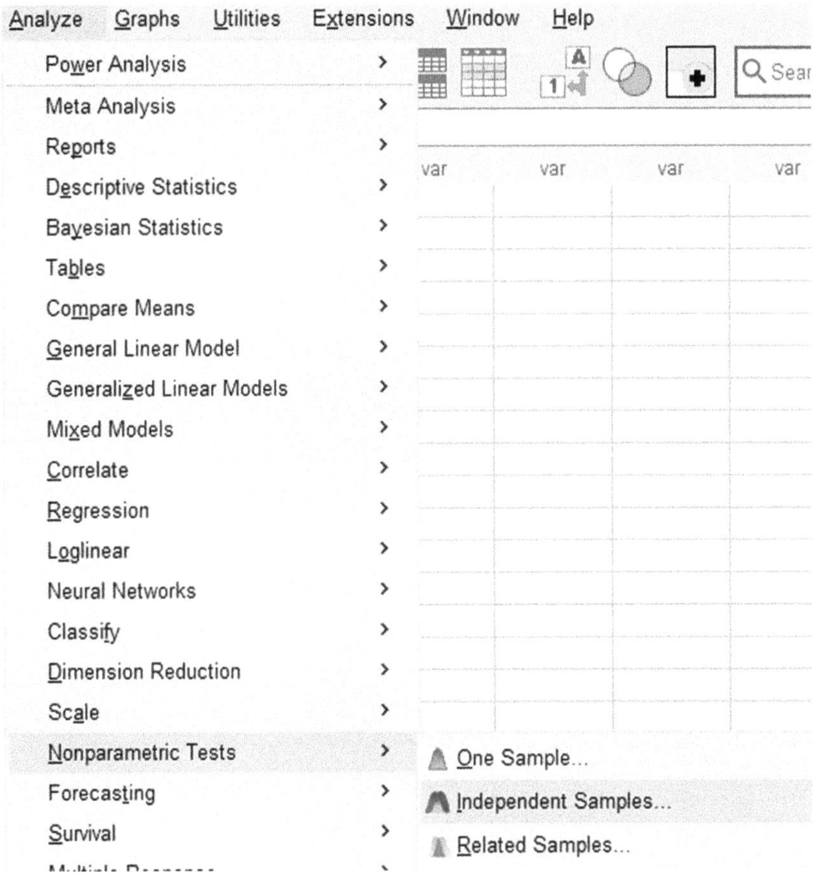

To run the command, click *Analyze*, then *Nonparametric Tests*, then *Independent Samples*. This will bring up the main **dialog box**, below.

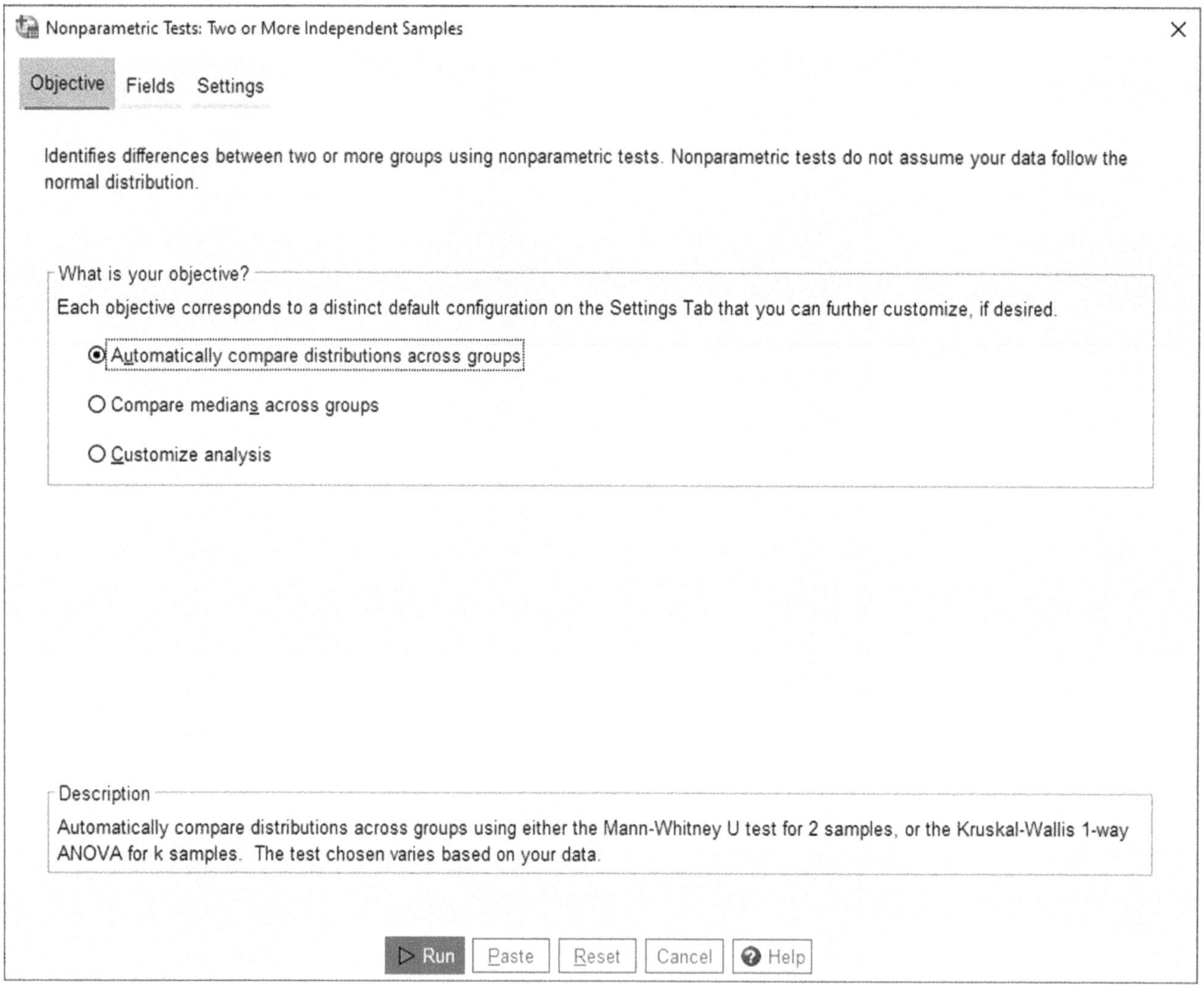

Leave the *Objective* the same and click the *Fields* tab. Move LONG to the *Test Fields* area and EXPERIENCE to the *Groups* area.

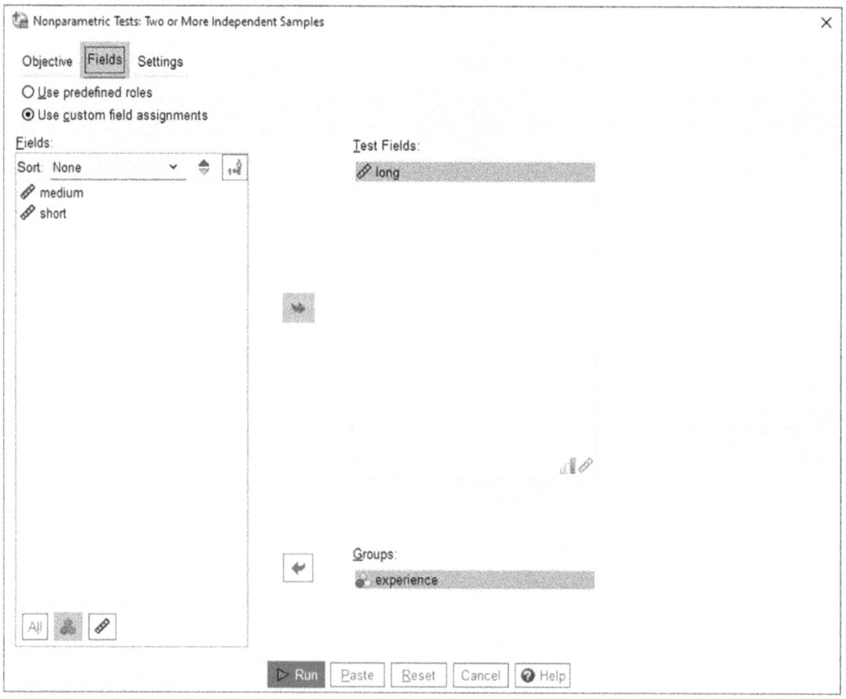

Select the *Settings* tab and *Customize tests* to select *Kruskal-Wallis 1-way ANOVA*. Also, select *Multiple comparisons* and choose *All pairwise*. Just as post-hoc tests are necessary with parametric one-way ANOVA, they are also necessary with non-parametric versions. This step will have SPSS compute them for us. Click *Run* to perform the analysis.

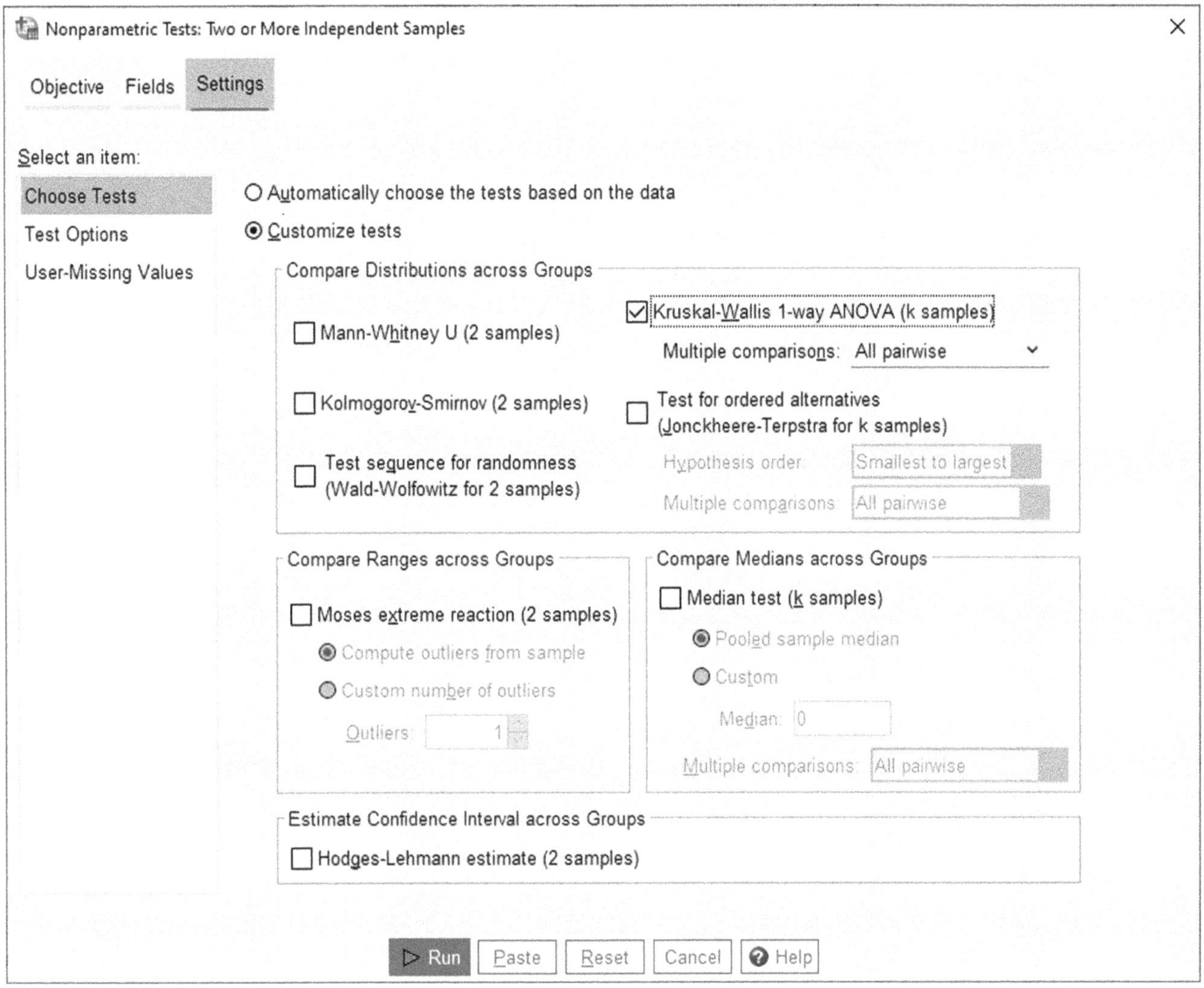

Reading the Output

The output will consist of the real-language Output. Here, it tells us that the distribution of the variable LONG is the same across all levels of EXPERIENCE.

*Output1 [Document1] - IBM SPSS Statistics Viewer

File Edit View Data Transform Insert Format Analyze Graphs Utilities Extensions Window Help

Output
- Nonparametric Tests
  - Title
  - Notes
  - Hypothesis Test Summary
  - Independent-Samples Krusk
    - Title
    - long across experience
      - Title
      - Independent-Sampl
      - Independent-Sampl
      - Pairwise Comparis
      - Pairwise Comparis
  - Continuous Field Information
  - Categorical Field Information

Nonparametric Tests

Hypothesis Test Summary

	Null Hypothesis	Test	Sig.[a,b]	Decision
1	The distribution of long is the same across categories of experience.	Independent-Samples Kruskal-Wallis Test	.007	Reject the null hypothesis.

a. The significance level is .050.

b. Asymptotic significance is displayed.

Independent-Samples Kruskal-Wallis Test

long across experience

Independent-Samples Kruskal-Wallis Test Summary

Total N	12
Test Statistic	9.846[a]
Degree Of Freedom	2
Asymptotic Sig.(2-sided test)	.007

a. The test statistic is adjusted for ties.

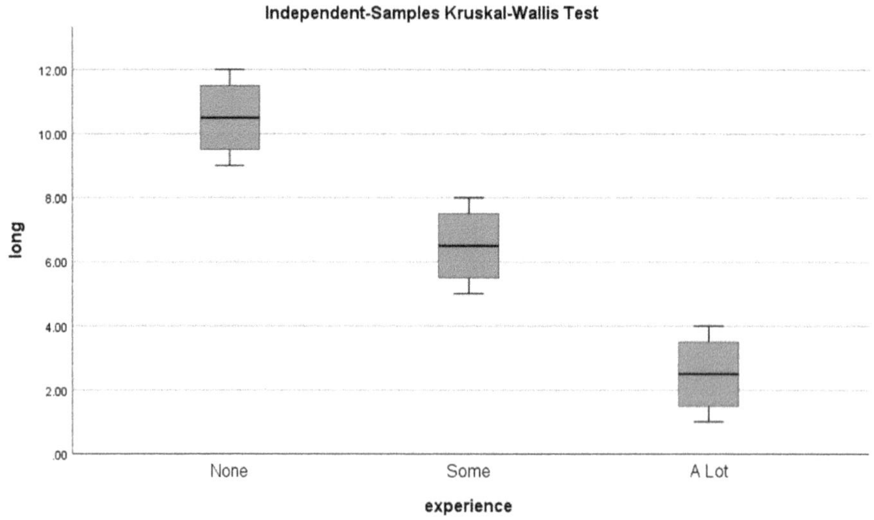

Pairwise Comparisons of experience

Sample 1-Sample 2	Test Statistic	Std. Error	Std. Test Statistic	Sig.	Adj. Sig.[a]
A Lot-Some	4.000	2.550	1.569	.117	.350
A Lot-None	8.000	2.550	3.138	.002	.005
Some-None	4.000	2.550	1.569	.117	.350

Each row tests the null hypothesis that the Sample 1 and Sample 2 distributions are the same.

Asymptotic significances (2-sided tests) are displayed. The significance level is .050.

a. Significance values have been adjusted by the Bonferroni correction for multiple tests.

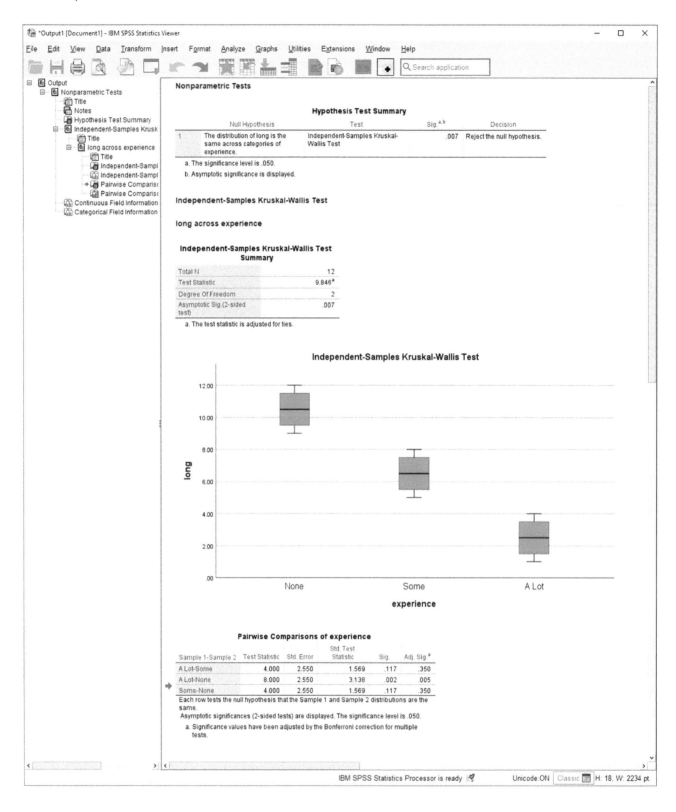

IBM SPSS Statistics Processor is ready Unicode:ON Classic H: 18, W: 2234 pt

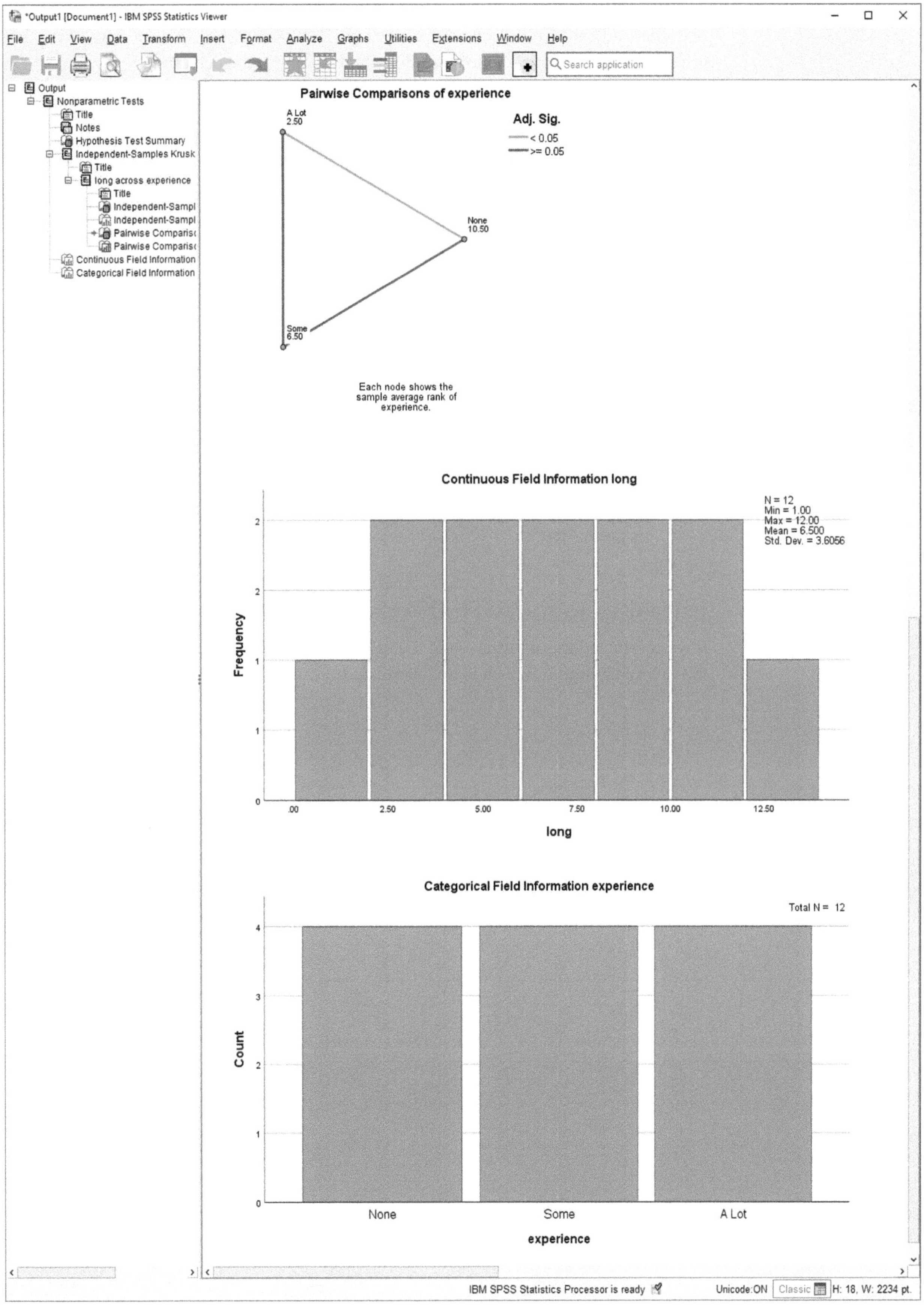
*Output1 [Document1] - IBM SPSS Statistics Viewer
File Edit View Data Transform Insert Format Analyze Graphs Utilities Extensions Window Help
Search application
Output
Nonparametric Tests
Title
Notes
Hypothesis Test Summary
Independent-Samples Krusk
Title
long across experience
Title
Independent-Sampl
Independent-Sampl
Pairwise Comparis
Pairwise Comparis
Continuous Field Information
Categorical Field Information
Pairwise Comparisons of experience
A Lot 2.50
Adj. Sig.
< 0.05
>= 0.05
None 10.50
Some 6.50
Each node shows the sample average rank of experience.
Continuous Field Information long
N = 12
Min = 1.00
Max = 12.00
Mean = 6.500
Std. Dev. = 3.6056
Frequency
.00
2.50
5.00
7.50
10.00
12.50
long
Categorical Field Information experience
Total N = 12
Count
None
Some
A Lot
experience
IBM SPSS Statistics Processor is ready
Unicode:ON
Classic
H: 18, W: 2234 pt

The Pairwise Comparisons of experience table show that EXPERIENCE 2 (Experienced) is significantly different from EXPERIENCE 0 (None). However, the other two comparisons are not significantly different from each other.

Drawing Conclusions

Like the one-way ANOVA, the Kruskal-Wallis test assumes that the groups are equal. Thus, a significant result indicates that at least one of the groups is different from at least one other group. The *Pairwise Comparisons* command will allow us to determine which groups are different.

Phrasing Results That Are Significant

The preceding example is significant, so we could state the following:

> A Kruskal-Wallis test was conducted comparing the outcome of a long-distance race for runners with varying levels of experience. A significant result was found (H (2) = 9.846, $p < .01$), indicating that the groups differed from each other. Follow-up pairwise comparisons indicated that runners with the greatest experience performed significantly better than runners with no experience.

Phrasing Results That Are Not Significant

If we conducted the analysis using the results of the short-distance race, we would get the following output, which is not significant.

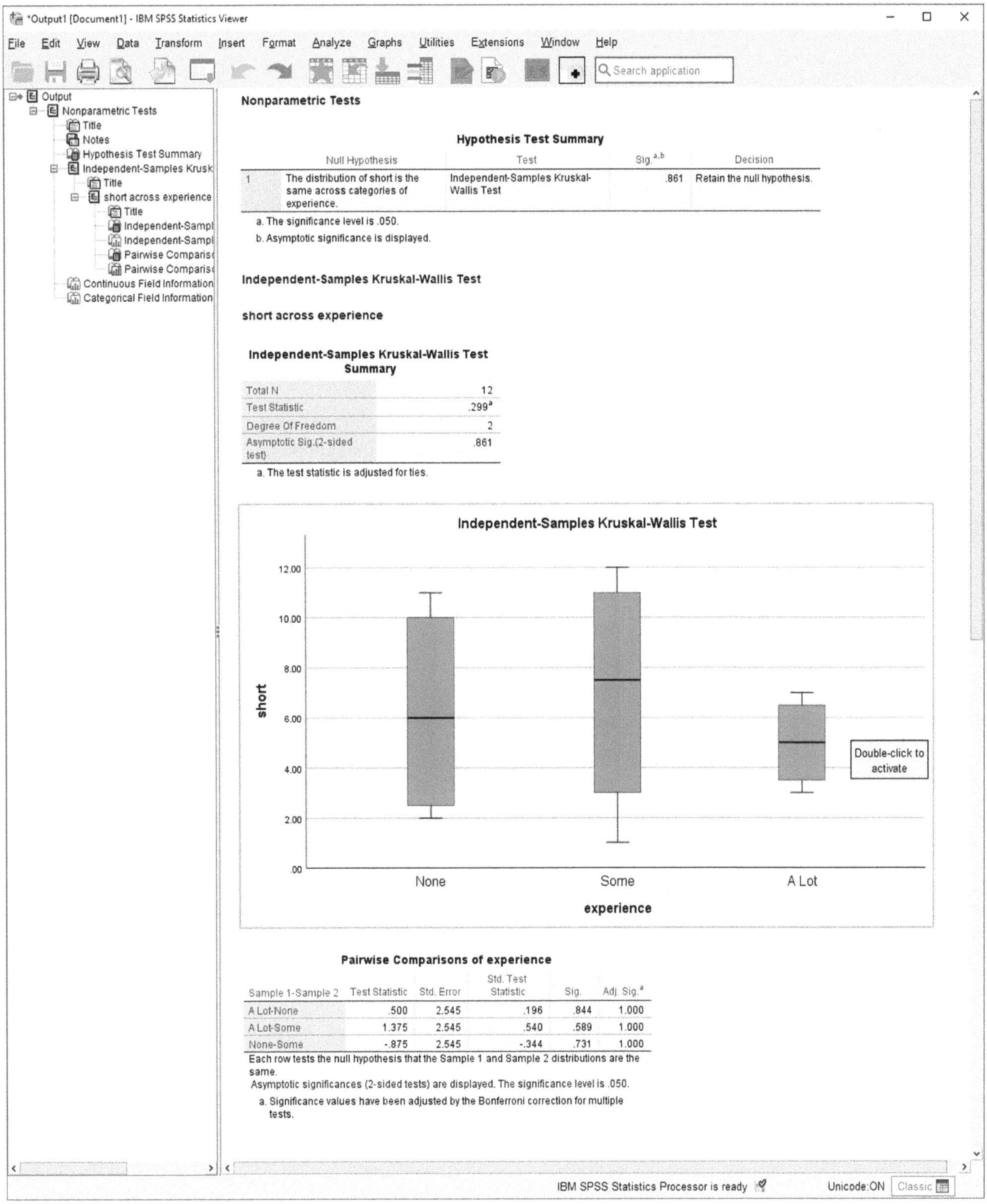

Nonparametric Tests

Hypothesis Test Summary

	Null Hypothesis	Test	Sig.[a,b]	Decision
1	The distribution of short is the same across categories of experience.	Independent-Samples Kruskal-Wallis Test	.861	Retain the null hypothesis.

a. The significance level is .050.

b. Asymptotic significance is displayed.

Independent-Samples Kruskal-Wallis Test

short across experience

Independent-Samples Kruskal-Wallis Test Summary

Total N	12
Test Statistic	.299[a]
Degree Of Freedom	2
Asymptotic Sig.(2-sided test)	.861

a. The test statistic is adjusted for ties.

Pairwise Comparisons of experience

Sample 1-Sample 2	Test Statistic	Std. Error	Std. Test Statistic	Sig.	Adj. Sig.[a]
A Lot-None	.500	2.545	.196	.844	1.000
A Lot-Some	1.375	2.545	.540	.589	1.000
None-Some	-.875	2.545	-.344	.731	1.000

Each row tests the null hypothesis that the Sample 1 and Sample 2 distributions are the same.
Asymptotic significances (2-sided tests) are displayed. The significance level is .050.

a. Significance values have been adjusted by the Bonferroni correction for multiple tests.

This result is not significant, so we could state the following:

> A Kruskal-Wallis test was conducted comparing the outcome of a short-distance race for runners with varying levels of experience. No significant difference was found (H (2) = 0.299, $p > .05$), indicating that the groups did not differ significantly from each other. Experience did not seem to influence the results of the short-distance race.

Section 8.6 Friedman Test

Description

The Friedman test is the nonparametric equivalent of a one-way repeated-measures ANOVA. It is used when you have more than two measurements from related participants.

Assumptions

The test uses the rankings of the variables, so the data must at least be **ordinal**. No other assumptions are required.

SPSS Data Format

SPSS requires at least three variables in the SPSS data file. Each variable represents the **dependent variable** at one of the **levels** of the **independent variable**.

Running the Command

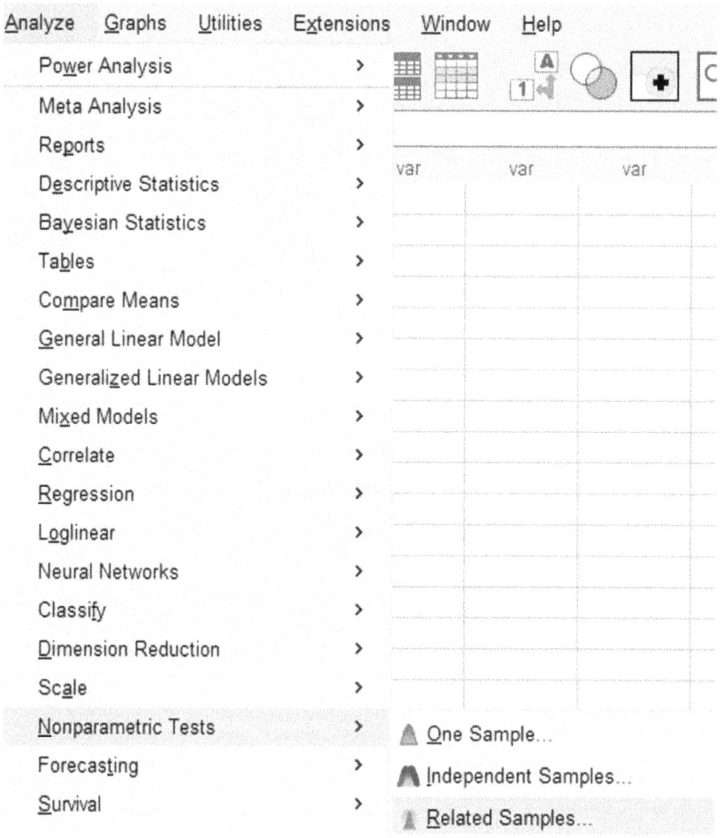

Locate the command by clicking *Analyze*, then *Nonparametric Tests*, then *Related Samples*. This will bring up the main **dialog box**. This example uses the RACE.sav data file.

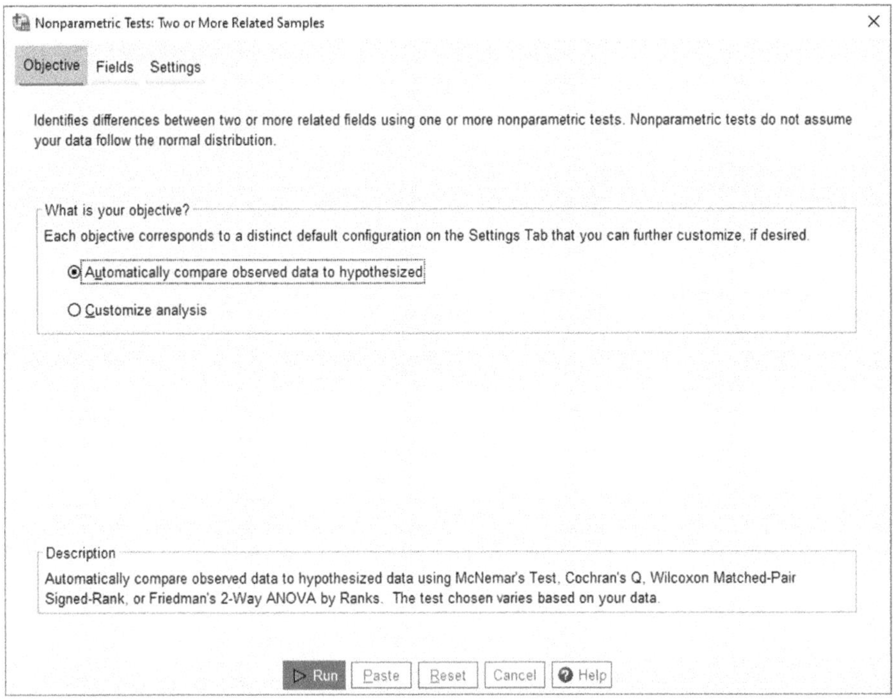

Leave the *Objective* the same, and select the *Fields* tab. Move all three variables representing race places over to the *Test Fields* area (LONG, MEDIUM, SHORT).

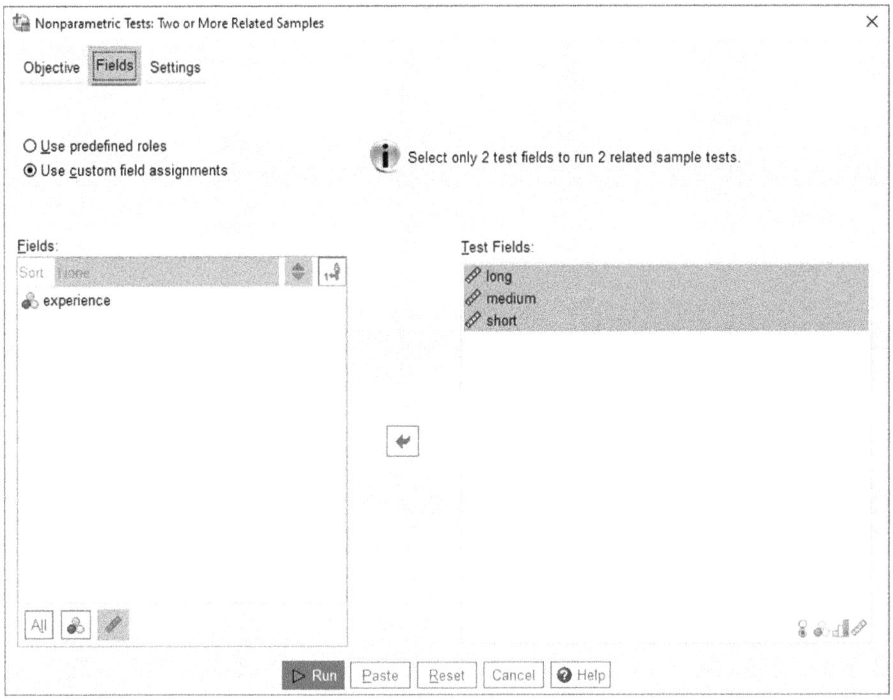

Under the *Settings* tab, select *Customize tests* and then select the *Friedman's 2-way ANOVA by ranks*. Select *Multiple comparisons* and select *All pairwise*. Run the test.

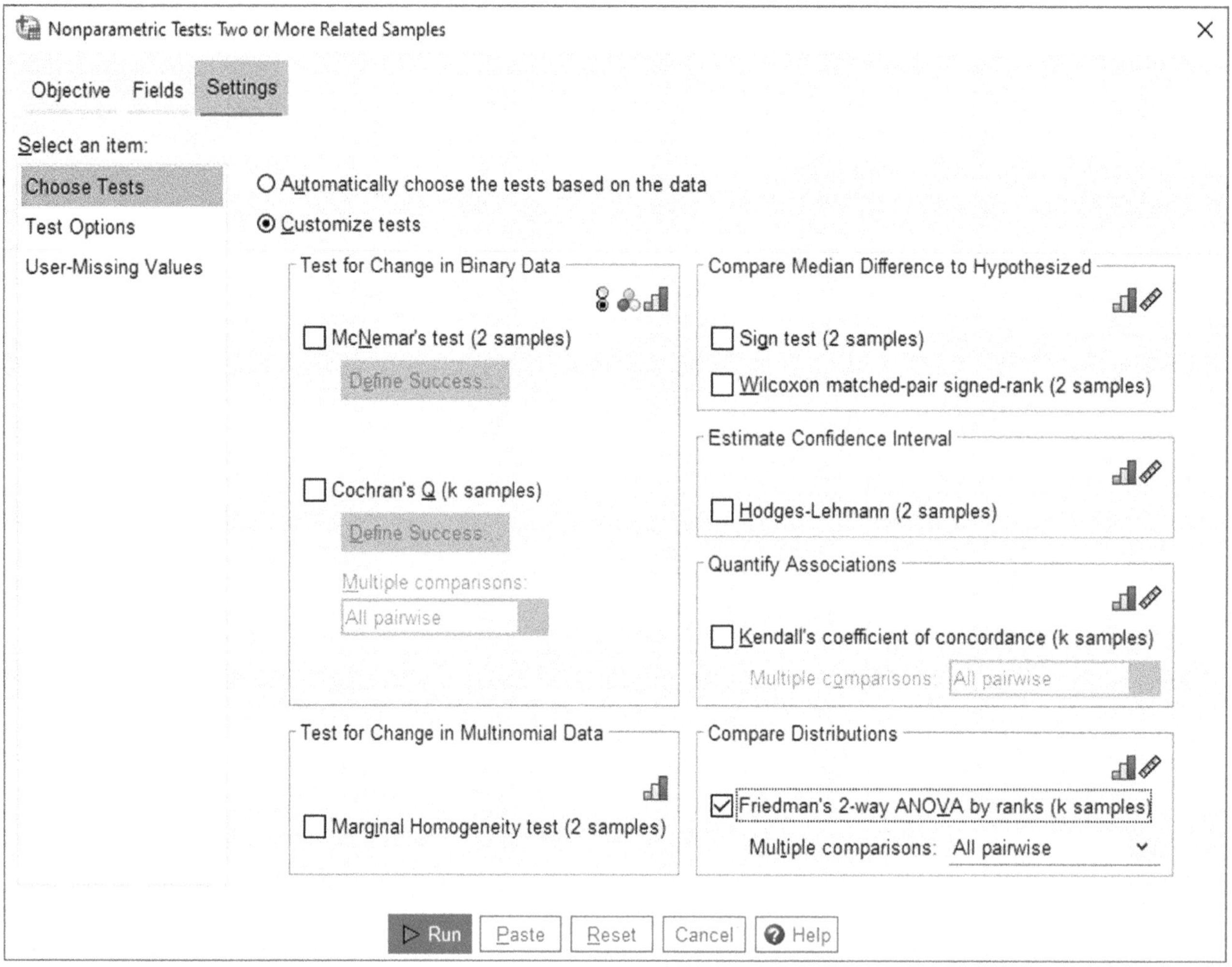

Reading the Output

The real-language Model Viewer will tell us that there was not a significant difference in the distributions of the three variables.

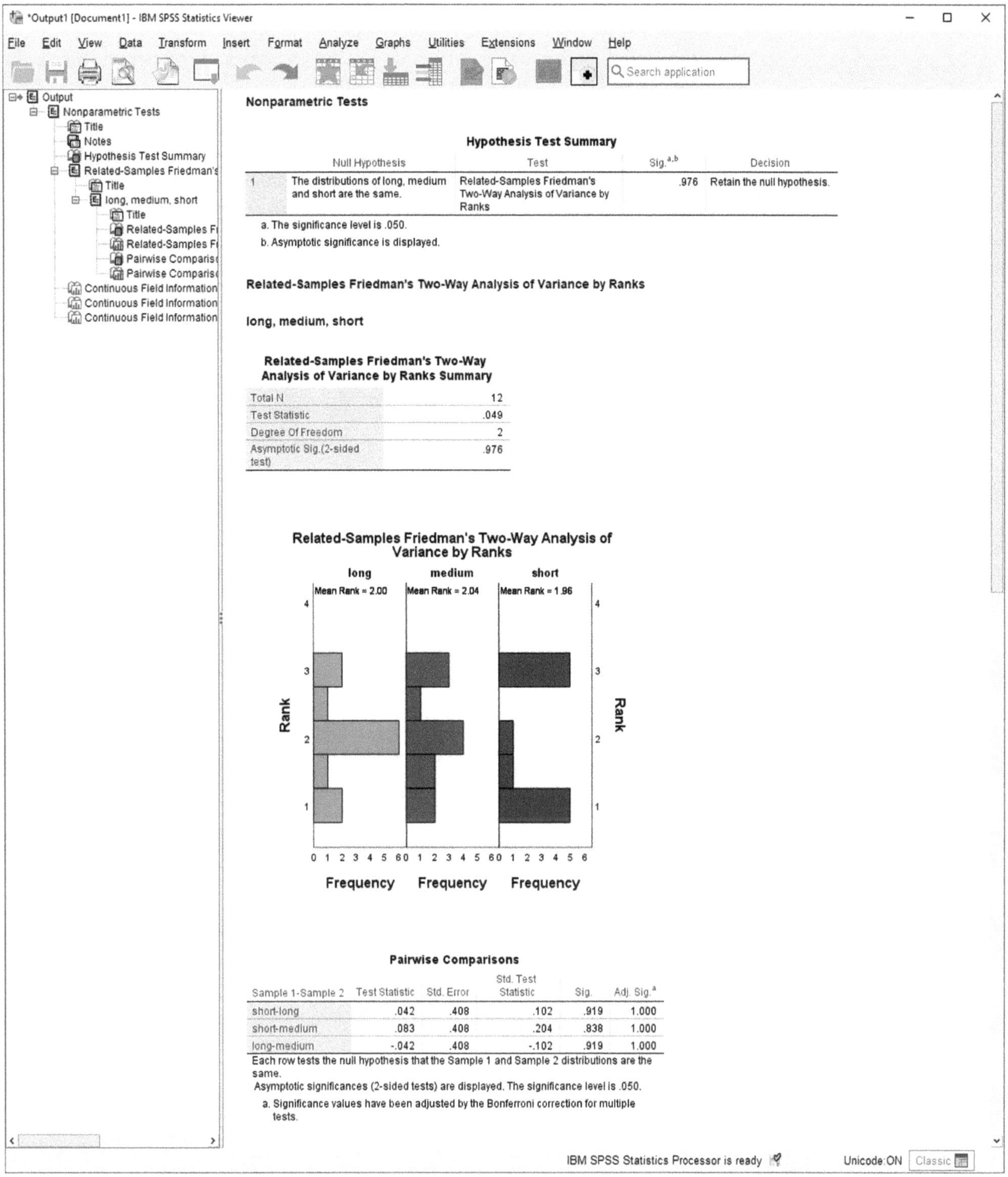

Drawing Conclusions

The Friedman test assumes that the three variables are from the same population. A significant value indicates that the variables are not equivalent.

Phrasing Results That Are Significant

If we obtained a significant result (e.g., using GRADES.sav and comparing PRETEST, MIDTERM, and FINAL), we could state the following:

> A Friedman test was conducted comparing the average scores on Pretest, Midterm, and Final exams. A significant difference was found (χ^2 (2) = 36.857, $p < .05$). Scores significantly changed over the course of the semester.

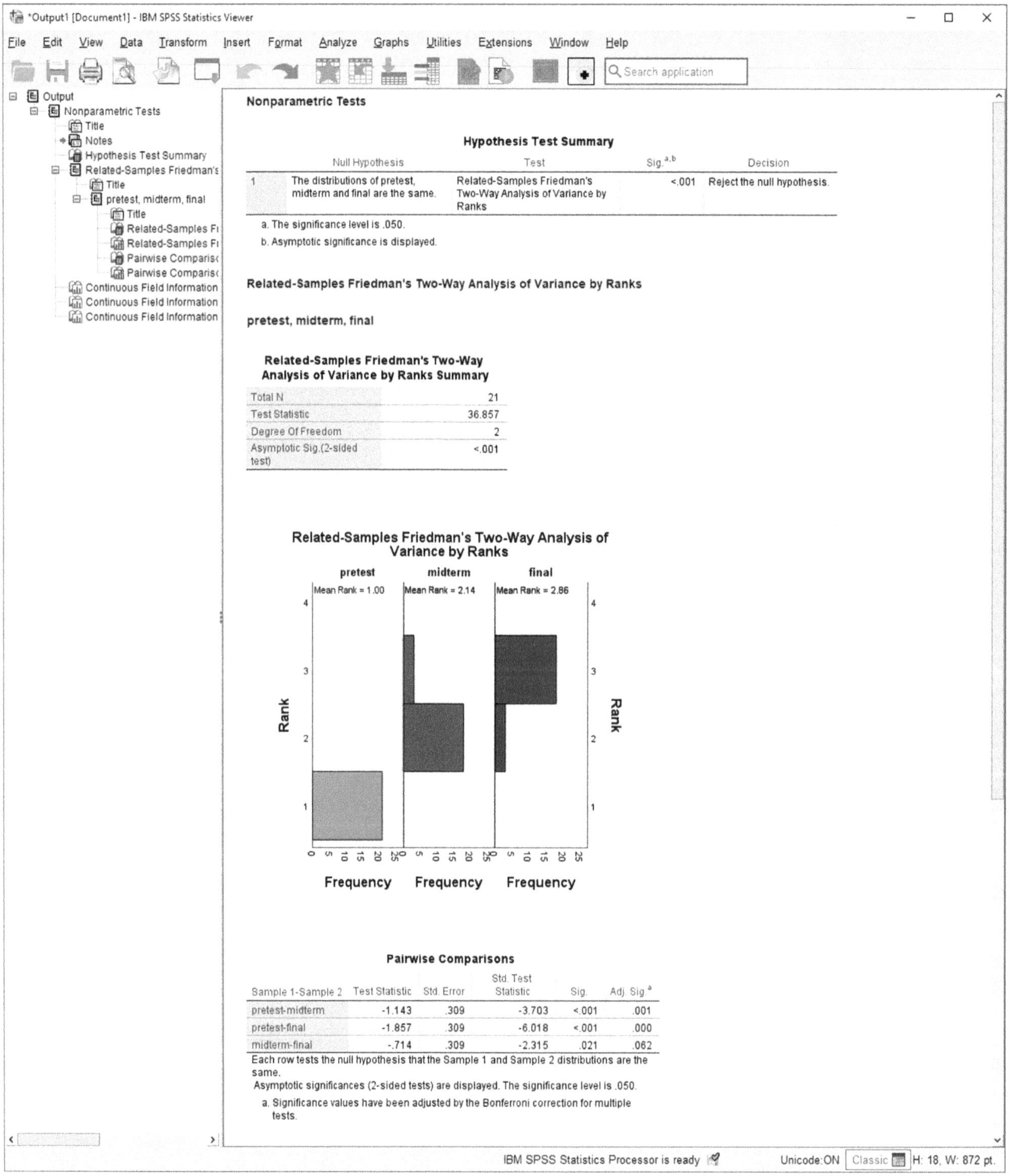

Phrasing Results That Are Not Significant

In fact, our example was not significant, so we could state the following:

> A Friedman test was conducted comparing the average place in a race of runners for short-distance, medium-distance, and long-distance races. No significant difference was found (χ^2 (2) = 0.839, $p > .05$). The length of the race did not significantly affect the results of the race.

Practice Exercise

Use the data in Practice Dataset 3 in Appendix B. If anxiety is measured on an **ordinal scale**, determine if anxiety levels changed over time. Phrase your results.

Chapter 9
Test Construction

Section 9.1 Item-Total Analysis

Description

Item-total analysis is a way to assess the **internal consistency** of a dataset. As such, it is one of many tests of **reliability**. Item-total analysis comprises a number of items that make up a scale or a test designed to measure a single construct (e.g., intelligence), and it determines the degree to which all of the items measure the same construct. It does not tell you if it is measuring the correct construct (that is a question of **validity**). Before a test can be valid, however, it must first be reliable.

Assumptions

All the items in the scale should be measured on an **interval** or **ratio scale**. In addition, each item should be normally distributed. If your items are **ordinal** in nature, you can conduct the analysis using the Spearman *rho* correlation instead of the Pearson *r* correlation.

SPSS Data Format

SPSS requires one variable for each item (or question) in the scale. In addition, you must have a variable representing the total score for the scale.

Conducting the Test

Item-total analysis uses the *Pearson Correlation* command. To conduct it, open the QUESTIONS.sav data file you created in Chapter 2. Click *Analyze*, then *Correlate*, then *Bivariate*.

DOI: 10.4324/9781003450467-9

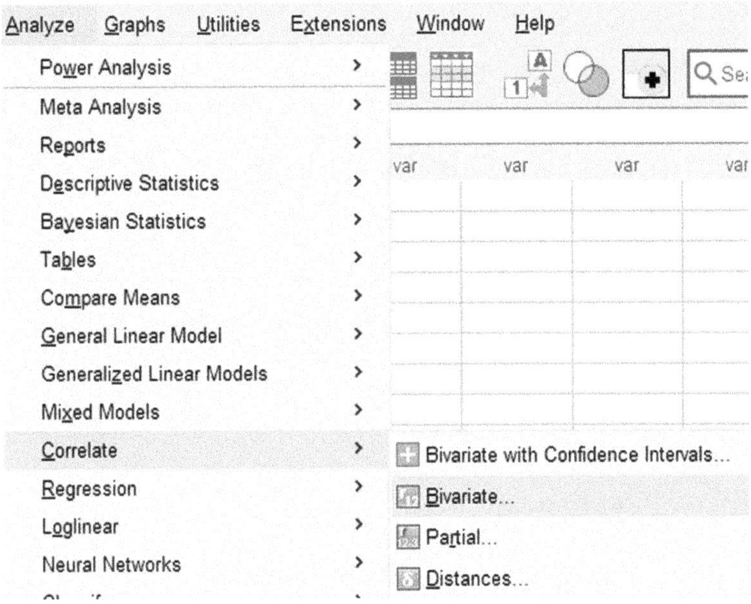

Place all questions and the total in the right-hand window, and click *OK*. (For more help on conducting correlations, see Chapter 5.) The total can be calculated with the techniques discussed in Chapter 2.

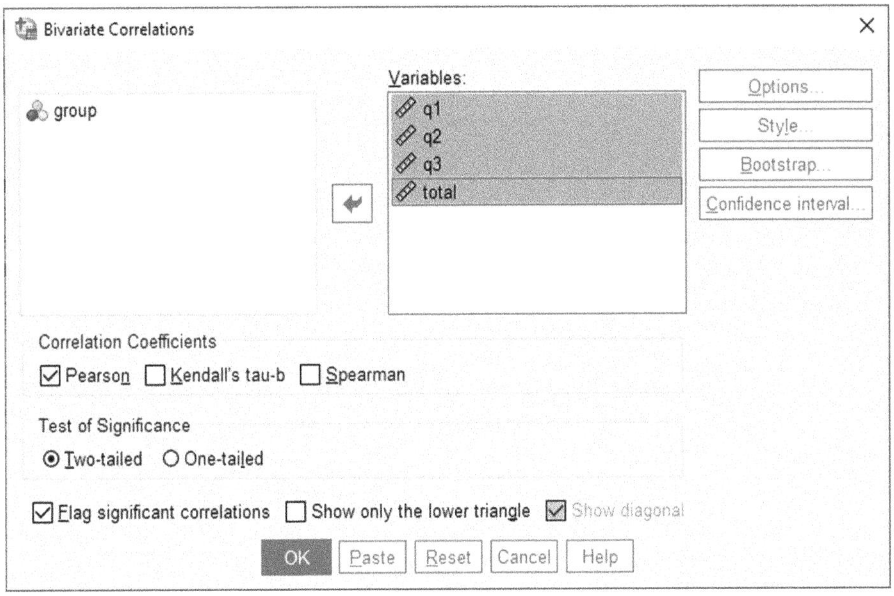

Reading the Output

The output consists of a **correlation matrix** containing all questions and the total. Use the column labeled TOTAL and locate the correlation between the total score and each question. In the example shown to the right, QUESTION 1 has a correlation of 0.873 with the total score. QUESTION 2 has a correlation of –.130 with the total score. QUESTION 3 has a correlation of .926 with the total score.

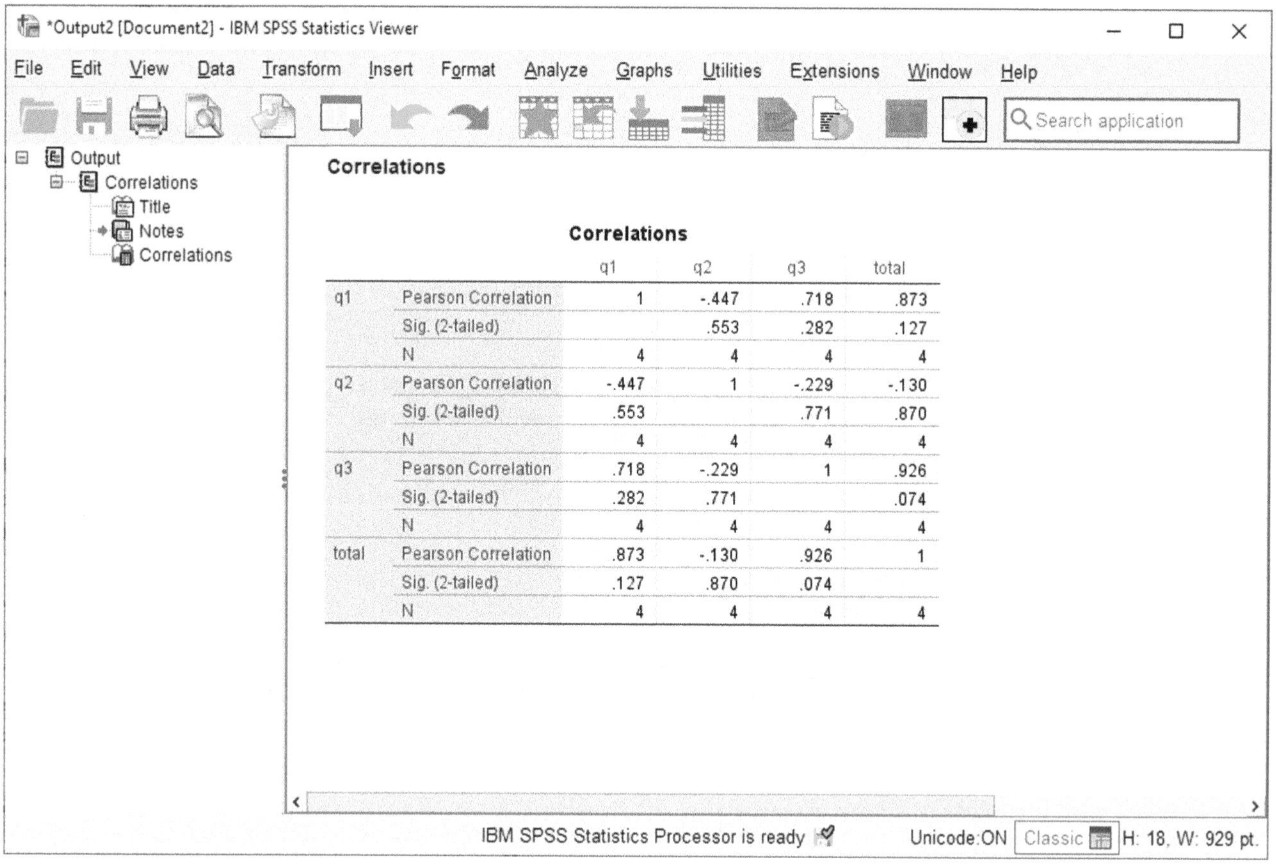

Correlations

Correlations

		q1	q2	q3	total
q1	Pearson Correlation	1	-.447	.718	.873
	Sig. (2-tailed)		.553	.282	.127
	N	4	4	4	4
q2	Pearson Correlation	-.447	1	-.229	-.130
	Sig. (2-tailed)	.553		.771	.870
	N	4	4	4	4
q3	Pearson Correlation	.718	-.229	1	.926
	Sig. (2-tailed)	.282	.771		.074
	N	4	4	4	4
total	Pearson Correlation	.873	-.130	.926	1
	Sig. (2-tailed)	.127	.870	.074	
	N	4	4	4	4

Interpreting the Output

Item-total correlations should always be positive. If you obtain a negative correlation, that question should be removed from the scale (or you may consider whether it should be reverse-keyed).

Generally, item-total correlations of greater than .7 are considered desirable. Those of less than .3 are considered weak. Any questions with correlations of less than .3 should be removed from the scale.

Normally, the worst question is removed, and then the total is recalculated. After the total is recalculated, the item-total analysis is repeated without the question that was removed. Then, if any questions have correlations of less than .3, the worst one is removed, and the process is repeated.

When all remaining correlations are greater than .3, the remaining items in the scale are considered to be those that are internally consistent.

Section 9.2 Cronbach's Alpha

Description

Cronbach's alpha is a measure of **internal consistency**. As such, it is one of many tests of **reliability**. Cronbach's alpha comprises a number of items that make up a scale designed to measure a single construct (e.g., intelligence), and it determines the degree to which all the items are measuring the same construct. It does not tell you if it is measuring the correct construct (that is a question of **validity**). As stated earlier, before a test can be valid, it must first be reliable.

Assumptions

All the items in the scale should be measured on an **interval** or **ratio scale**. In addition, each item should be normally distributed.

SPSS Data Format

SPSS requires one variable for each item (or question) in the scale.

Running the Command

We will continue to use the QUESTIONS.sav data file we first created in Chapter 2. Click *Analyze*, then *Scale*, then *Reliability Analysis*.

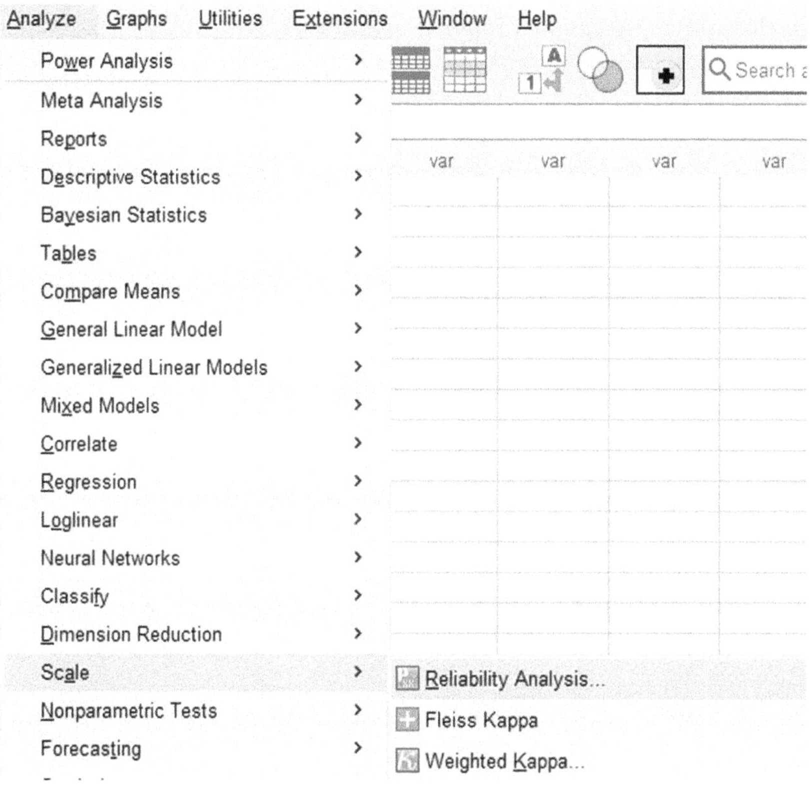

This will bring up the main **dialog box** for Reliability Analysis. Transfer the questions from your scale to the *Items* blank and click *OK*. Do not transfer any variables representing total scores.

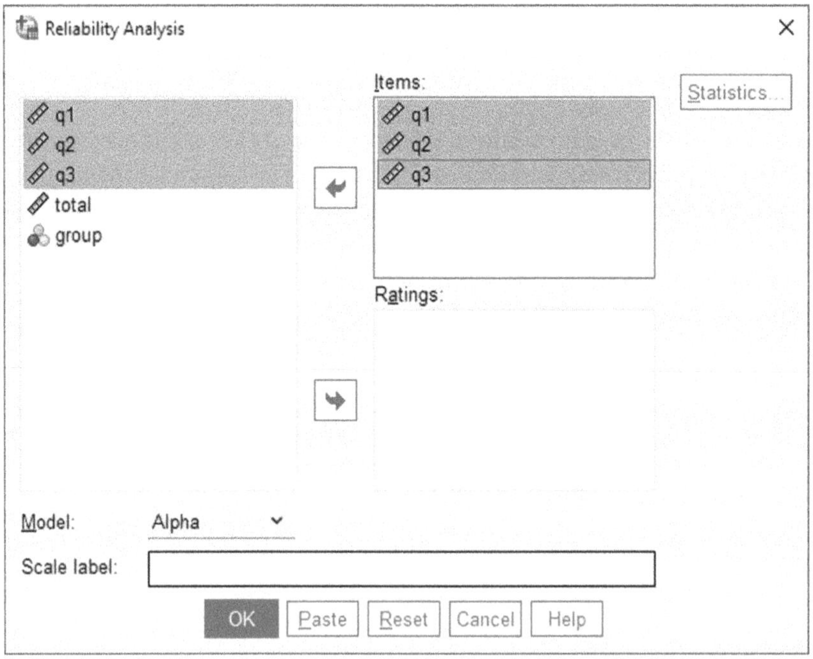

Note that if you change the options under *Model*, additional measures of **internal consistency** (e.g., split-half) can be calculated.

Reading the Output

In this example, the **reliability** coefficient is .407. Numbers close to 1 are very good, but numbers close to 0 represent poor **internal consistency**.

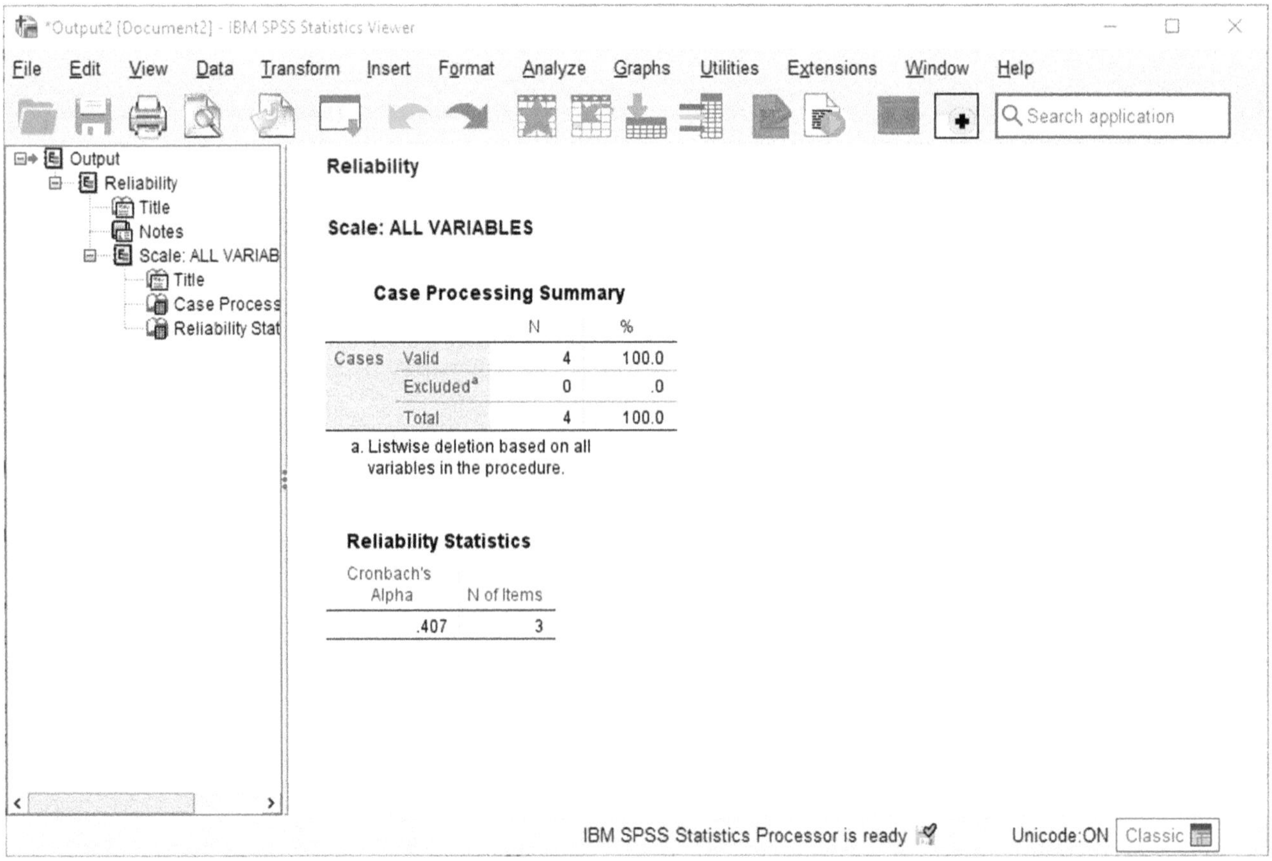

Section 9.3 Test-Retest Reliability

Description

Test-retest reliability is a measure of **temporal stability**. As such, it is a measure of **reliability**. Unlike measures of **internal consistency** that tell you the extent to which all of the questions that make up a scale measure the same construct, measures of **temporal stability** tell you whether or not the instrument is consistent over time or over multiple administrations.

Assumptions

The total score for the scale should be an **interval** or **ratio scale**. The scale scores should be normally distributed.

SPSS Data Format

SPSS requires a variable representing the total score for the scale at the time of first administration. A second variable representing the total score for the same participants at a different time (normally two weeks later) is also required.

Running the Command

The test-retest reliability coefficient is simply a Pearson correlation coefficient for the relationship between the total scores for the two administrations. To compute the coefficient, follow the directions for computing a Pearson correlation coefficient (Chapter 5, Section 5.1). Use the two variables representing the two administrations of the test.

Reading the Output

The correlation between the two scores is the test-retest reliability coefficient. It should be positive. Strong **reliability** is indicated by values close to 1. Weak **reliability** is indicated by values close to 0.

Section 9.4 Criterion-Related Validity

Description

Criterion-related validity determines the extent to which the scale you are testing correlates with a criterion. For instance, *ACT* scores should correlate highly with GPA. If they do, that is a measure of **validity** for *ACT* scores. If they do not, that indicates that *ACT* scores may not be valid for the intended purpose.

Assumptions

All of the same assumptions for the Pearson correlation coefficient apply to measures of criterion-related validity (**interval** or **ratio scales**, **normal distribution**, etc.).

SPSS Data Format

Two variables are required. One variable represents the total score for the scale you are testing. The other represents the criterion you are testing it against.

Running the Command

Calculating criterion-related validity involves determining the Pearson correlation value between the scale and the criterion. See Chapter 5, Section 5.1 for complete information.

Reading the Output

The correlation between the two scores is the criterion-related validity coefficient. It should be positive. Strong **validity** is indicated by values close to 1. Weak **validity** is indicated by values close to 0.

Section 9.5 Inter-Rater Reliability

Description

There are many ways to measure inter-rater reliability. It can be as simple as the correlation between two reviewer's scores if you have an interval or ratio scale (or even a Spearman *rho* if you have a nominal scale). A common measure with nominal scales, however, is Cohen's Kappa.

Assumptions

Cohen's Kappa works best with nominal scales with a limited number of categories.

SPSS Data Format

Two variables are required. One variable represents the rating of one person and the second variable represents the rating of the other rater. For example, you could have a data file using the data presented here.

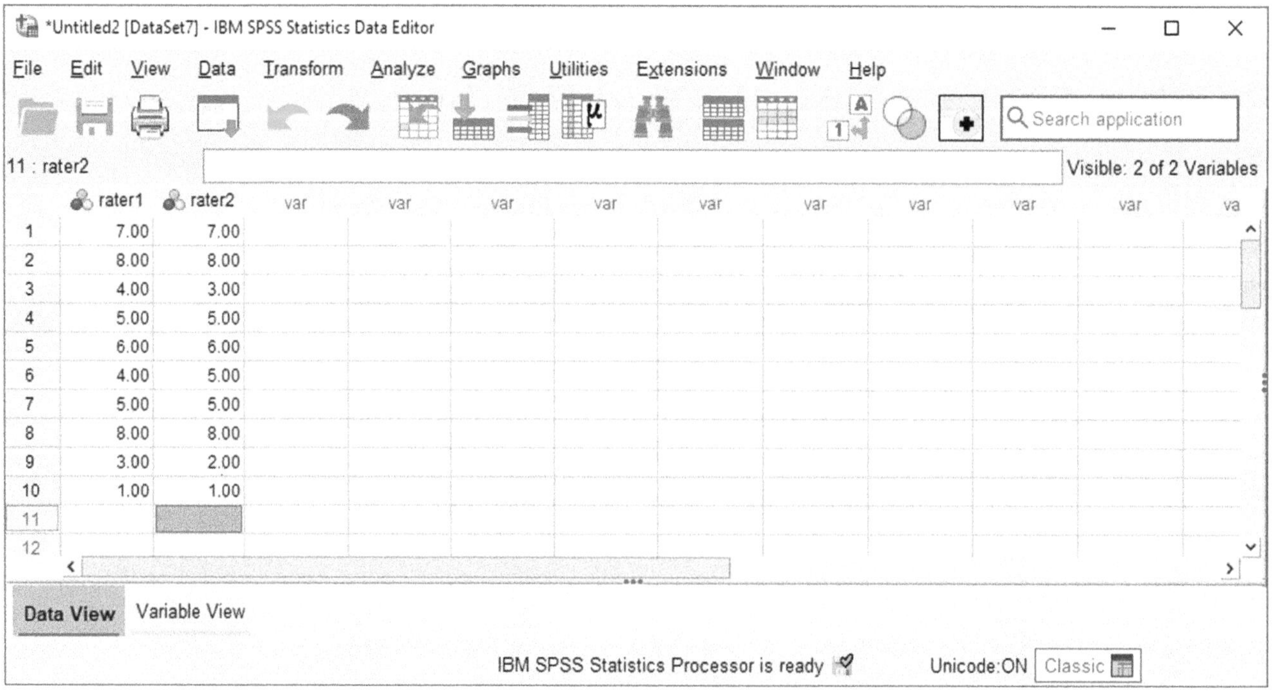

Running the Command

Calculating inter-rater reliability involves the crosstabs command. See Chapter 3, Section 3.2 for more information. Enter one rater's scores as the columns and the other rater's scores as the rows.

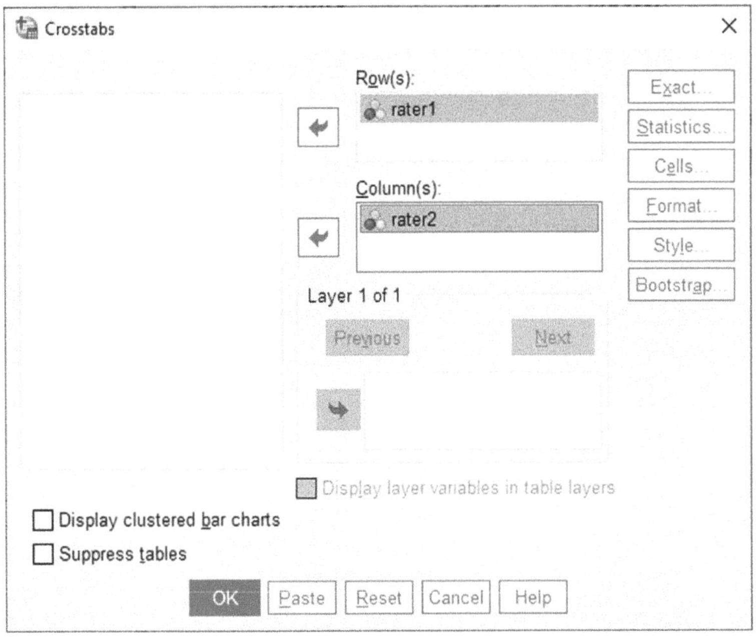

When running crosstabs, you select the *Statistics* option and then select *Kappa.*

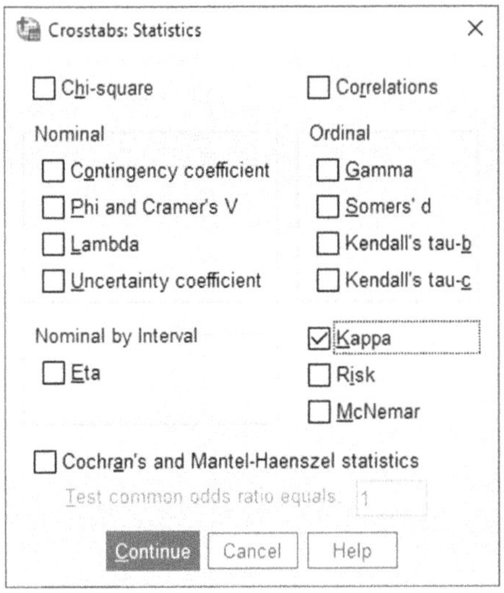

Reading the Output

The inter-rater reliability is represented at the bottom of the output as Kappa. In the example here, Kappa is equal to .651. While there is some disagreement on interpreting Kappa, Landis and Koch (1977) have suggested values <.2 as slight, .2–.4 as fair, .4–.6 as moderate, .5–.8 as strong, and >.8 as almost perfect.[1]

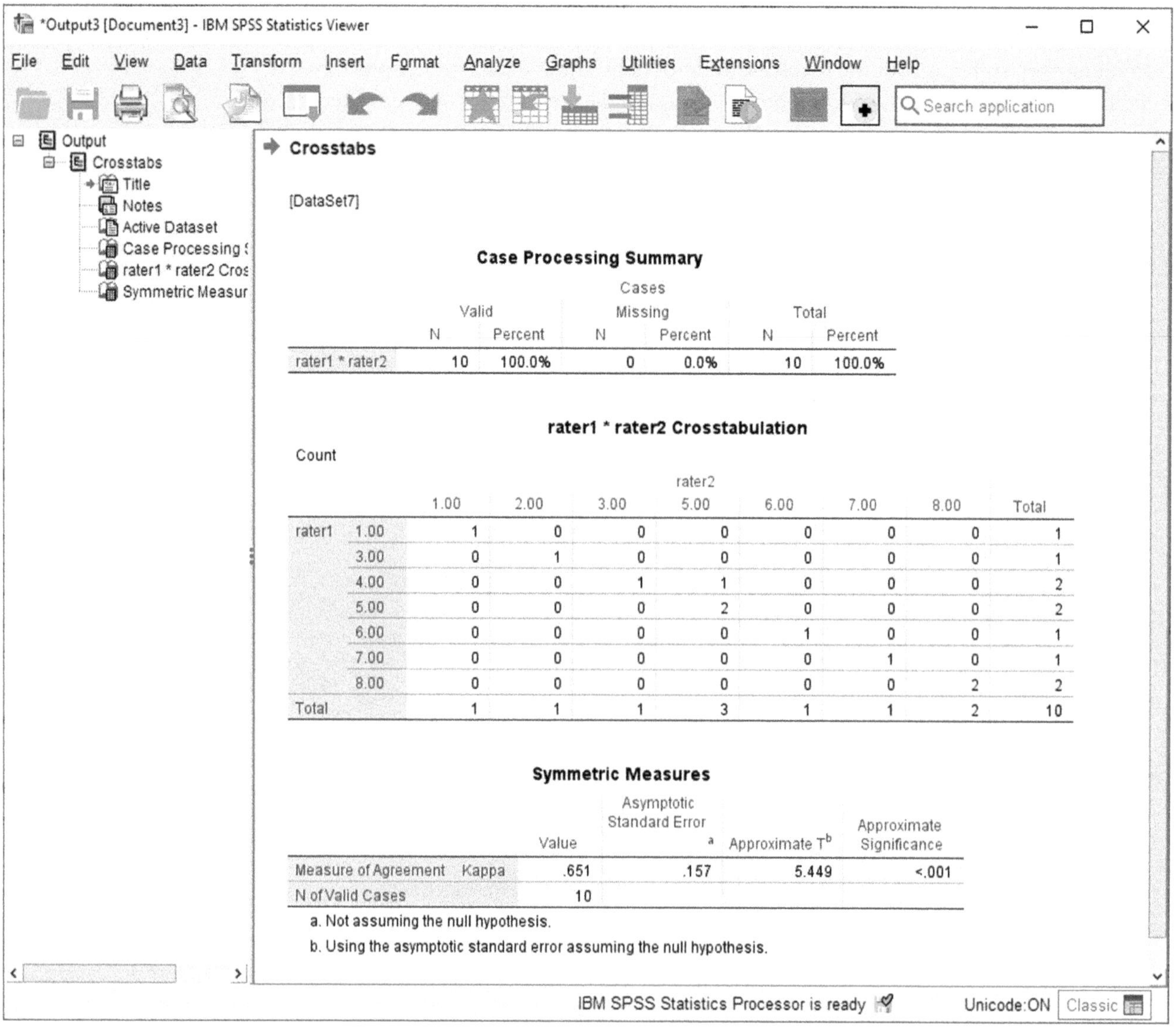

Crosstabs

[DataSet7]

Case Processing Summary

	Cases					
	Valid		Missing		Total	
	N	Percent	N	Percent	N	Percent
rater1 * rater2	10	100.0%	0	0.0%	10	100.0%

rater1 * rater2 Crosstabulation

Count

		rater2							
		1.00	2.00	3.00	5.00	6.00	7.00	8.00	Total
rater1	1.00	1	0	0	0	0	0	0	1
	3.00	0	1	0	0	0	0	0	1
	4.00	0	0	1	1	0	0	0	2
	5.00	0	0	0	2	0	0	0	2
	6.00	0	0	0	0	1	0	0	1
	7.00	0	0	0	0	0	1	0	1
	8.00	0	0	0	0	0	0	2	2
Total		1	1	1	3	1	1	2	10

Symmetric Measures

		Value	Asymptotic Standard Error[a]	Approximate T[b]	Approximate Significance
Measure of Agreement	Kappa	.651	.157	5.449	<.001
N of Valid Cases		10			

a. Not assuming the null hypothesis.

b. Using the asymptotic standard error assuming the null hypothesis.

Appendix A
Effect Size

Many disciplines are placing increased emphasis on reporting **effect size**. While statistical hypothesis testing provides a way to tell the odds that differences are real, effect sizes provide a way to judge the relative importance of those differences. That is, they tell us the size of the difference or relationship. They are also critical if you would like to estimate necessary sample sizes, conduct a power analysis, or conduct a meta-analysis. Many professional organizations (e.g., the American Psychological Association) are now requiring or strongly suggesting that effect sizes be reported in addition to the results of hypothesis tests.

Because there are at least 41 different types of effect sizes,[1] each with somewhat different properties, the purpose of this appendix is not to be a comprehensive resource on **effect size**, but rather to show you how to calculate some of the most common measures of **effect size** using SPSS. The most recent versions of SPSS now also include **effect size** calculations as options in some procedures.

Section A.1 Cohen's *d*

One of the simplest and most popular measures of **effect size** is **Cohen's *d*. Cohen's *d*** is a member of a class of measurements called *standardized mean differences*. In essence, *d* is the difference between the two means divided by the overall **standard deviation**.

It is not only a popular measure of **effect size,** but Cohen has also suggested a simple basis to interpret the value obtained. Cohen[2] suggests that effect sizes less than .2 are small, .2–.8 are medium, and greater than .8 are large.

We will discuss **Cohen's *d*** as the preferred measure of **effect size** for *t* tests. With the newest versions of SPSS, the software will now calculate **Cohen's *d*** for you. Just be sure the E*stimate effect sizes* box is checked in the appropriate *t* test dialog box.

One-Sample Effect Sizes

				95% Confidence Interval	
		Standardizer[a]	Point Estimate	Lower	Upper
GRADE	Cohen's d	1.41421	.707	-1.002	2.239
	Hedges' correction	.	.	.	.

a. The denominator used in estimating the effect sizes.
Cohen's d uses the sample standard deviation.
Hedges' correction uses the sample standard deviation, plus a correction factor.

Each is interpreted the same way. The Cohen's d row and Point Estimate column show the value. In the example here the answer is .707—a medium effect size.

All three tests (single sample, independent samples, paired samples) include **effect size** in SPSS 29 and above. Please refer to Chapter 6 as these effect sizes are now integrated into the chapter.

Section A.2 r^2 (Coefficient of Determination)

While **Cohen's *d*** is the appropriate measure of **effect size** for *t* tests, correlation and regression effect sizes should be determined by squaring the correlation coefficient. This squared correlation is called the **coefficient of determination.** Cohen[3] suggested that correlations of greater than .5 represented a large relationship, less than .1 a small relationship, and between .1 and .5 a moderate relationship. Those values squared yield **coefficients of determination** of > .25 and <.01, respectively. It would appear, therefore, that Cohen is suggesting that accounting for 25% of the variability represents a large effect and less than 1% a small effect.

Effect Size for Correlation

Nowhere is the effect of sample size on statistical power (and therefore **significance)** more apparent than with correlations. Given a large enough sample, *any* correlation can become significant. Thus, **effect size** becomes critically important in the interpretation of correlations.

The standard measure of **effect size** for correlations is the **coefficient of determination** (r^2) previously discussed. The coefficient should be interpreted as the proportion of **variance** in the **dependent variable** that can be accounted for by the relationship between the **independent** and **dependent variables.** While Cohen provided useful guidelines for interpretation, each problem should be interpreted in terms of its true practical **significance**. For instance, if a treatment is very expensive to implement, or has significant side effects, then a larger correlation should be required before the relationship becomes important. For treatments that are very inexpensive, a much smaller correlation can be considered important.

To calculate the **coefficient of determination**, simply take the *r* value that SPSS provides and square it.

Effect Size for Regression

The Model Summary section of the output reports R^2 for you. The example output here (from the HEIGHT.sav dataset for a regression to predict WEIGHT from HEIGHT) shows a **coefficient of determination** of .649, meaning that almost 65% (.649) of the variability in the **dependent variable** is accounted for by the relationship between the **dependent** and **independent variables**.

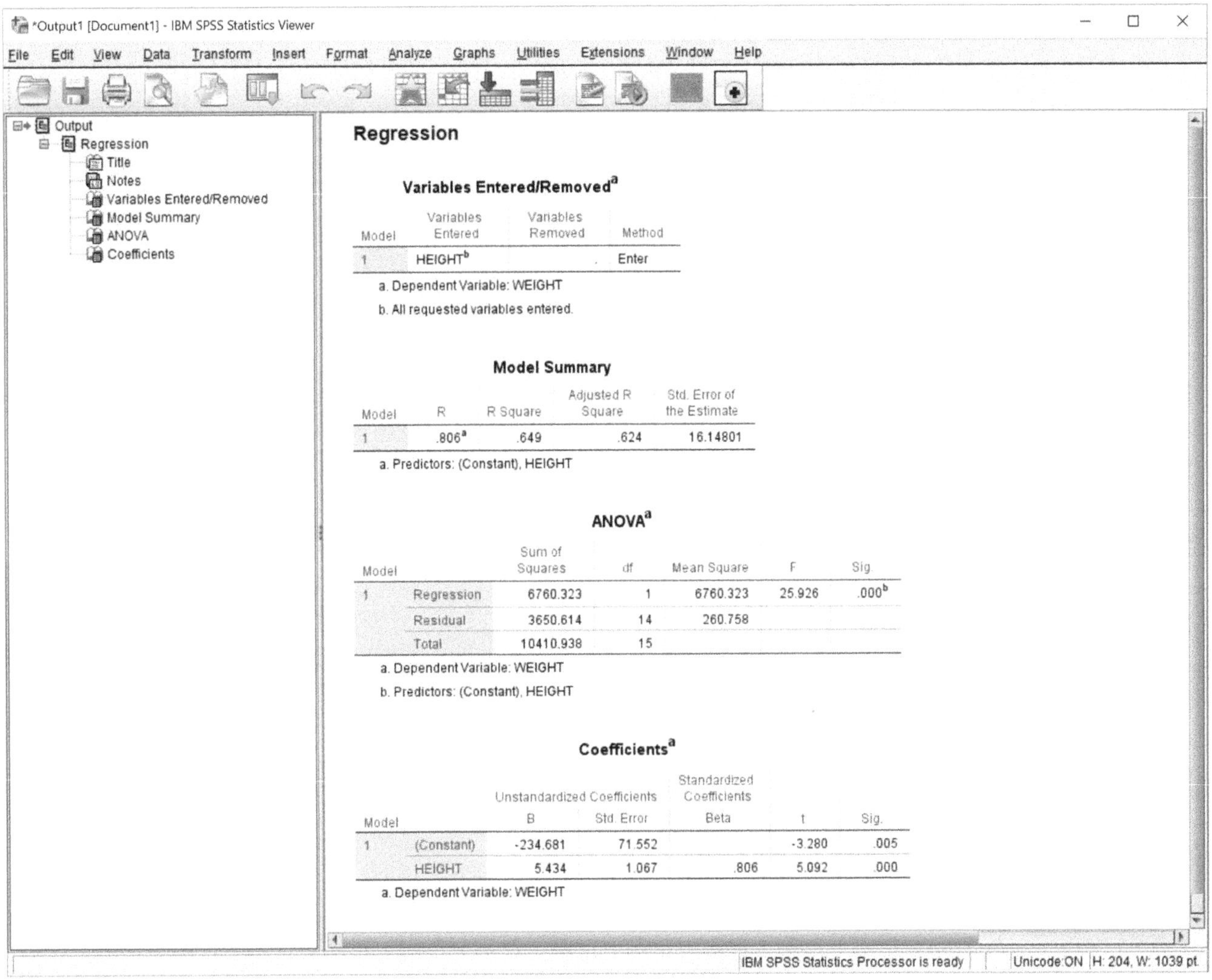

Regression

Variables Entered/Removed[a]

Model	Variables Entered	Variables Removed	Method
1	HEIGHT[b]	.	Enter

a. Dependent Variable: WEIGHT
b. All requested variables entered.

Model Summary

Model	R	R Square	Adjusted R Square	Std. Error of the Estimate
1	.806[a]	.649	.624	16.14801

a. Predictors: (Constant), HEIGHT

ANOVA[a]

Model		Sum of Squares	df	Mean Square	F	Sig.
1	Regression	6760.323	1	6760.323	25.926	.000[b]
	Residual	3650.614	14	260.758		
	Total	10410.938	15			

a. Dependent Variable: WEIGHT
b. Predictors: (Constant), HEIGHT

Coefficients[a]

Model		Unstandardized Coefficients		Standardized Coefficients	t	Sig.
		B	Std. Error	Beta		
1	(Constant)	-234.681	71.552		-3.280	.005
	HEIGHT	5.434	1.067	.806	5.092	.000

a. Dependent Variable: WEIGHT

Section A.3 Eta Squared (η^2)

A third measure of **effect size** is **Eta Squared (η^2). Eta Squared** is used for Analysis of Variance models. The *GLM* (General Linear Model) function in SPSS (the function that runs the procedures under *Analyze—General Linear Model*) will provide **Eta Squared (η^2)**.

In newer versions of SPSS the one-way command also now provides effect size. Please refer to Section 7.2 for information on **effect size** for one-way ANOVA.

$$\eta^2 = \frac{SS_{effect}}{SS_{effect} + SS_{error}}$$

Eta Squared has an interpretation similar to a squared correlation coefficient (r^2). It represents the proportion of the **variance** accounted for by the effect. Unlike r^2, however, which represents only linear relationships, η^2 can represent any type of relationship.

Effect Size for Analysis of Variance

For most Analysis of Variance problems, you should elect to report **Eta Squared** as your **effect size** measure. SPSS provides this calculation for you as part of the *General Linear Model* (*GLM*) command. To obtain **Eta Squared,** simply click on the *Options* box in the main **dialog box** for the *GLM* command you are running (this works for Univariate, Multivariate, and Repeated Measures versions of the command even though only the Univariate option is presented here).

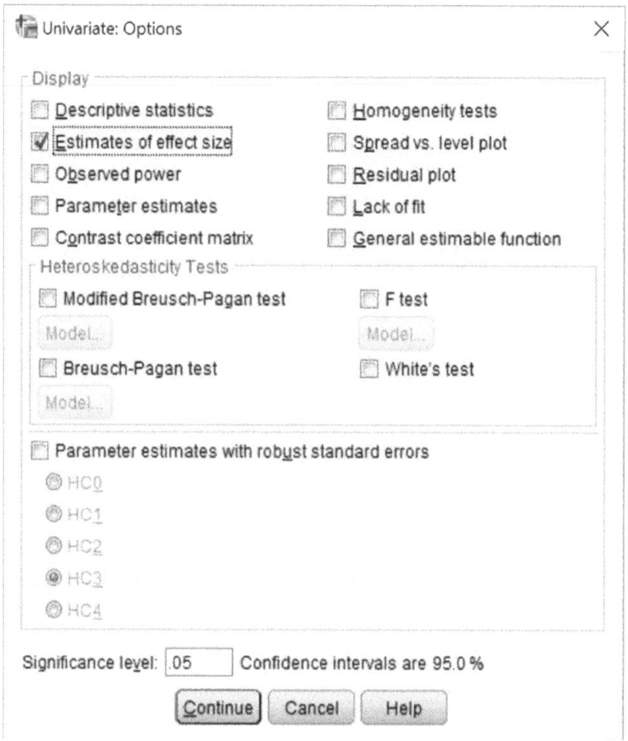

Once you have selected *Options,* a new **dialog box** will appear. One of the options in that box will be *Estimates of effect size.* When you select that box, SPSS will provide **Eta Squared** values as part of your output.

The example here uses the GRADES.sav dataset an ANOVA on FINAL grades and INSTRUCT.

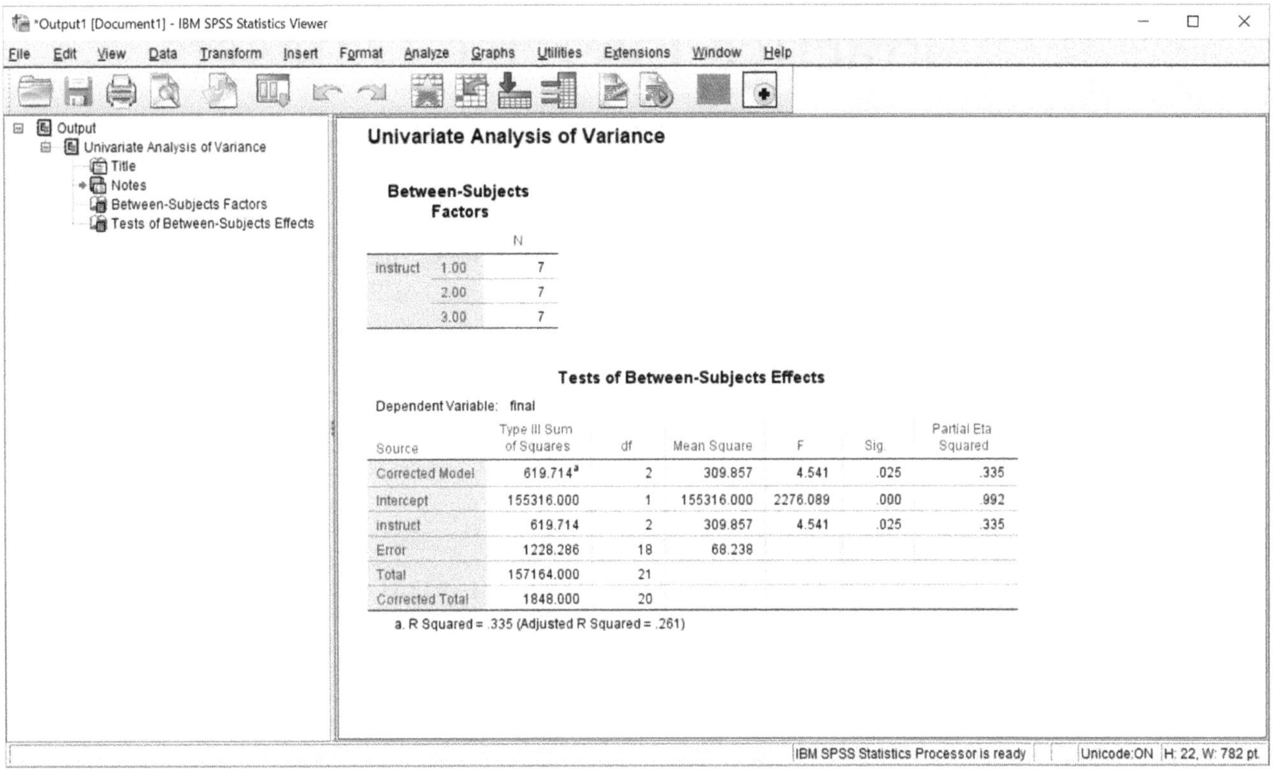

Univariate Analysis of Variance

Between-Subjects Factors

		N
instruct	1.00	7
	2.00	7
	3.00	7

Tests of Between-Subjects Effects

Dependent Variable: final

Source	Type III Sum of Squares	df	Mean Square	F	Sig.	Partial Eta Squared
Corrected Model	619.714[a]	2	309.857	4.541	.025	.335
Intercept	155316.000	1	155316.000	2276.089	.000	.992
instruct	619.714	2	309.857	4.541	.025	.335
Error	1228.286	18	68.238			
Total	157164.000	21				
Corrected Total	1848.000	20				

a. R Squared = .335 (Adjusted R Squared = .261)

In the above example, we obtained an **Eta Squared** of .335 for our main effect for Instructor. Because we interpret **Eta Squared** using the same guidelines as r^2, we would conclude that this represents a large **effect size** for instructor (>.25).

Appendix B
Practice Exercise Datasets

Practice Dataset 1

You have conducted a study in which you collected data from 23 participants. You asked each subject to indicate his or her sex (SEX), age (AGE), and marital status (MARITAL). You gave each subject a test to measure mathematics skills (SKILL), where the higher scores indicated a higher skill level. The data are presented below. Note that you will have to code the variables SEX and MARITAL and also indicate that they are measured on a **nominal scale**.

This example assumes students code SEX as 1 = M, 2 = F and MARITAL as 1 = Single, 2 = Married, 3 = Divorced.

SEX	AGE	MARITAL	SKILL
M	23	Single	34
F	35	Married	40
F	40	Divorced	38
M	19	Single	20
M	28	Married	30
F	35	Divorced	40
F	20	Single	38
F	29	Single	47
M	29	Married	26
M	40	Married	24
F	24	Single	45
M	23	Single	37
F	18	Single	44
M	21	Single	38
M	50	Divorced	32
F	25	Single	29
F	20	Single	38
M	24	Single	19
F	37	Married	29
M	42	Married	42
M	35	Married	59
M	23	Single	45
F	40	Divorced	20

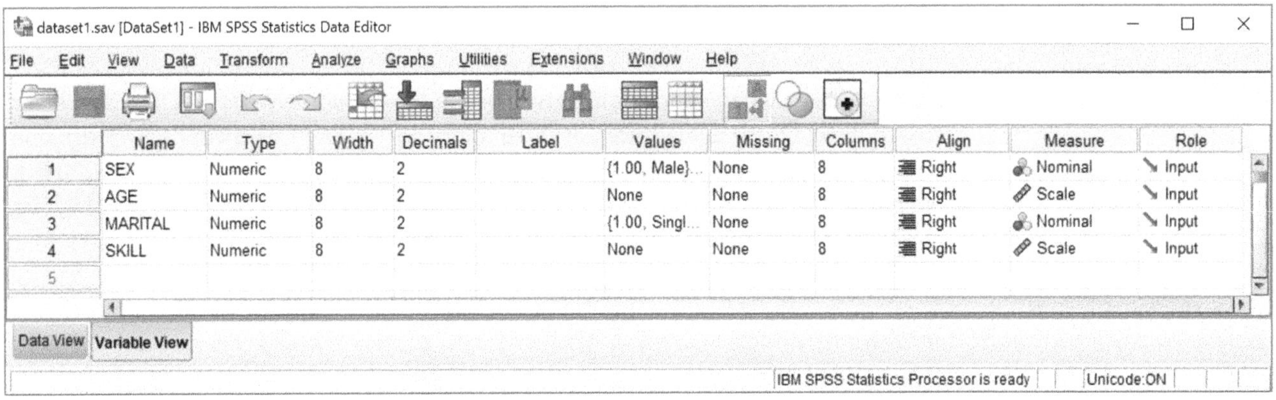

Practice Dataset 2

A survey of employees is conducted. Each employee provides the following information: Salary (SALARY), Years of Service (YOS), Sex (SEX), Job Classification (CLASSIFY), and Education Level (EDUC). Note that you will have to code SEX (Male = 1, Female = 2) and CLASSIFY (Clerical = 1, Technical = 2, Professional = 3) and indicate that they are measured on a **nominal scale**.

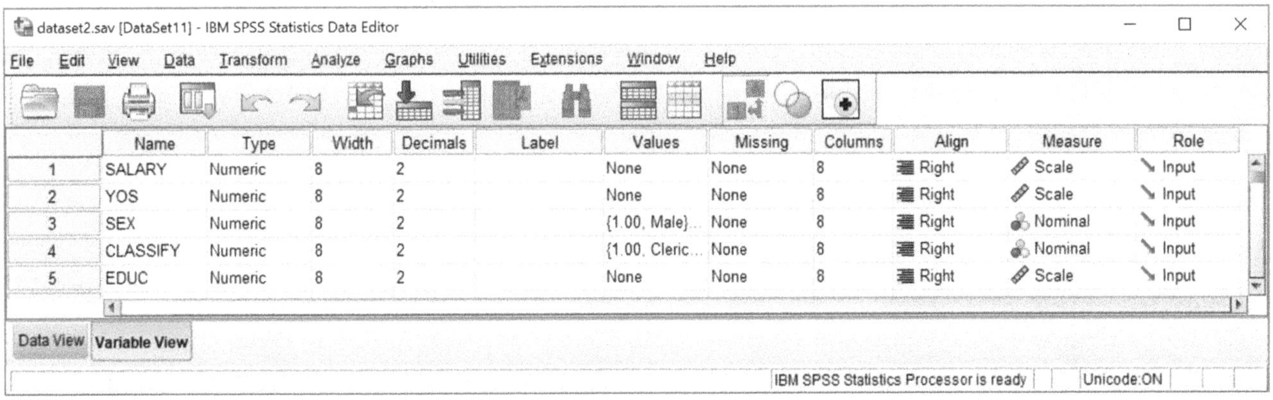

SALARY	YOS	SEX	CLASSIFY	EDUC
35,000	8	Male	Technical	14
18,000	4	Female	Clerical	10
20,000	1	Male	Professional	16
50,000	20	Female	Professional	16
38,000	6	Male	Professional	20
20,000	6	Female	Clerical	12
75,000	17	Male	Professional	20
40,000	4	Female	Technical	12
30,000	8	Male	Technical	14
22,000	15	Female	Clerical	12
23,000	16	Male	Clerical	12
45,000	2	Female	Professional	16

Practice Dataset 3

Participants who have phobias are given one of three treatments (CONDITION). Their anxiety level (1–10) is measured at three intervals—before treatment (ANXPRE), one hour after treatment (ANX1HR), and again four hours after treatment (ANX4HR). Note that you will have to code the variable CONDITION and indicate that they are measured on a **nominal scale**.

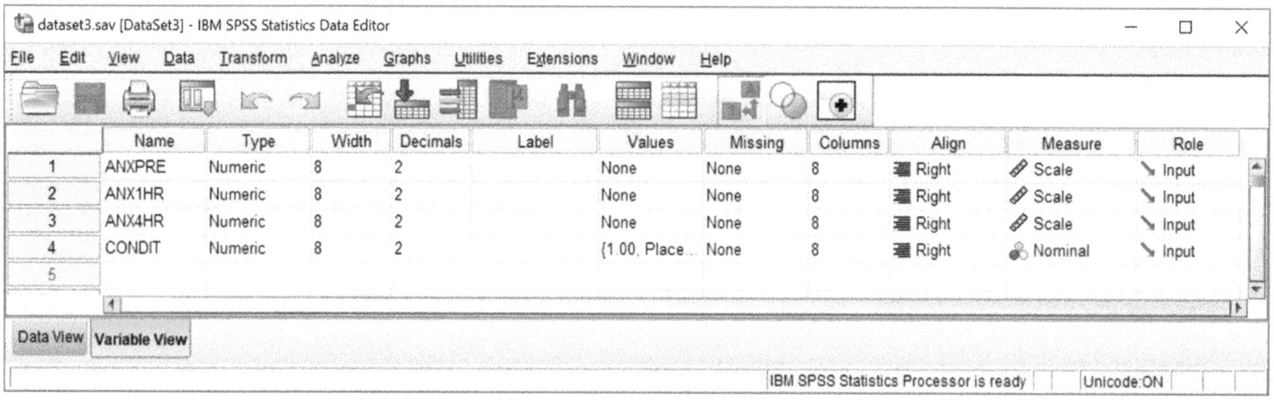

ANXPRE	ANX1HR	ANX4HR	CONDITION
8	7	7	Placebo
10	10	10	Placebo
9	7	8	Placebo
7	6	6	Placebo
7	7	7	Placebo
9	4	5	Valium®
10	6	8	Valium®
9	5	5	Valium®
8	3	5	Valium®
6	3	4	Valium®
8	5	3	Experimental Drug
6	5	2	Experimental Drug
9	8	4	Experimental Drug
10	9	4	Experimental Drug
7	6	3	Experimental Drug

Appendix C
Sample Data Files Used in Text

Practice Datasets 1, 2, and 3 Are in Appendix B

COINS.sav

Created in Section 8.1
Used also in Section 8.2

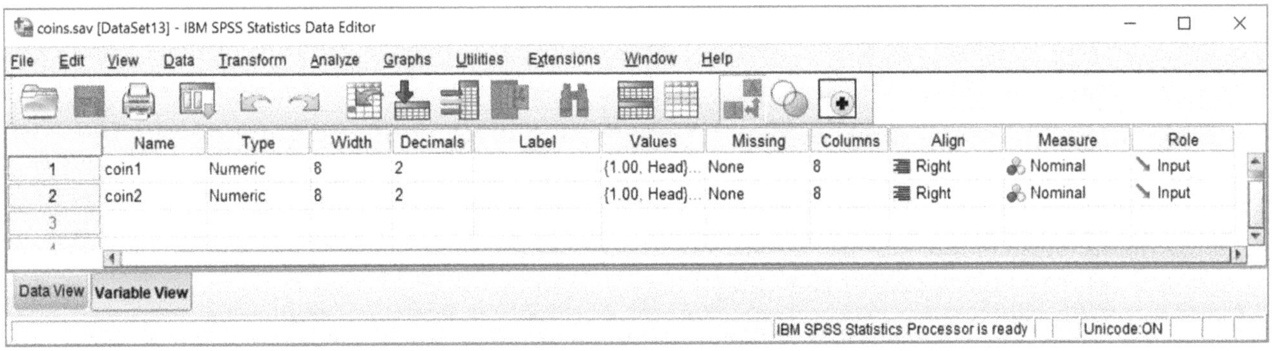

	Name	Type	Width	Decimals	Label	Values	Missing	Columns	Align	Measure	Role
1	coin1	Numeric	8	2		{1.00, Head}...	None	8	Right	Nominal	Input
2	coin2	Numeric	8	2		{1.00, Head}...	None	8	Right	Nominal	Input
3											

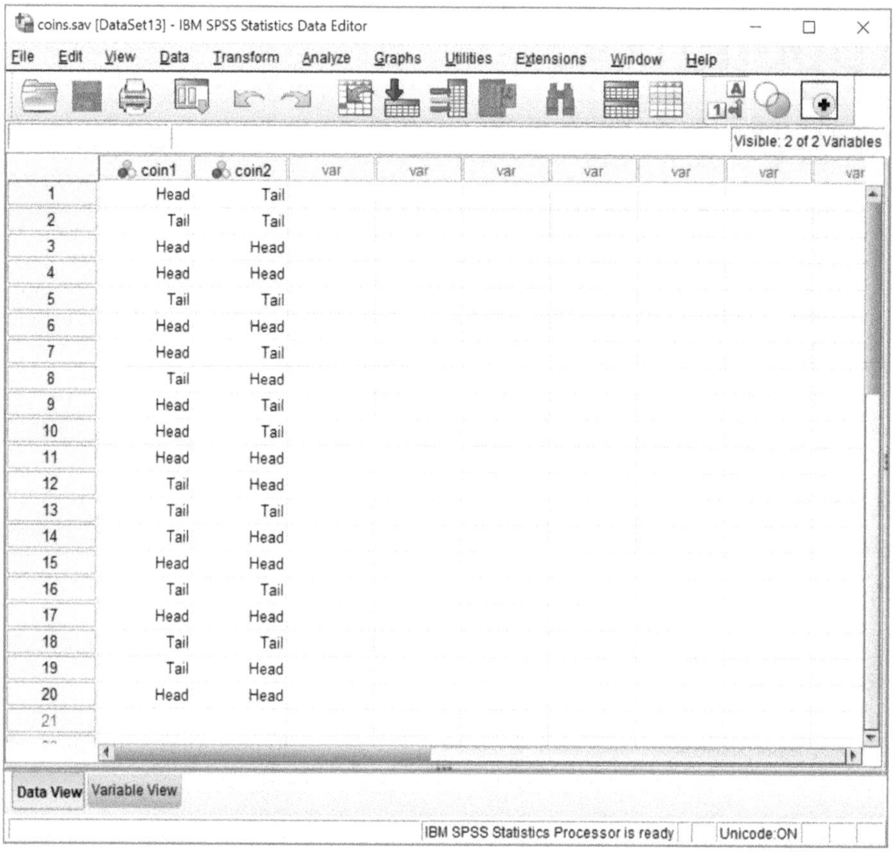

coins.sav [DataSet13] - IBM SPSS Statistics Data Editor

File Edit View Data Transform Analyze Graphs Utilities Extensions Window Help

Visible: 2 of 2 Variables

	coin1	coin2
1	Head	Tail
2	Tail	Tail
3	Head	Head
4	Head	Head
5	Tail	Tail
6	Head	Head
7	Head	Tail
8	Tail	Head
9	Head	Tail
10	Head	Tail
11	Head	Head
12	Tail	Head
13	Tail	Tail
14	Tail	Head
15	Head	Head
16	Tail	Tail
17	Head	Head
18	Tail	Tail
19	Tail	Head
20	Head	Head

Data View Variable View

IBM SPSS Statistics Processor is ready Unicode:ON

GRADES.sav

Created in Section 6.4
Used also in Sections 4.5, 7.2, 7.3, 7.4, and 7.5
Used also in Appendix A sections A.1 and A.3

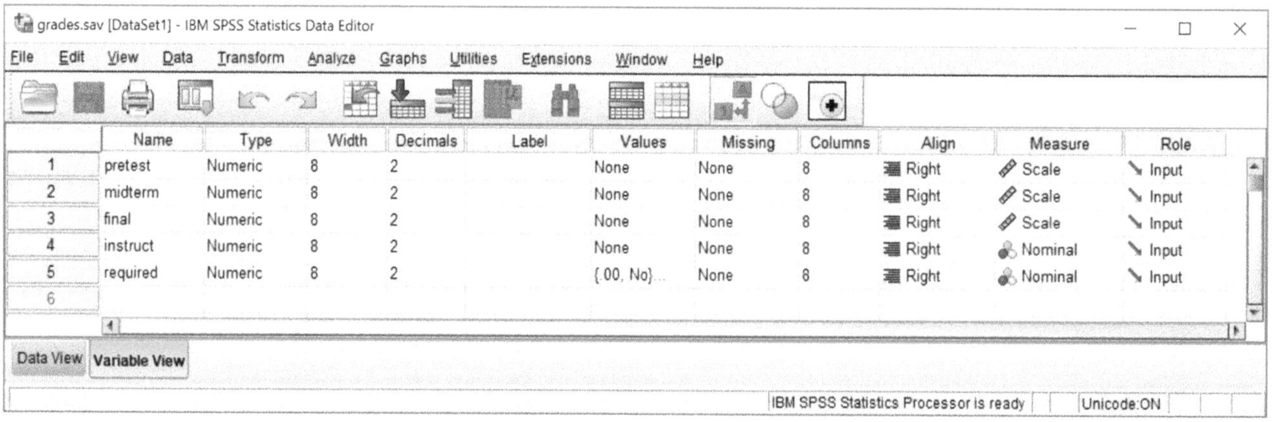

grades.sav [DataSet1] - IBM SPSS Statistics Data Editor

File Edit View Data Transform Analyze Graphs Utilities Extensions Window Help

	Name	Type	Width	Decimals	Label	Values	Missing	Columns	Align	Measure	Role
1	pretest	Numeric	8	2		None	None	8	Right	Scale	Input
2	midterm	Numeric	8	2		None	None	8	Right	Scale	Input
3	final	Numeric	8	2		None	None	8	Right	Scale	Input
4	instruct	Numeric	8	2		None	None	8	Right	Nominal	Input
5	required	Numeric	8	2		{.00, No}...	None	8	Right	Nominal	Input

Data View Variable View

IBM SPSS Statistics Processor is ready Unicode:ON

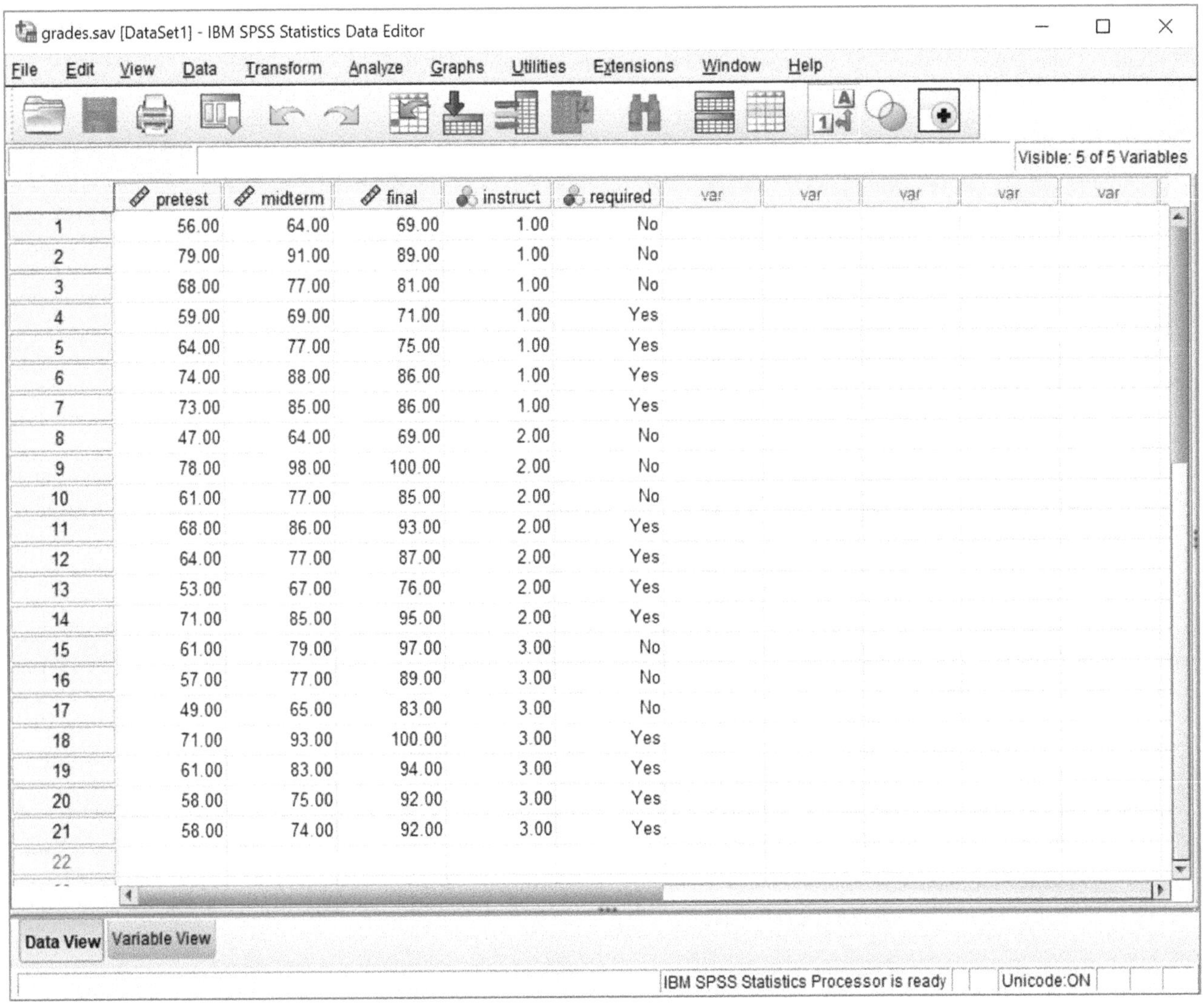

	pretest	midterm	final	instruct	required
1	56.00	64.00	69.00	1.00	No
2	79.00	91.00	89.00	1.00	No
3	68.00	77.00	81.00	1.00	No
4	59.00	69.00	71.00	1.00	Yes
5	64.00	77.00	75.00	1.00	Yes
6	74.00	88.00	86.00	1.00	Yes
7	73.00	85.00	86.00	1.00	Yes
8	47.00	64.00	69.00	2.00	No
9	78.00	98.00	100.00	2.00	No
10	61.00	77.00	85.00	2.00	No
11	68.00	86.00	93.00	2.00	Yes
12	64.00	77.00	87.00	2.00	Yes
13	53.00	67.00	76.00	2.00	Yes
14	71.00	85.00	95.00	2.00	Yes
15	61.00	79.00	97.00	3.00	No
16	57.00	77.00	89.00	3.00	No
17	49.00	65.00	83.00	3.00	No
18	71.00	93.00	100.00	3.00	Yes
19	61.00	83.00	94.00	3.00	Yes
20	58.00	75.00	92.00	3.00	Yes
21	58.00	74.00	92.00	3.00	Yes

HEIGHT.sav

Created in Section 4.2
Used also in Section 4.3
Used also throughout Chapter 5
Used also in Sections 6.2 and 6.9
Used also in Appendix A sections A.1 and A.2

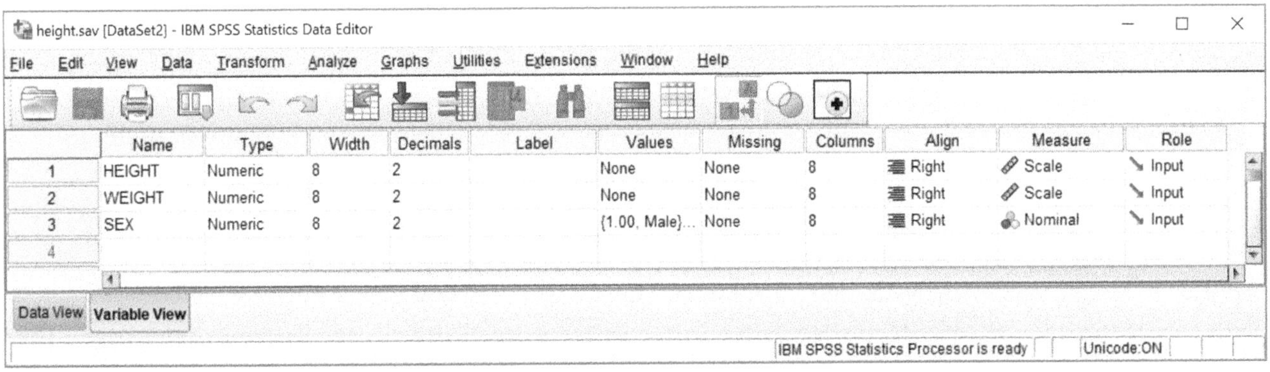

	Name	Type	Width	Decimals	Label	Values	Missing	Columns	Align	Measure	Role
1	HEIGHT	Numeric	8	2		None	None	8	Right	Scale	Input
2	WEIGHT	Numeric	8	2		None	None	8	Right	Scale	Input
3	SEX	Numeric	8	2		{1.00, Male}...	None	8	Right	Nominal	Input

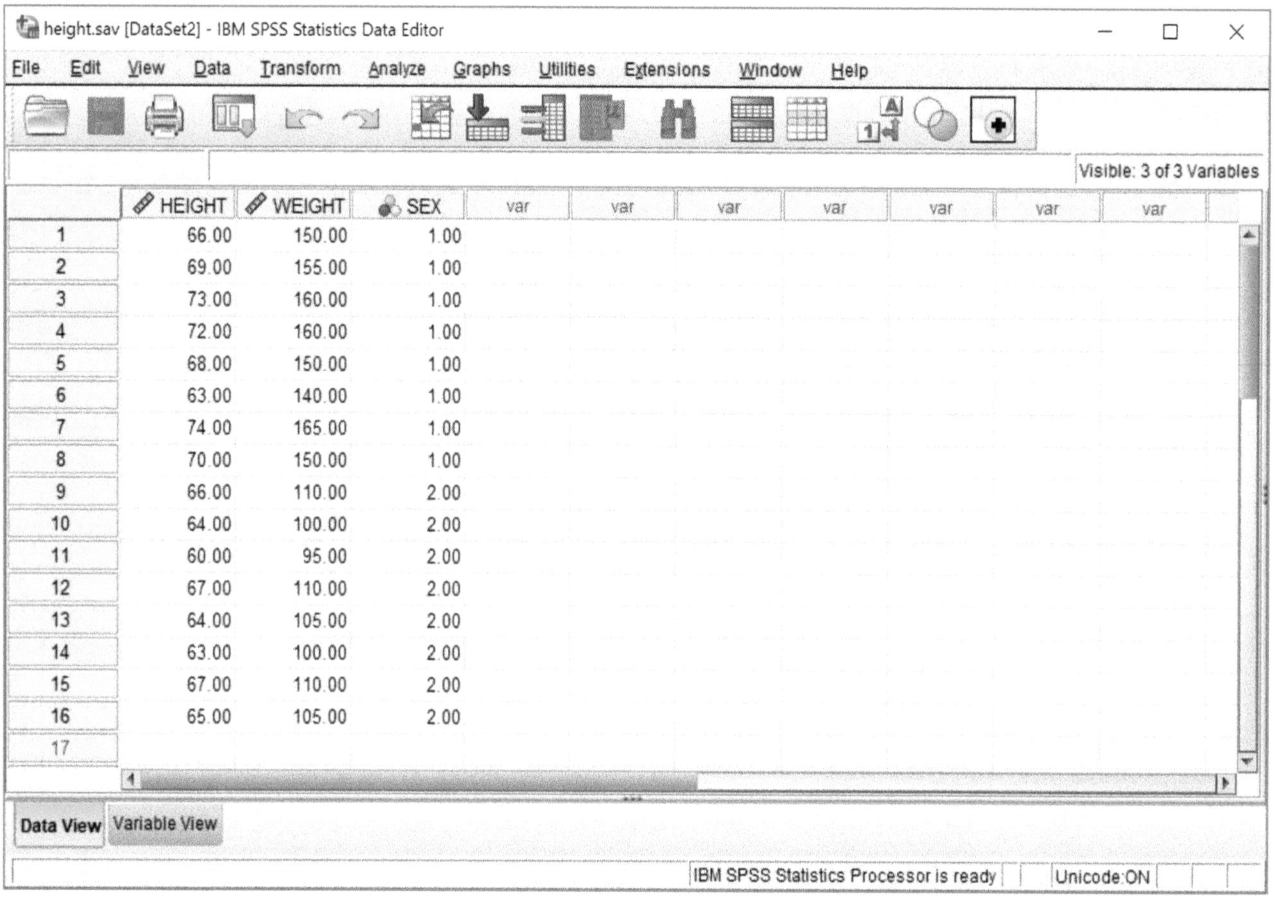

	HEIGHT	WEIGHT	SEX
1	66.00	150.00	1.00
2	69.00	155.00	1.00
3	73.00	160.00	1.00
4	72.00	160.00	1.00
5	68.00	150.00	1.00
6	63.00	140.00	1.00
7	74.00	165.00	1.00
8	70.00	150.00	1.00
9	66.00	110.00	2.00
10	64.00	100.00	2.00
11	60.00	95.00	2.00
12	67.00	110.00	2.00
13	64.00	105.00	2.00
14	63.00	100.00	2.00
15	67.00	110.00	2.00
16	65.00	105.00	2.00

QUESTIONS.sav

Created in Section 2.2

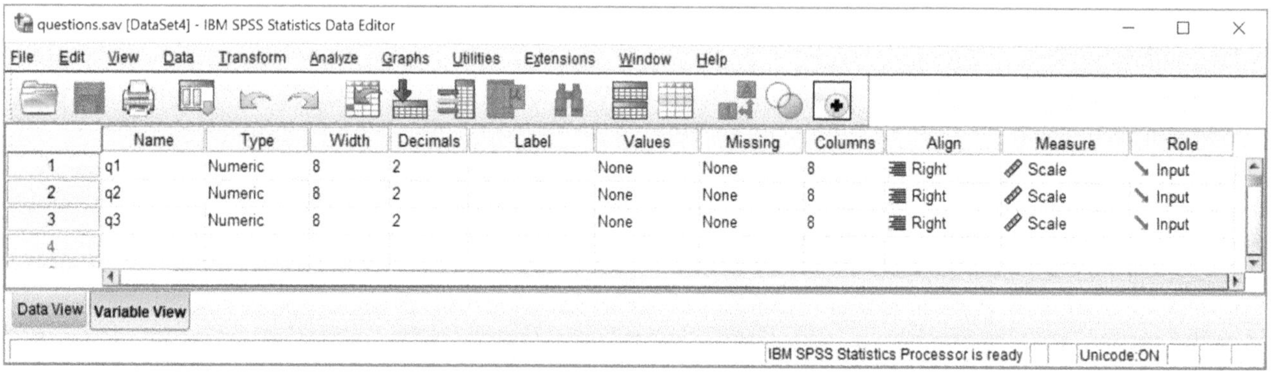

	Name	Type	Width	Decimals	Label	Values	Missing	Columns	Align	Measure	Role
1	q1	Numeric	8	2		None	None	8	Right	Scale	Input
2	q2	Numeric	8	2		None	None	8	Right	Scale	Input
3	q3	Numeric	8	2		None	None	8	Right	Scale	Input

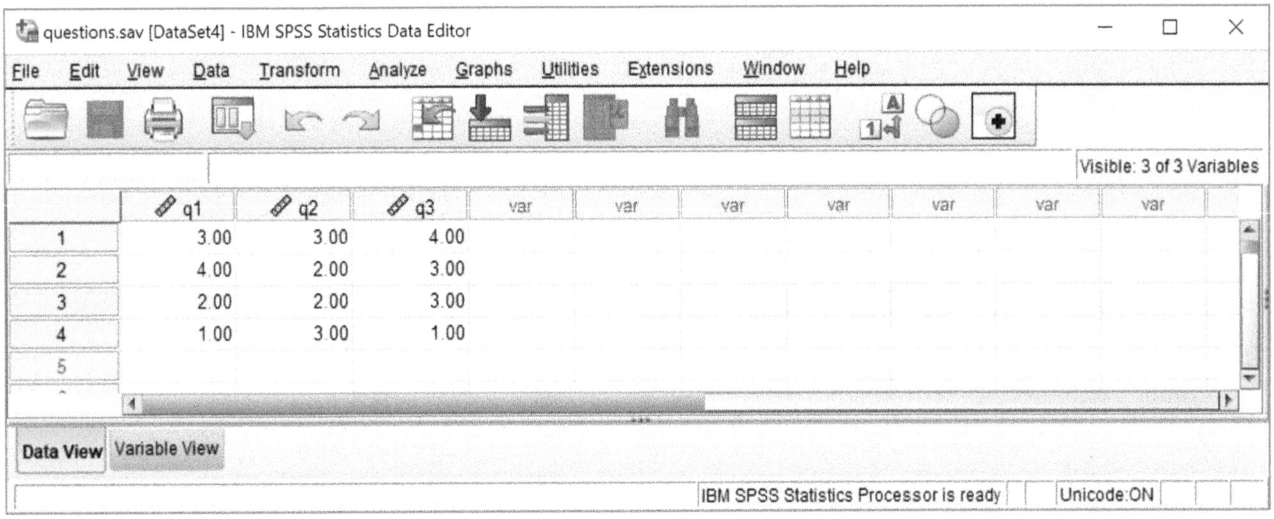

Modified in Section 2.2

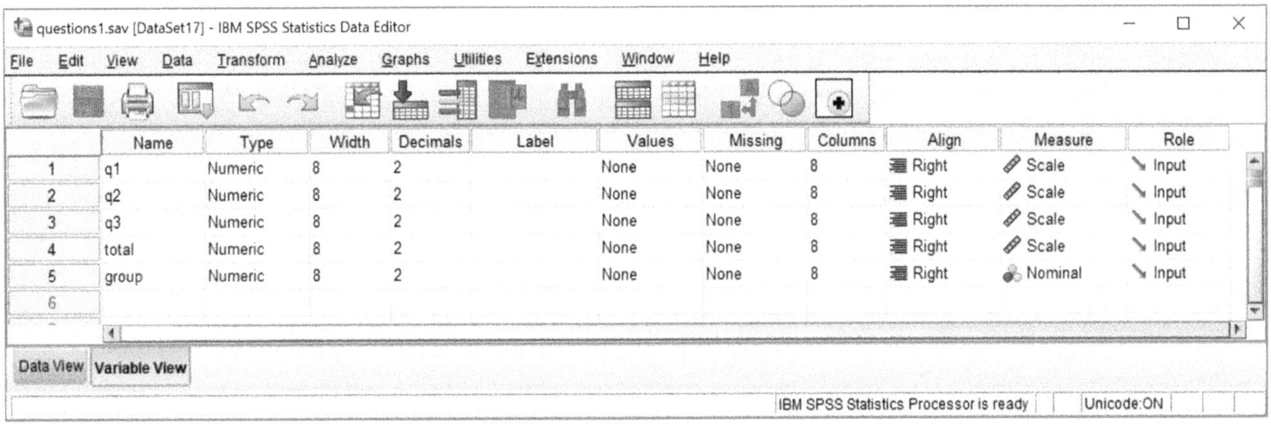

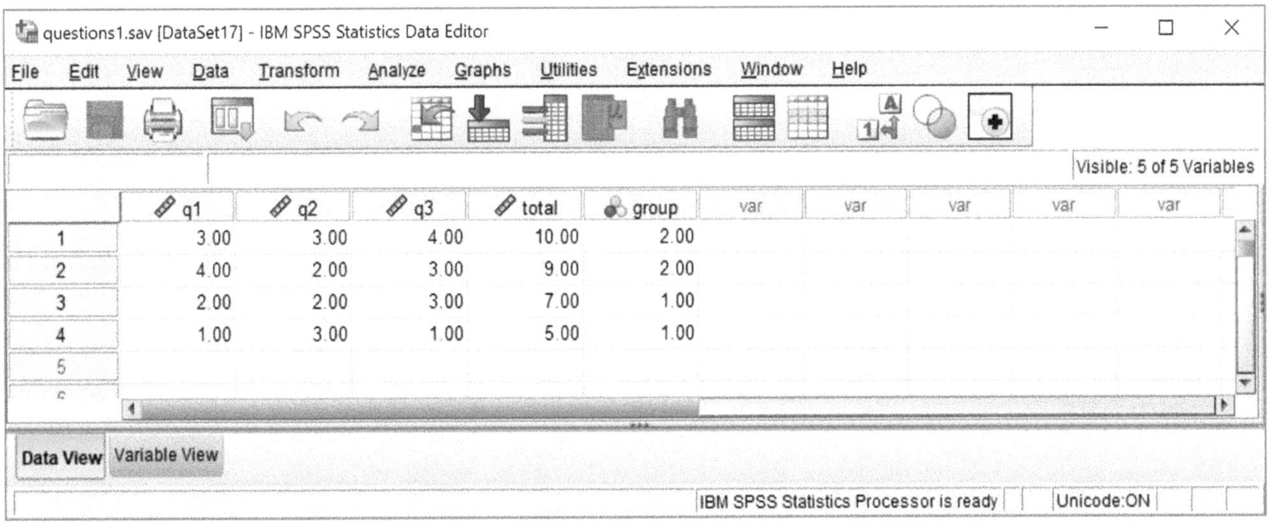

Used also in Sections 9.1 and 9.2

RACE.sav

Created in Section 8.3

Used also in Section 8.4

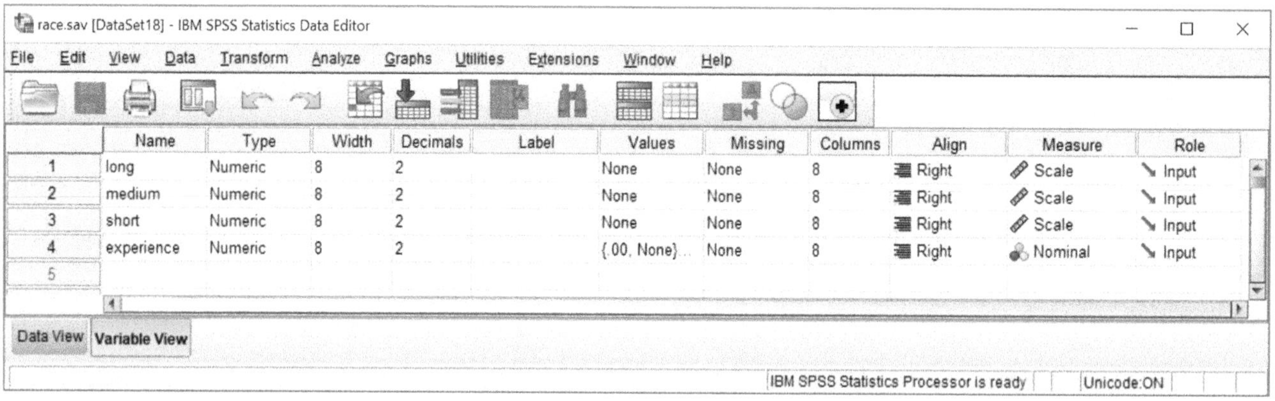

race.sav [DataSet18] - IBM SPSS Statistics Data Editor

	Name	Type	Width	Decimals	Label	Values	Missing	Columns	Align	Measure	Role
1	long	Numeric	8	2		None	None	8	Right	Scale	Input
2	medium	Numeric	8	2		None	None	8	Right	Scale	Input
3	short	Numeric	8	2		None	None	8	Right	Scale	Input
4	experience	Numeric	8	2		{.00, None}...	None	8	Right	Nominal	Input

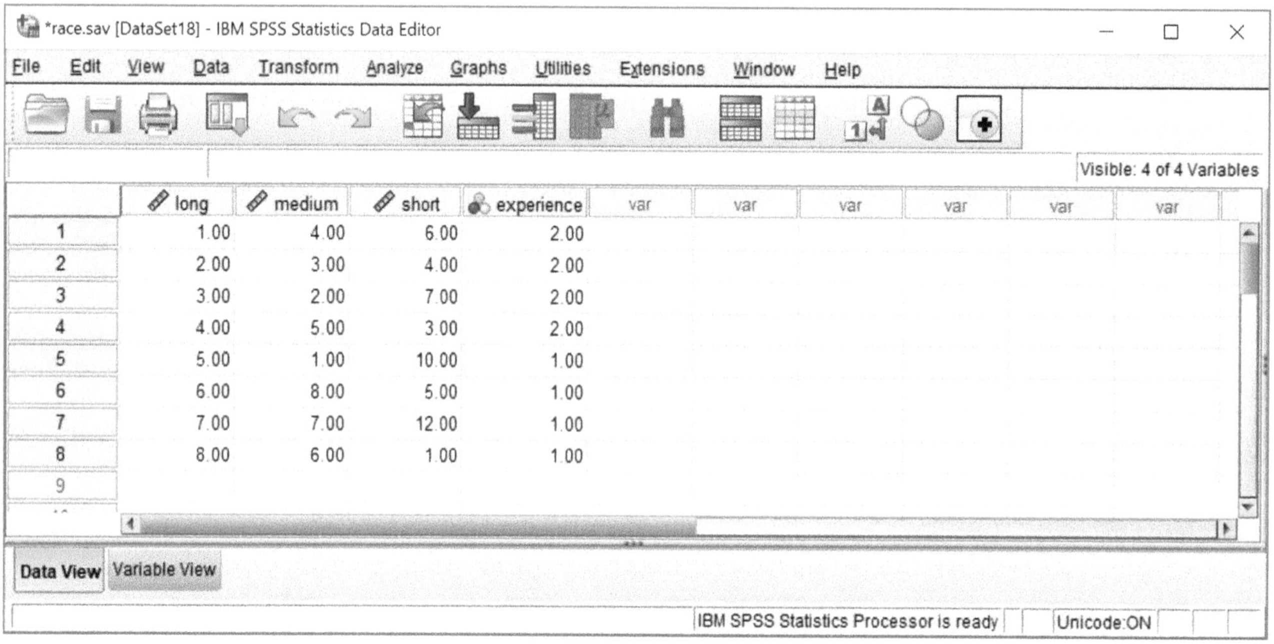

*race.sav [DataSet18] - IBM SPSS Statistics Data Editor

	long	medium	short	experience
1	1.00	4.00	6.00	2.00
2	2.00	3.00	4.00	2.00
3	3.00	2.00	7.00	2.00
4	4.00	5.00	3.00	2.00
5	5.00	1.00	10.00	1.00
6	6.00	8.00	5.00	1.00
7	7.00	7.00	12.00	1.00
8	8.00	6.00	1.00	1.00

Modified in Section 8.5 to be:

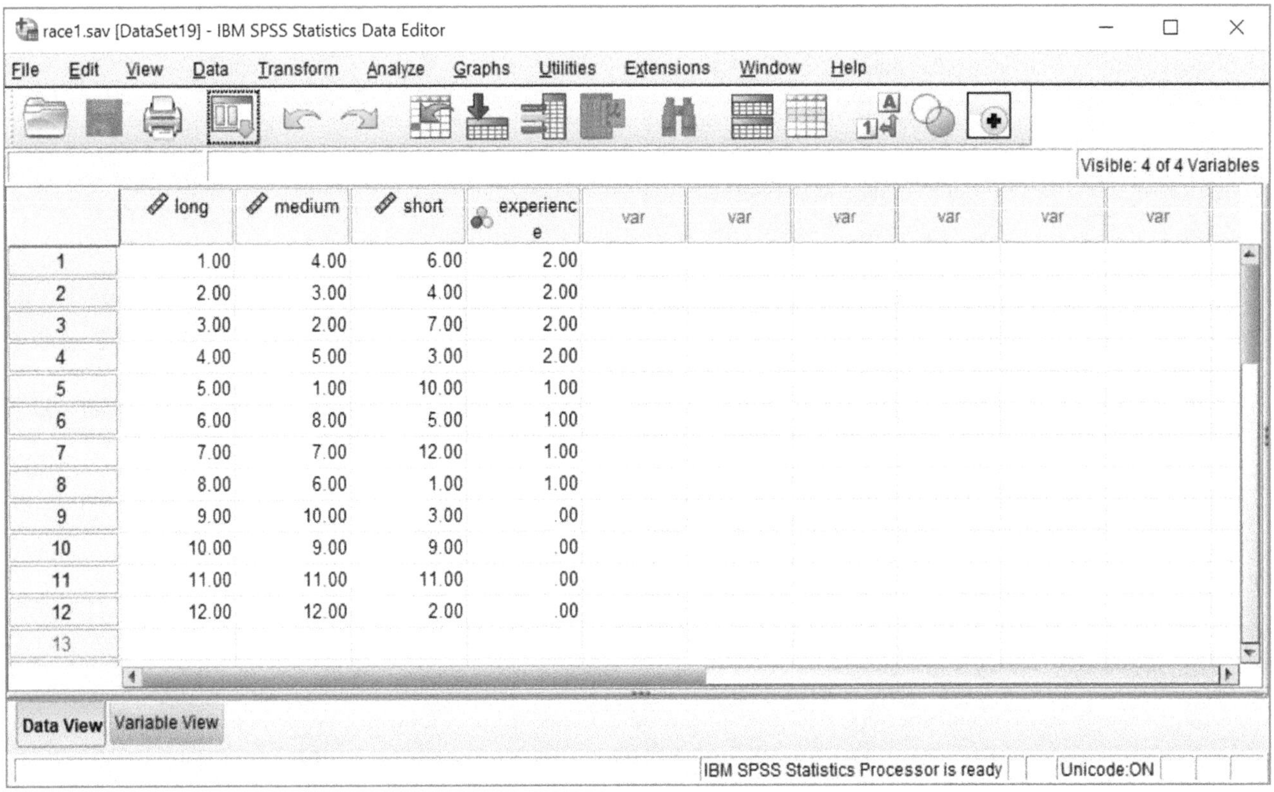

race1.sav [DataSet19] - IBM SPSS Statistics Data Editor

	long	medium	short	experience
1	1.00	4.00	6.00	2.00
2	2.00	3.00	4.00	2.00
3	3.00	2.00	7.00	2.00
4	4.00	5.00	3.00	2.00
5	5.00	1.00	10.00	1.00
6	6.00	8.00	5.00	1.00
7	7.00	7.00	12.00	1.00
8	8.00	6.00	1.00	1.00
9	9.00	10.00	3.00	.00
10	10.00	9.00	9.00	.00
11	11.00	11.00	11.00	.00
12	12.00	12.00	2.00	.00

Modified version also used in Section 8.6

RATER.sav

Created in Section 9.5

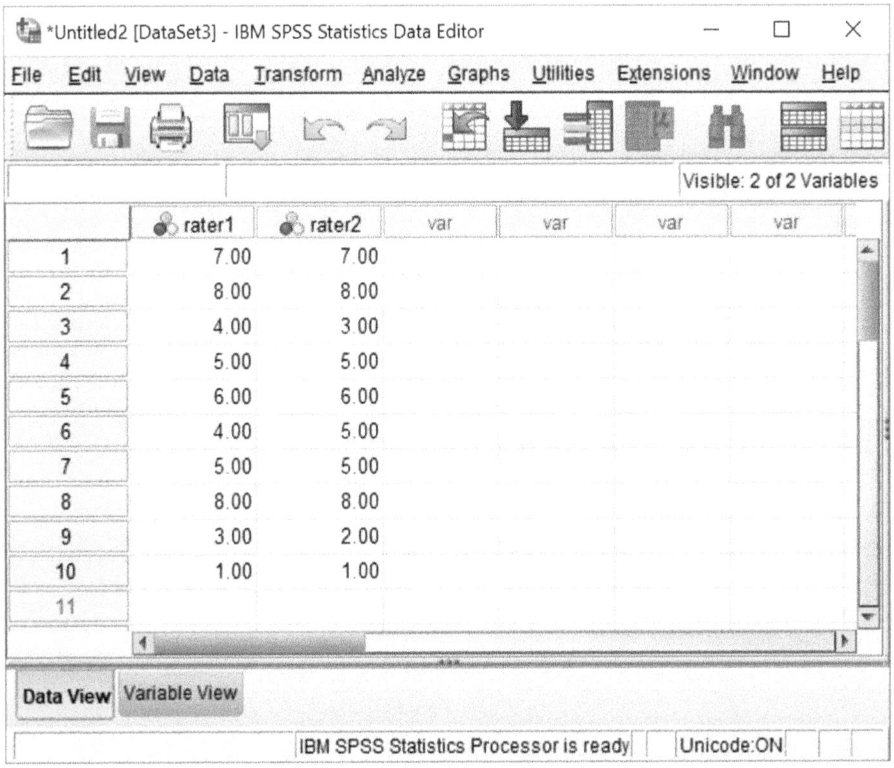

*Untitled2 [DataSet3] - IBM SPSS Statistics Data Editor

	rater1	rater2
1	7.00	7.00
2	8.00	8.00
3	4.00	3.00
4	5.00	5.00
5	6.00	6.00
6	4.00	5.00
7	5.00	5.00
8	8.00	8.00
9	3.00	2.00
10	1.00	1.00

SAMPLE.sav

Created in Section 1.4

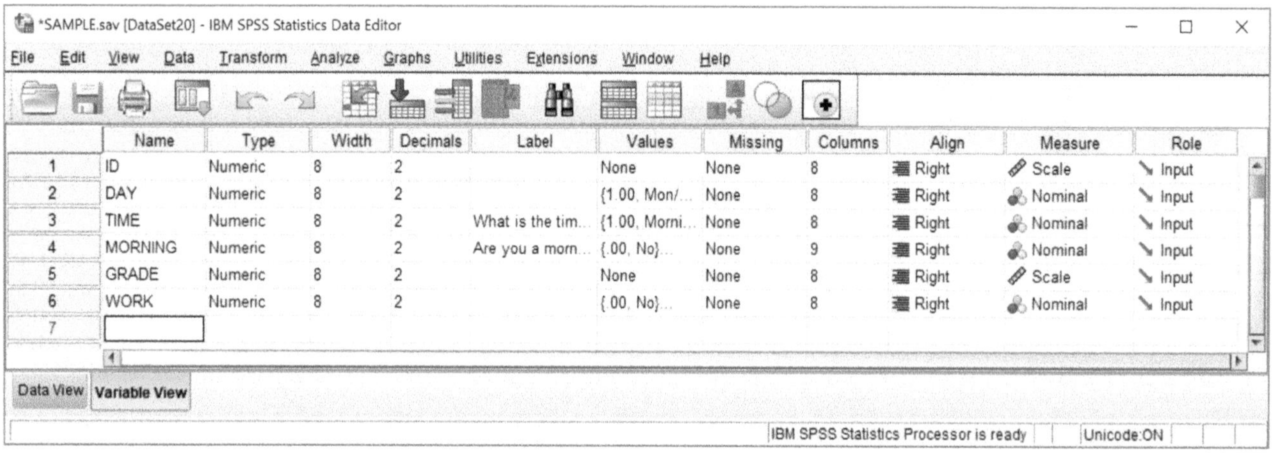

	Name	Type	Width	Decimals	Label	Values	Missing	Columns	Align	Measure	Role
1	ID	Numeric	8	2		None	None	8	Right	Scale	Input
2	DAY	Numeric	8	2		{1.00, Mon/...	None	8	Right	Nominal	Input
3	TIME	Numeric	8	2	What is the tim...	{1.00, Morni...	None	8	Right	Nominal	Input
4	MORNING	Numeric	8	2	Are you a morn...	{.00, No}...	None	9	Right	Nominal	Input
5	GRADE	Numeric	8	2		None	None	8	Right	Scale	Input
6	WORK	Numeric	8	2		{.00, No}...	None	8	Right	Nominal	Input
7											

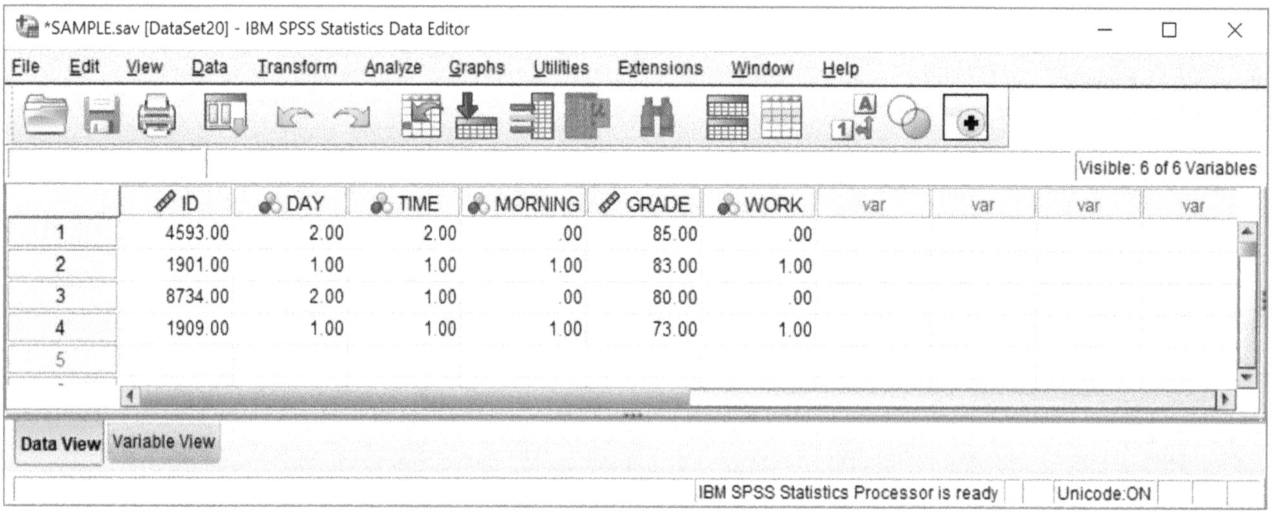

	ID	DAY	TIME	MORNING	GRADE	WORK	var	var	var	var
1	4593.00	2.00	2.00	.00	85.00	.00				
2	1901.00	1.00	1.00	1.00	83.00	1.00				
3	8734.00	2.00	1.00	.00	80.00	.00				
4	1909.00	1.00	1.00	1.00	73.00	1.00				
5										

Used throughout Chapter 1

Modified in Section 1.7 (TRAINING added)

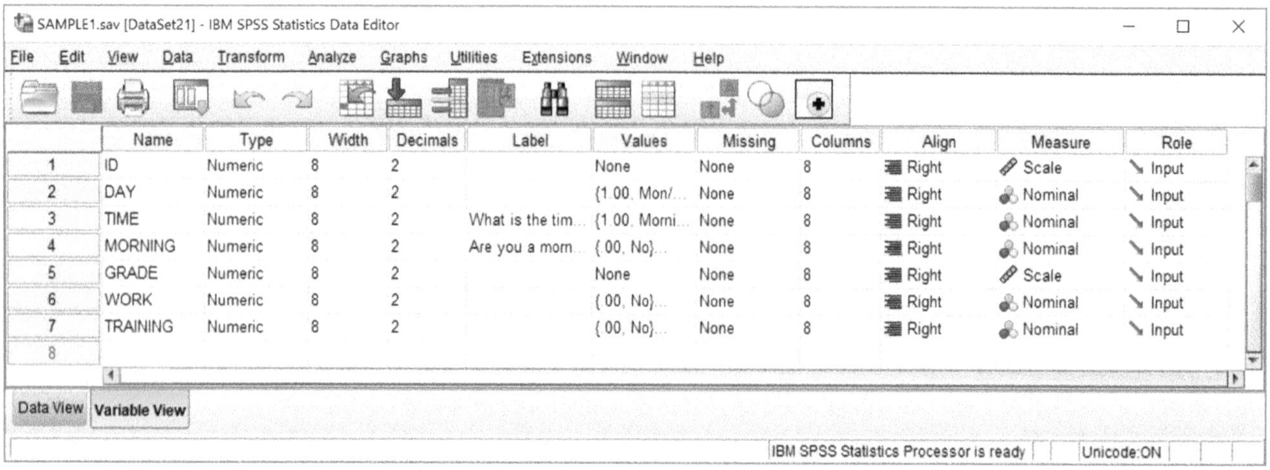

	Name	Type	Width	Decimals	Label	Values	Missing	Columns	Align	Measure	Role
1	ID	Numeric	8	2		None	None	8	Right	Scale	Input
2	DAY	Numeric	8	2		{1.00, Mon/...	None	8	Right	Nominal	Input
3	TIME	Numeric	8	2	What is the tim...	{1.00, Morni...	None	8	Right	Nominal	Input
4	MORNING	Numeric	8	2	Are you a morn...	{.00, No}...	None	8	Right	Nominal	Input
5	GRADE	Numeric	8	2		None	None	8	Right	Scale	Input
6	WORK	Numeric	8	2		{.00, No}...	None	8	Right	Nominal	Input
7	TRAINING	Numeric	8	2		{.00, No}...	None	8	Right	Nominal	Input
8											

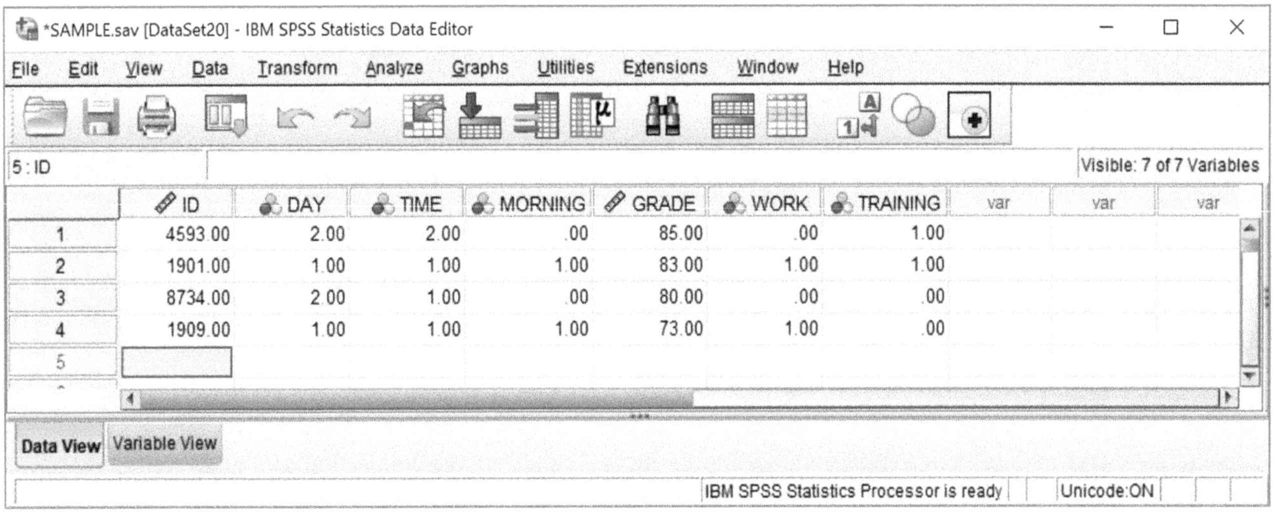

	ID	DAY	TIME	MORNING	GRADE	WORK	TRAINING	var	var	var
1	4593.00	2.00	2.00	.00	85.00	.00	1.00			
2	1901.00	1.00	1.00	1.00	83.00	1.00	1.00			
3	8734.00	2.00	1.00	.00	80.00	.00	.00			
4	1909.00	1.00	1.00	1.00	73.00	1.00	.00			
5										

Used also in Sections 2.1 and 2.2
Used also throughout Chapter 3
Used also in Section 6.3

SAT.sav

Created in Section 7.7

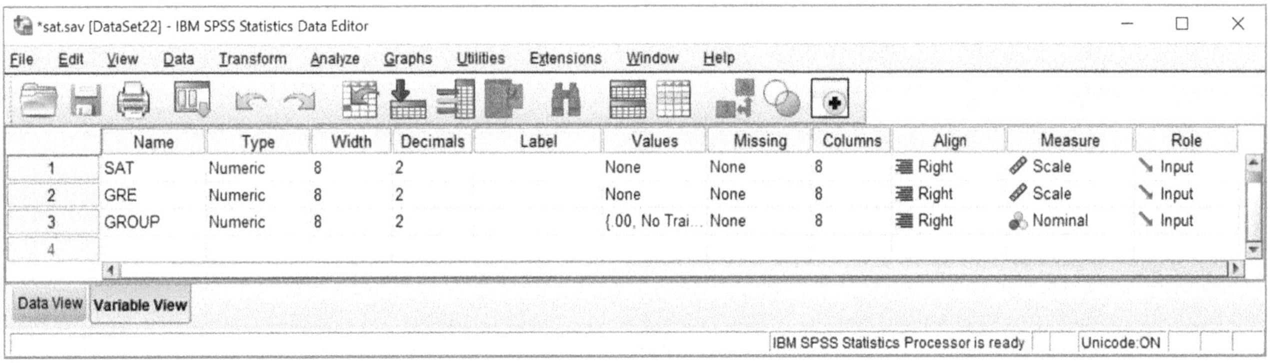

	Name	Type	Width	Decimals	Label	Values	Missing	Columns	Align	Measure	Role
1	SAT	Numeric	8	2		None	None	8	Right	Scale	Input
2	GRE	Numeric	8	2		None	None	8	Right	Scale	Input
3	GROUP	Numeric	8	2		{.00, No Trai...	None	8	Right	Nominal	Input
4											

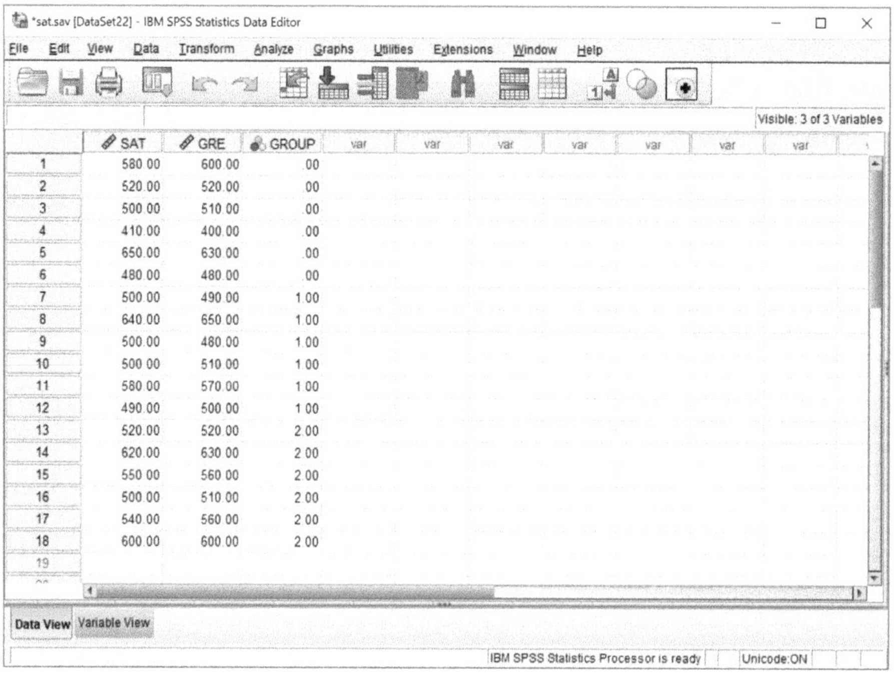

*sat.sav [DataSet22] - IBM SPSS Statistics Data Editor

	SAT	GRE	GROUP
1	580.00	600.00	.00
2	520.00	520.00	.00
3	500.00	510.00	.00
4	410.00	400.00	.00
5	650.00	630.00	.00
6	480.00	480.00	.00
7	500.00	490.00	1.00
8	640.00	650.00	1.00
9	500.00	480.00	1.00
10	500.00	510.00	1.00
11	580.00	570.00	1.00
12	490.00	500.00	1.00
13	520.00	520.00	2.00
14	620.00	630.00	2.00
15	550.00	560.00	2.00
16	500.00	510.00	2.00
17	540.00	560.00	2.00
18	600.00	600.00	2.00

TIME.sav

Created in Section 7.4

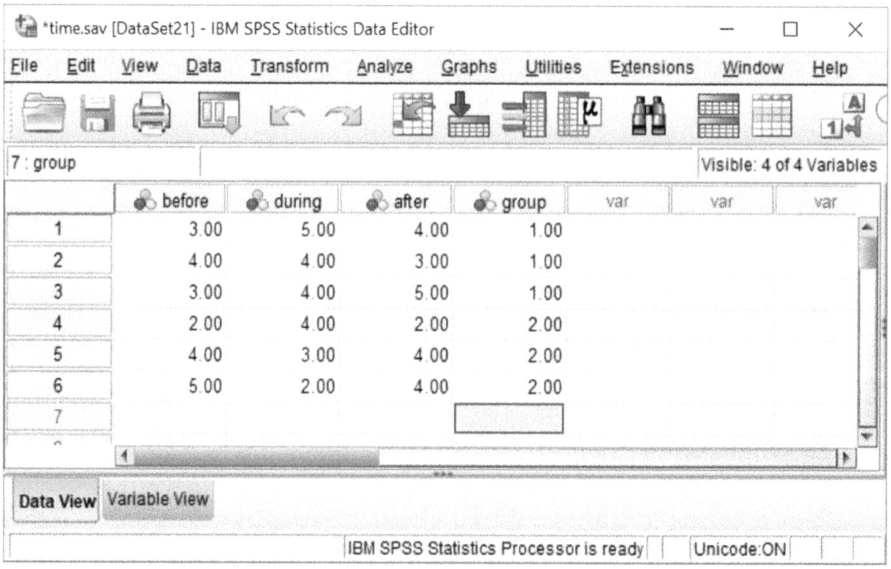

*time.sav [DataSet21] - IBM SPSS Statistics Data Editor

	before	during	after	group
1	3.00	5.00	4.00	1.00
2	4.00	4.00	3.00	1.00
3	3.00	4.00	5.00	1.00
4	2.00	4.00	2.00	2.00
5	4.00	3.00	4.00	2.00
6	5.00	2.00	4.00	2.00

Appendix D
SPSS Syntax Basics

Introduction to Syntax

IBM SPSS Statistics has been easy to use with its point and click interface for several decades. Before that, however, SPSS required users to write programs to conduct their analyses. In fact, all the point and click interface does is to create syntax for you that SPSS runs.

Using syntax can be very useful if you want to save an analysis and run it later. Understanding it is an important part of being able to understand your output as well.

This text will not cover how to create SPSS syntax from scratch. Rather, it will simply cover how to get SPSS to create it for you, and how to read the syntax it creates.

Having SPSS Create Syntax

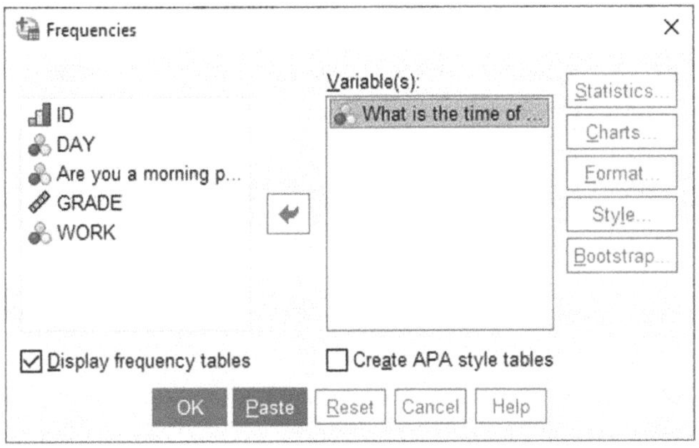

Most SPSS Dialog Boxes will have a *Paste* option in them. Clicking *Paste* instead of *OK* will put the relevant SPSS Syntax in a Syntax Window.

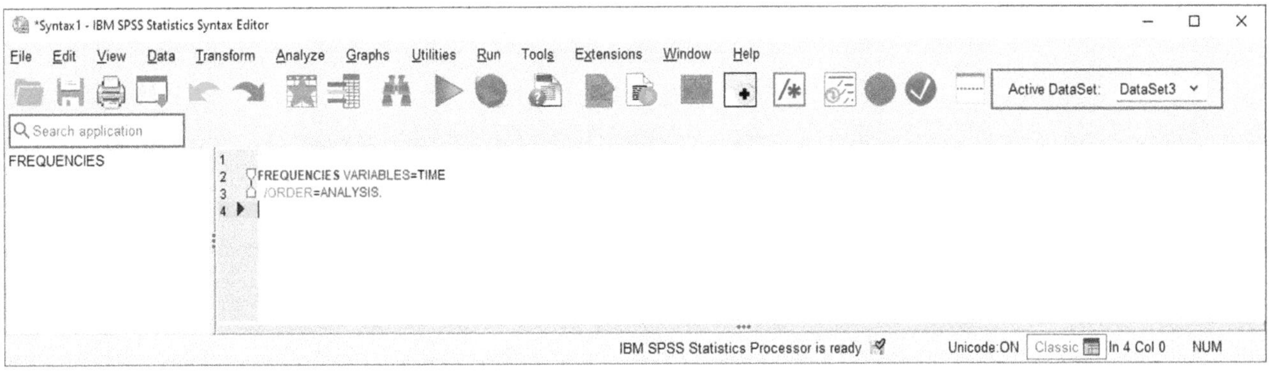

The syntax window is highlighted and color coded to help you understand what the commands are asking SPSS to do. In the example above, the command that was run was the *Frequencies* command. It was run on the variable TIME.

Many SPSS commands have options that are added to them but are not necessarily important at the level of this text. For example, the /ORDER=ANALYSIS option above.

SPSS Syntax Format and SPSS Help

SPSS syntax is written like sentences. In fact, each one ends with a period ("."). The number of lines it is on does not matter—so it is often best to use several lines to make it easier to read.

Rather than provide extensive coverage in this text, I suggest that if you are interested in SPSS Syntax you use the excellent help system provided within SPSS itself.

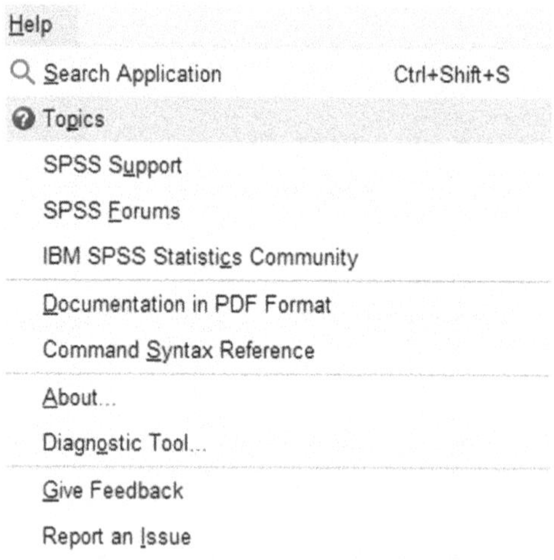

To access that help system, click *Help* then select *Topics.* This will open the SPSS Help web site. Select *Core System*, then *Working with Command Syntax*. You can also select *Help* and then *Command Syntax Reference.*

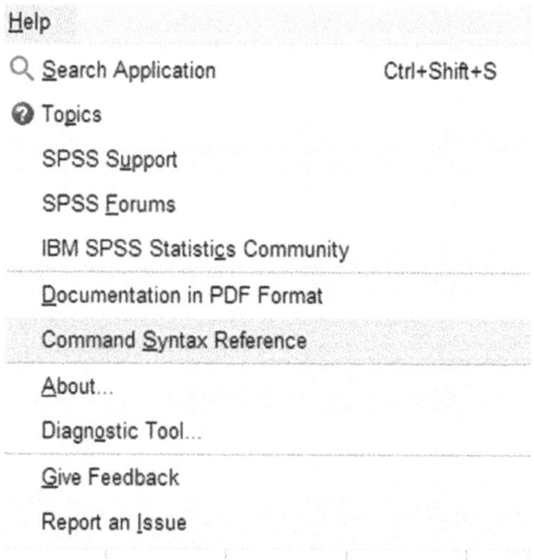

Running SPSS Syntax

Whether SPSS creates syntax for you, whether you create it yourself, or whether you modify syntax that SPSS has created, at some point you are going to want to use that syntax.

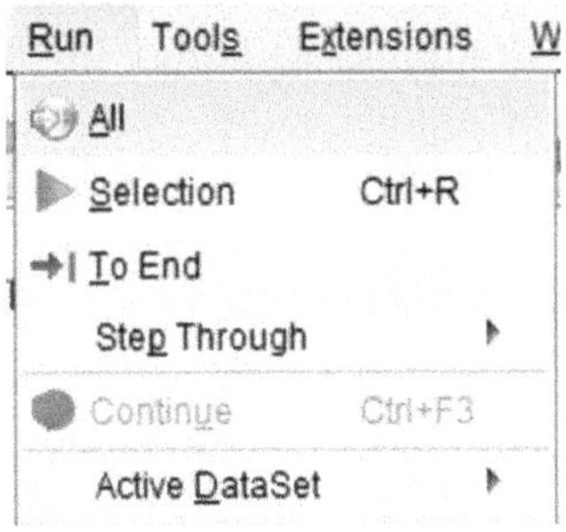

From a Syntax Window simply select *Run* and then *All*. Alternatively, you can select just some of the text in the window with your mouse and then click *Selection* instead of *All*.

Saving and Loading SPSS Syntax

Syntax files have the extension ".sps" and can be saved and loaded just like other files. To load a file select *File—Open—Syntax*.

Appendix E
Glossary

All Inclusive A set of events that encompasses every possible outcome. Also referred to as *exhaustive*.

Alternative Hypothesis The opposite of the null hypothesis, normally showing that there is a true difference. Generally, this is the statement that the researcher would like to support.

Case Processing Summary A section of SPSS output that lists the number of subjects used in the analysis.

Coefficient of Determination The value of the correlation, squared. It provides the proportion of variance accounted for by the relationship.

Cohen's *d* A common and simple measure of effect size that standardizes the difference between groups.

Correlation Matrix A section of SPSS output in which correlation coefficients are reported for all pairs of variables.

Covariate A variable known to be related to the dependent variable but not treated as an independent variable. Used in ANCOVA as a statistical control technique.

Data Window The SPSS window that contains the data in a spreadsheet format. This is the window used for running most commands.

Dependent Variable An outcome or response variable. The dependent variable is normally dependent on the independent variable.

Descriptive Statistics Statistical procedures that organize and summarize data.

Dialog Box A window that allows you to enter information that SPSS will use in a command.

Dichotomous Variables Variables with only two levels (e.g., gender).

Discrete Variable A variable that can have only certain values (i.e., values between which there is no score, like A, B, C, D, F).

Effect Size A measure that allows one to judge the relative importance of a difference or relationship by reporting the size of a difference.

Eta Squared (η^2) A measure of effect size used in Analysis of Variance models.

Grouping Variable In SPSS, the variable used to represent group membership. SPSS often refers to independent variables as grouping variables; SPSS sometimes refers to grouping variables as independent variables.

Independent Events Two events are independent if information about one event gives no information about the second event (e.g., two flips of a coin).

Independent Variable The variable whose levels (values) determine the group to which a subject belongs. A true independent variable is manipulated by the researcher. See Grouping Variable.

Inferential Statistics Statistical procedures designed to allow the researcher to draw inferences about a population on the basis of a sample.

Interaction With more than one independent variable, an interaction occurs when a level of one independent variable affects the influence of another independent variable.

Internal Consistency A reliability measure that assesses the extent to which all of the items in an instrument measure the same construct.

Interval Scale A measurement scale where items are placed in mutually exclusive categories, with equal intervals between values. Appropriate transformations include counting, sorting, and addition/subtraction.

Levels The values that a variable can have. A variable with three levels has three possible values.

Mean A measure of central tendency where the sum of the deviation scores equals zero.

Median A measure of central tendency representing the middle of a distribution when the data are sorted from low to high. Fifty percent of the cases are below the median.

Mode A measure of central tendency representing the value (or values) with the most subjects (the score with the greatest frequency).

Mutually Exclusive Two events are mutually exclusive when they cannot occur simultaneously.

Nominal Scale A measurement scale where items are placed in mutually exclusive categories. Differentiation is by name only (e.g., race, sex). Appropriate categories include "same" or "different." Appropriate transformations include counting.

Normal Distribution A symmetric, unimodal, bell-shaped curve.

Null Hypothesis The hypothesis to be tested, normally in which there is no true difference. It is mutually exclusive of the alternative hypothesis.

Ordinal Scale A measurement scale where items are placed in mutually exclusive categories, in order. Appropriate categories include "same," "less," and "more." Appropriate transformations include counting and sorting.

Outliers Extreme scores in a distribution. Scores that are very distant from the mean and the rest of the scores in the distribution.

Output Window The SPSS window that contains the results of an analysis. The left side summarizes the results in an outline. The right side contains the actual results.

Percentiles (Percentile Ranks) A relative score that gives the percentage of subjects who scored at the same value or lower.

Pooled Standard Deviation A single value that represents the standard deviation of two groups of scores.

Protected Dependent *t* Tests *T* tests where the acceptable Type I error rate has been adjusted to prevent the inflation of a Type I error.

Quartiles The points that define a distribution into four equal parts:The scores at the 25th, 50th, and 75th percentile ranks.

Random Assignment A procedure for assigning subjects to conditions in which each subject has an equal chance of being assigned to any condition.

Range A measure of dispersion representing the number of points from the highest score through the lowest score.

Ratio Scale A measurement scale where items are placed in mutually exclusive categories, with equal intervals between values and a true zero. Appropriate transformations include counting, sorting, addition/subtraction, and multiplication/division.

Reliability An indication of the consistency of a scale. A reliable scale is internally consistent and stable over time.

Robust A test is said to be robust if it continues to provide accurate results even after the violation of some assumptions.

Significance A difference is said to be significant if the probability of making a Type I error is less than the accepted limit (normally 5%). If a difference is significant, the null hypothesis is rejected.

Skew The extent to which a distribution is not symmetrical. A positive skew has outliers on the positive (right) side of the distribution. A negative skew has outliers on the negative (left) side of the distribution.

Source Table The results of ANOVA are normally presented in the form of a source table in which rows represent a source of variability and the columns represent different amounts of the variability that can be attributed to that source.

Standard Deviation A measure of dispersion representing a special type of average deviation from the mean.

Standard Error of Estimate The equivalent of the standard deviation for a regression line. The data points will be normally distributed around the regression line with a standard deviation equal to the standard error of the estimate.

Standard Normal Distribution A normal distribution with a mean of 0.0 and a standard deviation of 1.0.

Temporal Stability This is achieved when reliability measures have determined that scores remain stable over multiple administrations of the instrument.

Tukey's *HSD* A post-hoc comparison purported to reveal an "honestly significant difference" (*HSD*).

Type I Error A Type I error occurs when the researcher erroneously rejects the null hypothesis.

Type II Error A Type II error occurs when the researcher erroneously fails to reject the null hypothesis.

Valid Data Data that SPSS will use in its analyses.

Validity An indication of the accuracy of a scale.

Variance A measure of dispersion equal to the squared standard deviation.

Appendix F Selecting the Appropriate Inferential Test

Being able to select the appropriate inferential test for SPSS to run is one of the most important skills you can learn. Unfortunately, the way most statistics classes and statistics texts are set up, each statistical procedure is presented in isolation. There is rarely an opportunity to synthesize the information and understand when each test is appropriate.

This text contains a decision tree on the inside cover. However, novices sometimes have trouble using that tree because they do not understand the terms used. This appendix expands on the decision tree and provides a comprehensive way to determine which test to run.

To help you use this section of the text, there will be a reference to a "branch" of the decision tree.

Branch 1

The first step in determining what type of statistical test to run is deciding what your main question is.

Sometimes you want to know if a group is "significantly different" from another. This can involve means or shapes of distributions. The basic question, however, is the same. You want to know if the things you are comparing are the same as each other or different from each other. For example, you want to know if a group that received a new treatment had a significantly shorter hospital stay than a group that did not receive the new treatment. If this is what you are trying to do, go to Branch 2.

Other times you are more interested in knowing whether or not there is a relationship between two variables, and if so, how strong that relationship is. These are known as correlations. Here, rather than looking for differences, you are simply looking for stable relationships. For example, is height related to weight? Do students who miss class more often do worse academically? Often, the variables here are continuous and have many levels rather than just a few. If this is what you are interested in, go to Branch 3.

More powerful than association is prediction. If you would like to be able to predict or guess the value of a variable from some other information that you have, then you are interested in regression. For example, if an insurance company wants to predict how much you will cost them in claims based on your driving history, that is a problem for regression. If these are the types of questions you are interested in, go to Branch 6.

Branch 2

Your Decisions So Far:

- Looking for Differences in Groups

The next step is to determine how many Dependent Variables you have. Dependent Variables are the variables that are different from each other. Most of the time you will have only a single Dependent Variable (although you could have one or more Independent Variables).

If you get Independent and Dependent Variables confused, a good way to help clarify is to rephrase the problem as "I am looking for the effect of the (Independent Variable) on the (Dependent Variable)."

- If you have one Dependent Variable, see Branch 9.
- If you have more than one Dependent Variable, see Branch 10.

Branch 3

Your Decisions So Far:

- Measuring Strength of Association

In general, correlations measure the strength of a relationship or association.

This text covers what are known as bivariate correlations—that is, correlations between only two variables at a time. In addition, this text covers only linear correlations—or correlations that assume the relationship being measured generally looks like a line.

In deciding which correlation is appropriate, first you must determine if you can do a parametric or nonparametric correlation. Parametric Tests generally require that the Dependent Variables are measured on an Interval or Ratio measurement scale and that the variables are approximately normally distributed. Nonparametric Tests do not make that assumption.

If either of the variables in your relationship is ordinal (e.g., ranks) you must do a nonparametric correlation. If one of your variables is nominal, you are just out of luck with correlations.

- If you can do a parametric correlation, see Branch 4.
- If you must do a nonparametric correlation, see Branch 5.

Branch 4

Your Decisions So Far:

- Measuring Strength of Association
- You can do a parametric correlation

The appropriate test to do is a Pearson *r*. This test is covered in detail in Section 5.1 of this text. It is accessed in SPSS at Analyze → Correlate → Bivariate.

Answers range from –1 to +1. Values close to 0 mean there is a weak relationship. Values closer to +1 or –1 mean a stronger relationship. Positive values mean as one variable gets larger, the other variable also gets larger. Negative values mean as one variable gets larger, the other variable gets smaller.

Branch 5

Your Decisions So Far:

- Measuring Strength of Association
- You must do a Nonparametric Correlation

The appropriate test to do is a Spearman *rho.* This test is covered in detail in Section 5.2 of this text. It is accessed in SPSS at Analyze → Correlate → Bivariate and then selecting Spearman in the dialog box that comes up.

Answers range from –1 to +1. Values close to 0 mean there is a weak relationship. Values closer to +1 or –1 mean a stronger relationship. Positive values mean as one variable gets larger, the other variable also gets larger. Negative values mean as one variable gets larger, the other variable gets smaller.

Branch 6

Your Decisions So Far:

- Making Predictions

Predictions are a powerful use of statistics. In this text, we only cover prediction models that assume the variables are related to each other in linear ways. If you have nonlinear relationships, they are beyond the scope of this text.

This text uses the term "Independent Variable" when, in fact, sometimes prediction is done when there is not a true Independent Variable. If you have a design where there is not a true Independent Variable, the variables you use to make a prediction should be considered Independent Variables for purposes of this decision process.

- If you have a single Independent Variable, see Branch 7.
- If you have multiple Independent Variables that you want to use to predict the value of a single Dependent Variable (e.g., predict grades from a standardized test score, number of classes missed, and hours spent studying), see Branch 8.

Branch 7

Your Decisions So Far:

- Making Predictions
- Single Independent Variable

The appropriate test to do is a Simple Linear Regression. This test is covered in detail in Section 5.3 of this text. It is accessed in SPSS at Analyze → Regression → Linear.

The output will have a B column that provides the answers. The B for Constant is your Y-Intercept, and the B next to the name of your Independent Variable is the slope of the line.

Branch 8

Your Decisions So Far:

- Making Predictions
- Multiple Independent Variables

The appropriate test to do is a Multiple Linear Regression. This test is covered in detail in Section 5.4 of this text. It is accessed in SPSS at Analyze → Regression → Linear and entering more than one Independent Variable.

The output will have a B column that provides the answers. The B for Constant is your Y-Intercept, and the Bs next to the names of your Independent Variables are the weights for each of them.

Branch 9

Your Decisions So Far:

- Looking for Differences in Groups
- One Dependent Variable

The next question to ask is whether you can do a Parametric Test or whether you have to do a Nonparametric Test. Parametric Tests generally require that the Dependent Variables are measured on an Interval or Ratio measurement scale, and the variables are approximately normally distributed. With Nonparametric Tests, you do not make that assumption.

- If you can do a Parametric Test, see Branch 11.
- If you must do a Nonparametric Test, see Branch 26.

Branch 10

Your Decisions So Far:

- Looking for Differences in Groups
- More than One Dependent Variable

The appropriate test to do is a MANOVA (Multivariate Analysis of Variance). This test is covered in detail in Section 7.7 of this text. It is accessed in SPSS at Analyze → General Linear Model → Multivariate.

This text does not cover tests for Nonparametric equivalents of MANOVA. Also, please note that the Advanced Module of IBM SPSS Statistics is required to run this command.

Branch 11

Your Decisions So Far:

- Looking for Differences in Groups
- One Dependent Variable
- Parametric Test

Sometimes when a researcher cannot implement appropriate procedural controls (e.g., random assignment of participants to groups), the overall analysis can be improved by "covarying out" variables that the researcher knows may have an effect on the Dependent Variable.

For example, a researcher may be comparing two different sections of a class, but was unable to assign students to the sections (the students self-selected). The researcher knows that overall GPA is likely to have an effect on the grade in the course. The researcher can, therefore, "covary out" the effect of cumulative GPA from the grades in the two sections to make a better comparison as to whether or not the two sections were different in some other way (e.g., because of a new teaching method).

- If you would like to analyze your data with a Covariate, see Branch 12.
- If you are not going to use a Covariate, see Branch 13.

Branch 12

Your Decisions So Far:

- Looking for Differences in Groups
- One Dependent Variable
- Parametric Test
- Covariate

The appropriate test to do is ANCOVA (Analysis of Covariance). This test is covered in detail in Section 7.6 of this text. It is accessed in SPSS at Analyze → General Linear Model → Univariate and by entering a variable in the Covariate space.

Branch 13

Your Decisions So Far:

- Looking for Differences in Groups
- One Dependent Variable
- Parametric Test
- No Covariate

Next, it is time to make a decision about your Independent Variables. The easiest analyses have a single Independent Variable (e.g., treatment); however, more complex analyses can have multiple Independent Variables that are examined at the same time (e.g., treatment and time).

- If you are simply comparing a single sample to a known population, see Branch 14.
- If you have a single Independent Variable, see Branch 15.
- If you have multiple Independent Variables, see Branch 22.

Branch 14

Your Decisions So Far:

- Looking for Differences in Groups
- One Dependent Variable
- Parametric Test
- No Covariate
- Comparison of a Sample to a Population

The appropriate test to do is a single-sample *t* Test. This test is covered in detail in Section 6.2 of this text. It is accessed in SPSS at Analyze → Compare Means → One-Sample T Test.

Branch 15

Your Decisions So Far:

- Looking for Differences in Groups
- One Dependent Variable
- Parametric Test

- No Covariate
- Single Independent Variable

The number of levels of your Independent Variable also impacts which test is appropriate. Students often get levels of a variable confused with the variable itself. Levels of a variable refer to the distinct and discrete values that it can take on. For example, the variable "sex" has two levels (male and female). If you were comparing three different instructors, the variable "instructor" would have three levels.

- If you have two levels of your Independent Variable, see Branch 16.
- If you have more than two levels of your Independent Variable, see Branch 17.

Branch 16

Your Decisions So Far:

- Looking for Differences in Groups
- One Dependent Variable
- Parametric Test
- No Covariate
- Single Independent Variable
- Two Levels of the Independent Variable

Next, you need to determine what kind of design produced the two levels of your Independent Variable.

A Repeated-Measures Design is where the data in the various conditions are in some way related to each other. For example, a pre-test/post-test design or a design utilizing twins would fall into this category. These designs are sometimes called Correlated Groups Designs or Paired Samples Designs.

In an Independent Groups Design, the data in each group are independent of each other. This is the classic Experimental Group/Control Group Design where each participant is in one (and only one) group.

- If you have an Independent Groups Design, see Branch 20.
- If you have a Repeated-Measures Design, see Branch 21.

Branch 17

Your Decisions So Far:

- Looking for Differences in Groups
- One Dependent Variable
- Parametric Test
- No Covariate
- Single Independent Variable
- More Than Two Levels of the Independent Variable

Next, you need to determine what kind of design produced the levels of your Independent Variable.

A Repeated-Measures Design is where the data in the various conditions are in some way related to each other. For example, comparing grades of the same students when they were freshmen, sophomores, juniors, and then seniors would fall into this category. These designs are sometimes called Correlated Groups Designs, Within-Subjects Designs, or Paired Samples Designs.

In an Independent Groups Design, the data in each group are independent of each other. This is the classic Experimental Group/Control Group Design where each participant is in one (and only one) group. For example, if you were comparing four different PSY101 instructors, you would use an Independent Groups Design. Sometimes this is called a Between-Subjects Design.

- If your Independent Variable represents Independent Groups, see Branch 18.
- If your Independent Variable represents Repeated Measures, see Branch 19.

Branch 18

Your Decisions So Far:

- Looking for Differences in Groups
- One Dependent Variable
- Parametric Test
- No Covariate
- Single Independent Variable
- More Than Two Levels of the Independent Variable
- Independent Groups

The appropriate test to do is a one-way ANOVA (technically a one-way between-subjects ANOVA). This test is covered in detail in Section 7.2 of this text. It is accessed in SPSS at Analyze → Compare Means → One-Way ANOVA.

Branch 19

Your Decisions So Far:

- Looking for Differences in Groups
- One Dependent Variable
- Parametric Test
- No Covariate
- Single Independent Variable
- More Than Two Levels of the Independent Variable
- Repeated Measures

The appropriate test to do is a Repeated-Measures ANOVA. This test is covered in detail in Section 7.4 of this text. It is accessed in SPSS at Analyze → General Linear Model → Repeated Measures.

Note: This procedure requires the Advanced Module in SPSS.

Branch 20

Your Decisions So Far:

- Looking for Differences in Groups
- One Dependent Variable
- Parametric Test
- No Covariate
- Single Independent Variable

- Two Levels of the Independent Variable
- Independent Groups

The appropriate test to do is an Independent-Samples *t* Test. This test is covered in detail in Section 6.3 of this text. It is accessed in SPSS at Analyze → Compare Means → Independent-Samples *t* Test.

Branch 21

Your Decisions So Far:

- Looking for Differences in Groups
- One Dependent Variable
- Parametric Test
- No Covariate
- Single Independent Variable
- Two Levels of the Independent Variable
- Repeated Measures

The appropriate test to do is a Paired-Samples *t* test (sometimes called a Dependent *t* Test). This test is covered in detail in Section 6.4 of this text. It is accessed in SPSS at Analyze → Compare Means → Paired-Samples T Test.

Branch 22

Your Decisions So Far:

- Looking for Differences in Groups
- One Dependent Variable
- Parametric Test
- No Covariate
- Multiple Independent Variables

Next, you need to determine what kind of design produced the levels of your Independent Variables.

A Repeated-Measures Design (also called Within-Subjects) is where the data in the various conditions are in some way related to each other. For example, comparing grades of the same students when they were freshmen, sophomores, juniors, and then seniors would fall into this category. These designs are sometimes called Correlated Groups Designs or Paired Samples Designs.

In an Independent Groups Design (also called Between-Subjects), the data in each group are independent of each other. This is the classic Experimental Group/Control Group Design where each participant is in one (and only one) group. For example, if you were comparing four different PSY101 instructors, you would use an Independent Groups Design.

Because you have more than one Independent Variable, your Independent Variables can all be Repeated Measures, all Independent Groups, or a mix of the two.

- If all of your Independent Variables represent Independent Groups, see Branch 23.
- If all of your Independent Variables represent Repeated Measures, see Branch 24.
- If your Independent Variables are a mix of Independent Groups and Repeated Measures, see Branch 25.

Branch 23

Your Decisions So Far:

- Looking for Differences in Groups
- One Dependent Variable
- Parametric Test
- No Covariate
- Multiple Independent Variables
- Independent Measures

The appropriate test to do is a Factorial ANOVA. This test is covered in detail in Section 7.3 of this text. It is accessed in SPSS at Analyze → General Linear Model → Univariate.

Branch 24

Your Decisions So Far:

- Looking for Differences in Groups
- One Dependent Variable
- Parametric Test
- No Covariate
- Multiple Independent Variables
- Repeated Measures

The appropriate test to do is a Repeated-Measures ANOVA. This test is covered in detail in Section 7.4 of this text. It is accessed in SPSS at Analyze → General Linear Model → Repeated Measures. Note: This test requires the Advanced Module to be installed.

Branch 25

Your Decisions So Far:

- Looking for Differences in Groups
- One Dependent Variable
- Parametric Test
- No Covariate
- Multiple Independent Variables
- Both Independent and Repeated Measures

The appropriate test to do is a Mixed-Design ANOVA. This test is covered in detail in Section 7.4 of this text. It is accessed in SPSS at Analyze → General Linear Model → Repeated Measures.

Note: This test requires the Advanced Module to be installed.

Branch 26

Your Decisions So Far:

- Looking for Differences in Groups
- One Dependent Variable
- Nonparametric Test

While Parametric Tests look for differences in Means, the data on which Nonparametric Tests are calculated often do not lend themselves to the calculation of a Mean (e.g., they are nominal or ordinal).

Thus, the first question in a Nonparametric Test is whether you are interested in differences in proportions of values (used with nominal data) or differences in the ranks of values (ordinal data).

- If you are looking for differences in proportions, see Branch 27.
- If you are looking for differences in ranks, see Branch 30.

Branch 27

Your Decisions So Far:

- Looking for Differences in Groups
- One Dependent Variable
- Nonparametric Test
- Difference in Proportions

Like their Parametric Equivalents, the number of Independent Variables you are examining for Nonparametric Tests must also be determined.

- If you have a single Independent Variable, see Branch 28.
- If you have multiple Independent Variables, see Branch 29.

Branch 28

Your Decisions So Far:

- Looking for Differences in Groups
- One Dependent Variable
- Nonparametric Test
- Difference in Proportions
- One Independent Variable

The appropriate test to do is a Chi-Square Goodness of Fit. This test is covered in detail in Section 8.1 of this text. It is accessed in SPSS at Analyze → Nonparametric Tests → One Sample.

Branch 29

Your Decisions So Far:

- Looking for Differences in Groups
- One Dependent Variable
- Nonparametric Test
- Difference in Proportions
- Multiple Independent Variables

The appropriate test to do is a Chi-Square Test of Independence. This test is covered in detail in Section 8.2 of this text. It is accessed in SPSS at Analyze → Descriptive Statistics → Crosstabs and then checking Chi-Square in the Statistics option.

Branch 30

Your Decisions So Far:

- Looking for Differences in Groups
- One Dependent Variable
- Nonparametric Test
- Difference in Ranks
- If you have more than one Independent Variable, this text does not cover the appropriate Nonparametric Test.
- If you have a single Independent Variable, however, the number of levels of that variable determines the appropriate test.
- If your Independent Variable has two levels, see Branch 31.
- If your Independent Variable has more than two levels, see Branch 32.

Branch 31

Your Decisions So Far:

- Looking for Differences in Groups
- One Dependent Variable
- Nonparametric Test
- Difference in Ranks
- Two Levels of the Independent Variable

Next, you need to determine what kind of design produced the levels of your Independent Variables.

A Repeated-Measures Design (also called Within-Subjects) is where the data in the various conditions are in some way related to each other. For example, comparing class rank of the same students when they were in high school to college would fall into this category. These designs are sometimes called Correlated Groups Designs or Paired Samples Designs.

In an Independent Groups Design (also called Between-Subjects), the data in each group are independent of each other. This is the classic Experimental Group/Control Group Design where each participant is in one (and only one) group. For example, if you were comparing two different people and their performances in various marathons, this would be an Independent Groups Design.

- If you have Independent Groups, see Branch 33.
- If you have Repeated Measures, see Branch 34.

Branch 32

Your Decisions So Far:

- Looking for Differences in Groups
- One Dependent Variable
- Nonparametric Test
- Difference in Ranks
- More Than Two Levels of the Independent Variable

Next, you need to determine what kind of design produced the levels of your Independent Variables.

A Repeated-Measures Design (also called Within-Subjects) is where the data in the various conditions are in some way related to each other. For example, comparing class rank of the same students when they were freshmen, sophomores, juniors, and then seniors would fall into this category. These designs are sometimes called Correlated Groups Designs or Paired Samples Designs.

In an Independent Groups Design (also called Between-Subjects), the data in each group are independent of each other. This is the classic Experimental Group/Control Group Design where each participant is in one (and only one) group. For example, if you were comparing three different people and their performances in various marathons, this would be an Independent Groups Design.

- If you have Independent Groups, see Branch 35.
- If you have Repeated-Measures Design, see Branch 36.

Branch 33

Your Decisions So Far:

- Looking for Differences in Groups
- One Dependent Variable
- Nonparametric Test
- Difference in Ranks
- Two Levels of the Independent Variable
- Independent Groups

The appropriate test to do is a Mann-Whitney *U*. This test is covered in detail in Section 8.3 of this text. It is accessed in SPSS at Analyze → Nonparametric Tests → Independent Samples.

Branch 34

Your Decisions So Far:

- Looking for Differences in Groups
- One Dependent Variable
- Nonparametric Test
- Difference in Ranks
- Two Levels of the Independent Variable
- Correlated Groups

The appropriate test to do is a Wilcoxon. This test is covered in detail in Section 8.4 of this text. It is accessed in SPSS at Analyze → Nonparametric Tests → Related Samples.

Branch 35

Your Decisions So Far:

- Looking for Differences in Groups
- One Dependent Variable
- Nonparametric Test
- Difference in Ranks

- Two Levels of the Independent Variable
- Independent Groups

The appropriate test to do is a Kruskal-Wallis *H* Test. This test is covered in detail in Section 8.5 of this text. It is accessed in SPSS at Analyze → Nonparametric Tests → Independent Samples.

Branch 36

Your Decisions So Far:

- Looking for Differences in Groups
- One Dependent Variable
- Nonparametric Test
- Difference in Ranks
- Two Levels of the Independent Variable
- Correlated Groups

The appropriate test to do is a Friedman. This test is covered in detail in Section 8.6 of this text. It is accessed in SPSS at Analyze → Nonparametric Tests → Related Samples.

Appendix G
Answer Key

Section 1.6

Load the sample data file you created earlier (SAMPLE.sav). Run the *Descriptives* command for the variable GRADE and view the output. Next, select the **data window** and print it.

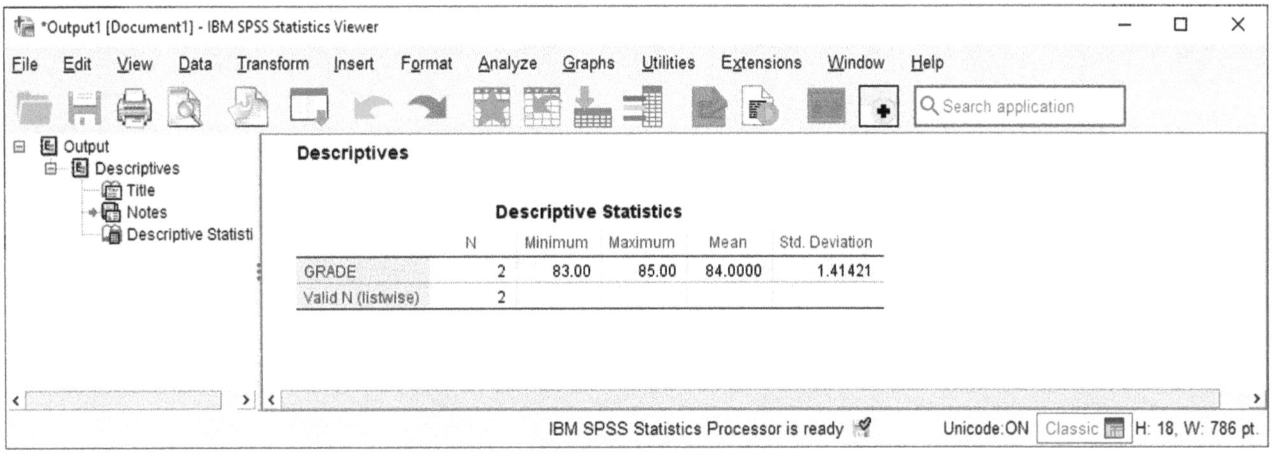

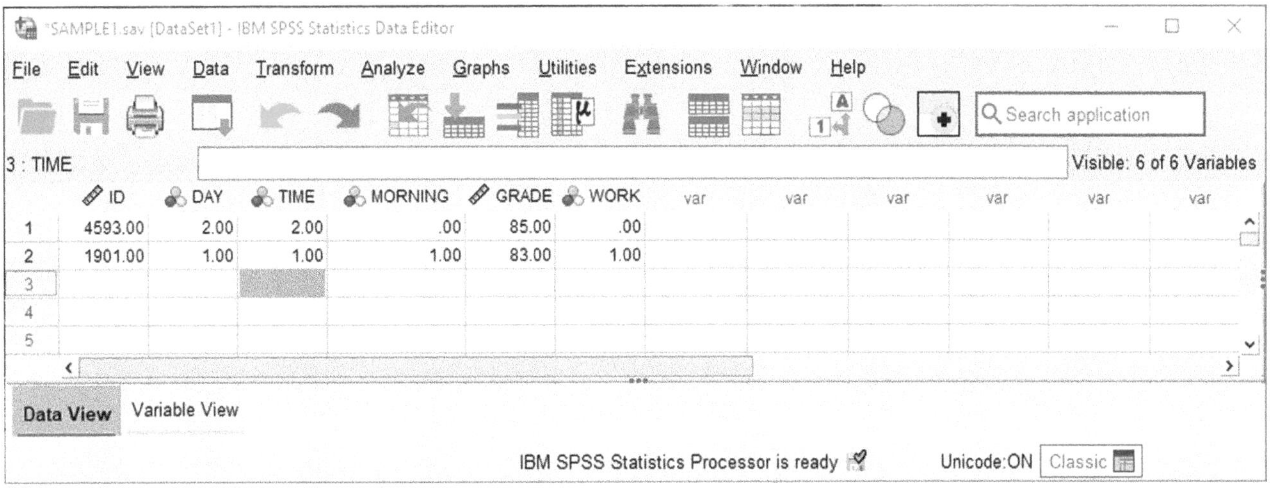

Section 1.7

Follow the previous example (where TRAINING is the new variable). Make the modifications to your SAMPLE.sav data file and save it.

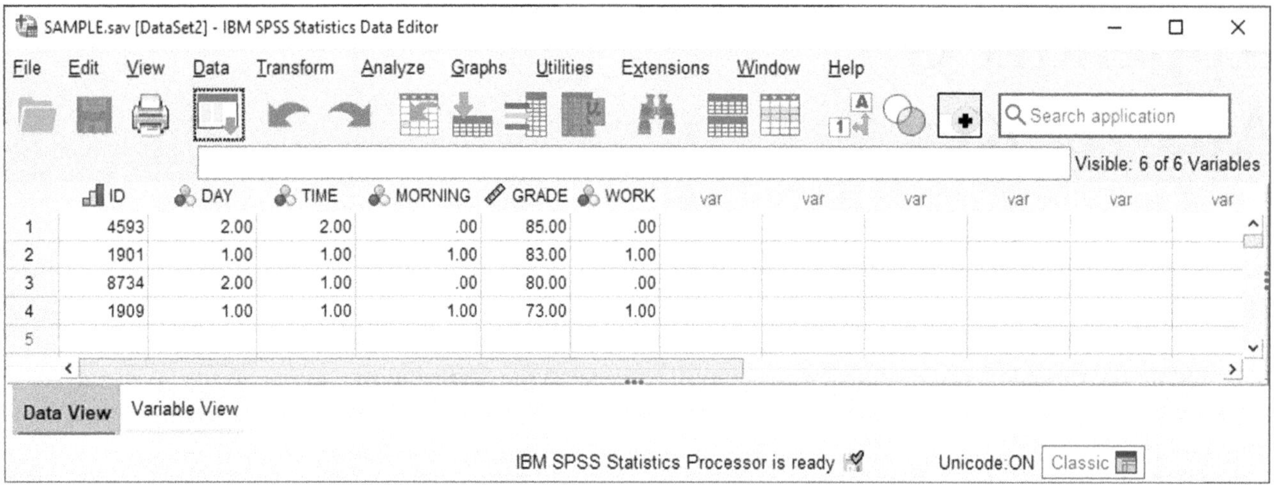

	ID	DAY	TIME	MORNING	GRADE	WORK
1	4593	2.00	2.00	.00	85.00	.00
2	1901	1.00	1.00	1.00	83.00	1.00
3	8734	2.00	1.00	.00	80.00	.00
4	1909	1.00	1.00	1.00	73.00	1.00
5						

Section 3.1

Use Practice Data Set 1 in Appendix B. Create a frequency distribution table for the mathematics skills scores. Determine the mathematics skills score at which the 60th percentile lies.

ANSWER: 38

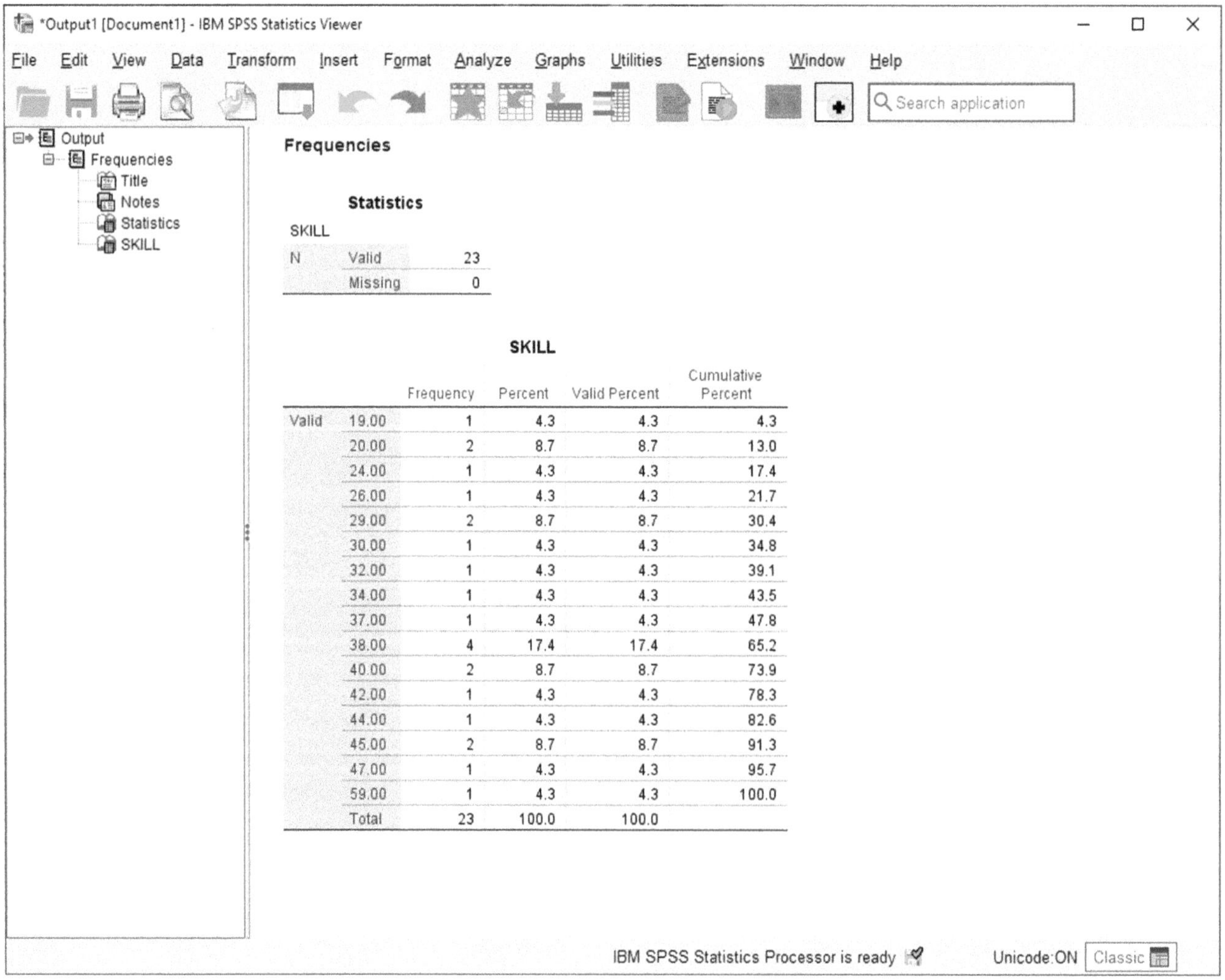

Frequencies

Statistics

SKILL

N	Valid	23
	Missing	0

SKILL

		Frequency	Percent	Valid Percent	Cumulative Percent
Valid	19.00	1	4.3	4.3	4.3
	20.00	2	8.7	8.7	13.0
	24.00	1	4.3	4.3	17.4
	26.00	1	4.3	4.3	21.7
	29.00	2	8.7	8.7	30.4
	30.00	1	4.3	4.3	34.8
	32.00	1	4.3	4.3	39.1
	34.00	1	4.3	4.3	43.5
	37.00	1	4.3	4.3	47.8
	38.00	4	17.4	17.4	65.2
	40.00	2	8.7	8.7	73.9
	42.00	1	4.3	4.3	78.3
	44.00	1	4.3	4.3	82.6
	45.00	2	8.7	8.7	91.3
	47.00	1	4.3	4.3	95.7
	59.00	1	4.3	4.3	100.0
	Total	23	100.0	100.0	

Section 3.2

Use Practice Data Set 1 in Appendix B. Create a contingency table using the *Crosstabs* command. Determine the number of participants in each combination of the variables SEX and MARITAL. What percentage of participants are married? (ANSWER: 30.4%). What percentage of participants are male and married? (ANSWER: 21.7%).

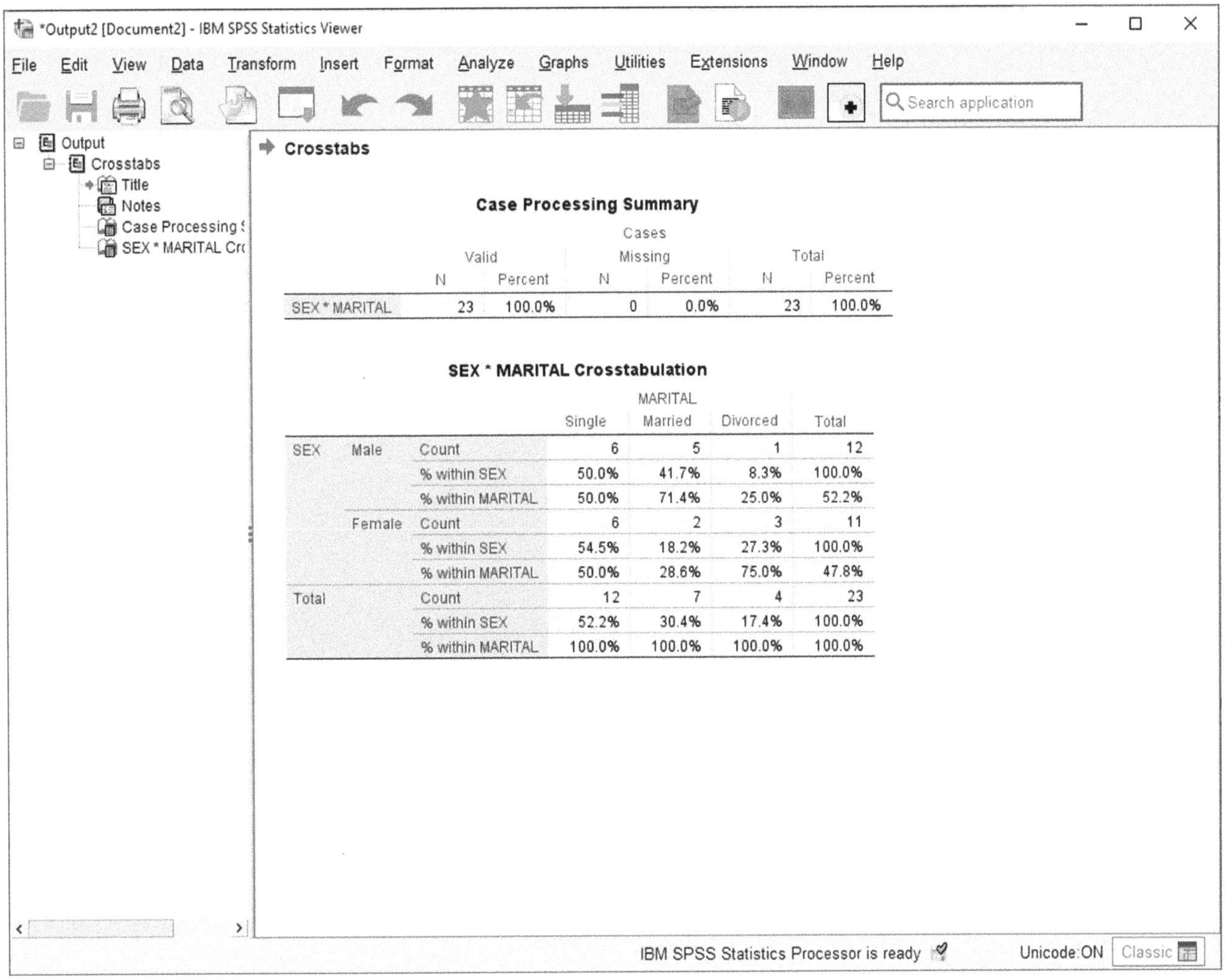

Case Processing Summary

	Cases					
	Valid		Missing		Total	
	N	Percent	N	Percent	N	Percent
SEX * MARITAL	23	100.0%	0	0.0%	23	100.0%

SEX * MARITAL Crosstabulation

			MARITAL			
			Single	Married	Divorced	Total
SEX	Male	Count	6	5	1	12
		% within SEX	50.0%	41.7%	8.3%	100.0%
		% within MARITAL	50.0%	71.4%	25.0%	52.2%
	Female	Count	6	2	3	11
		% within SEX	54.5%	18.2%	27.3%	100.0%
		% within MARITAL	50.0%	28.6%	75.0%	47.8%
Total		Count	12	7	4	23
		% within SEX	52.2%	30.4%	17.4%	100.0%
		% within MARITAL	100.0%	100.0%	100.0%	100.0%

Section 3.3

Use Practice Data Set 1 in Appendix B. Obtain the **descriptive statistics** for the age of the participants. What is the **mean**? (ANSWER: 29.5652). The **median**? (ANSWER: 28.0). The **mode**? (ANSWER: 23.0). What is the **standard deviation**? (ANSWER: 8.9434). Minimum? (ANSWER: 18.0). Maximum? (ANSWER: 50.0). The **range**? (ANSWER: 50 – 18=32).

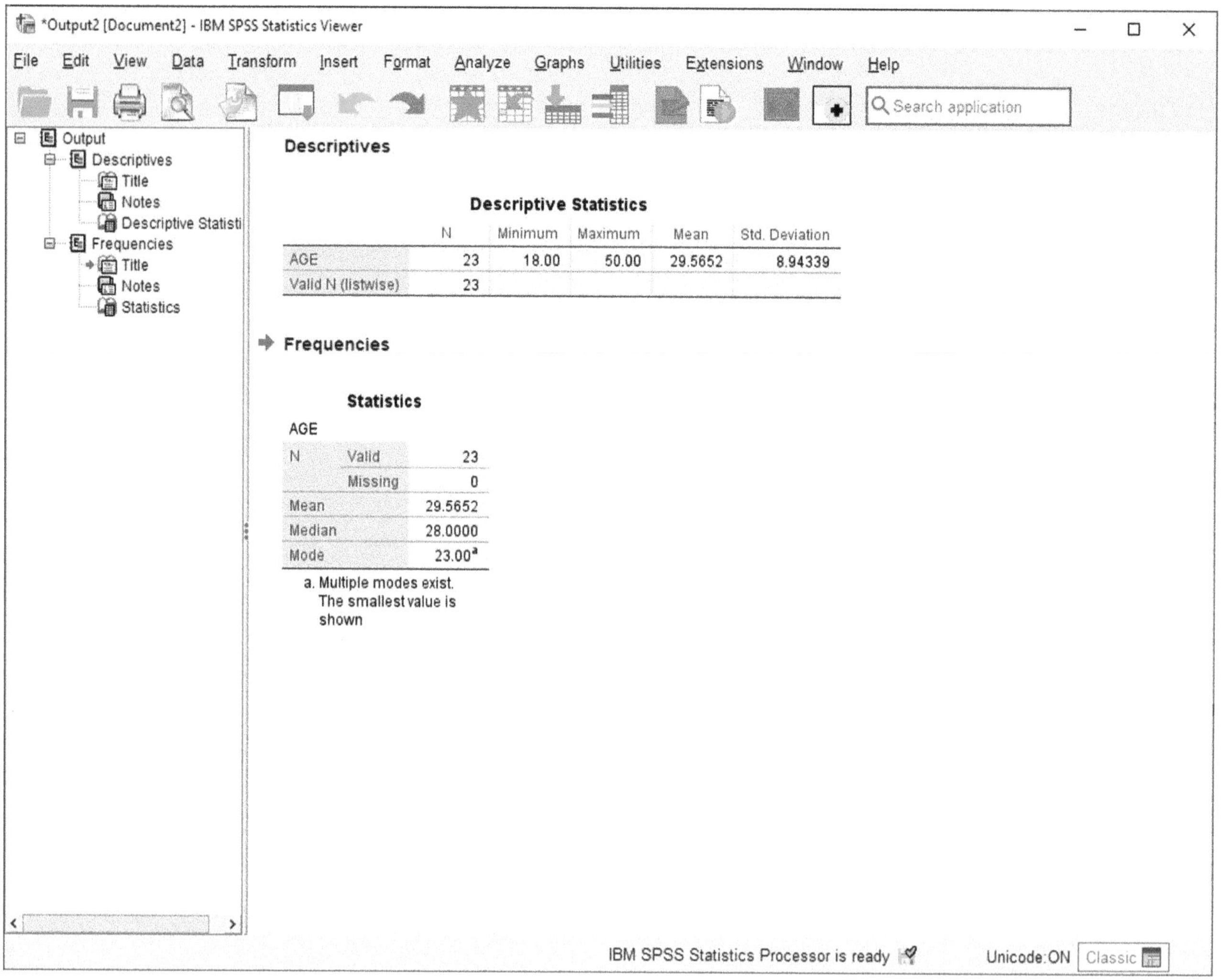

Descriptive Statistics

	N	Minimum	Maximum	Mean	Std. Deviation
AGE	23	18.00	50.00	29.5652	8.94339
Valid N (listwise)	23				

Statistics

AGE

N	Valid	23
	Missing	0
Mean		29.5652
Median		28.0000
Mode		23.00[a]

a. Multiple modes exist. The smallest value is shown

By using the *Descriptives* command, you can get everything except the **median** and **mode**. Using the *Frequencies* command, you can get all of the information.

Section 3.4

Use Practice Data Set 1 in Appendix B. Compute the **mean** and **standard deviation** of ages for each value of marital status. What is the average age of the married participants? (ANSWER: 35.1429). The single participants? (ANSWER: 22.4167). The divorced participants? (ANSWER: 41.25).

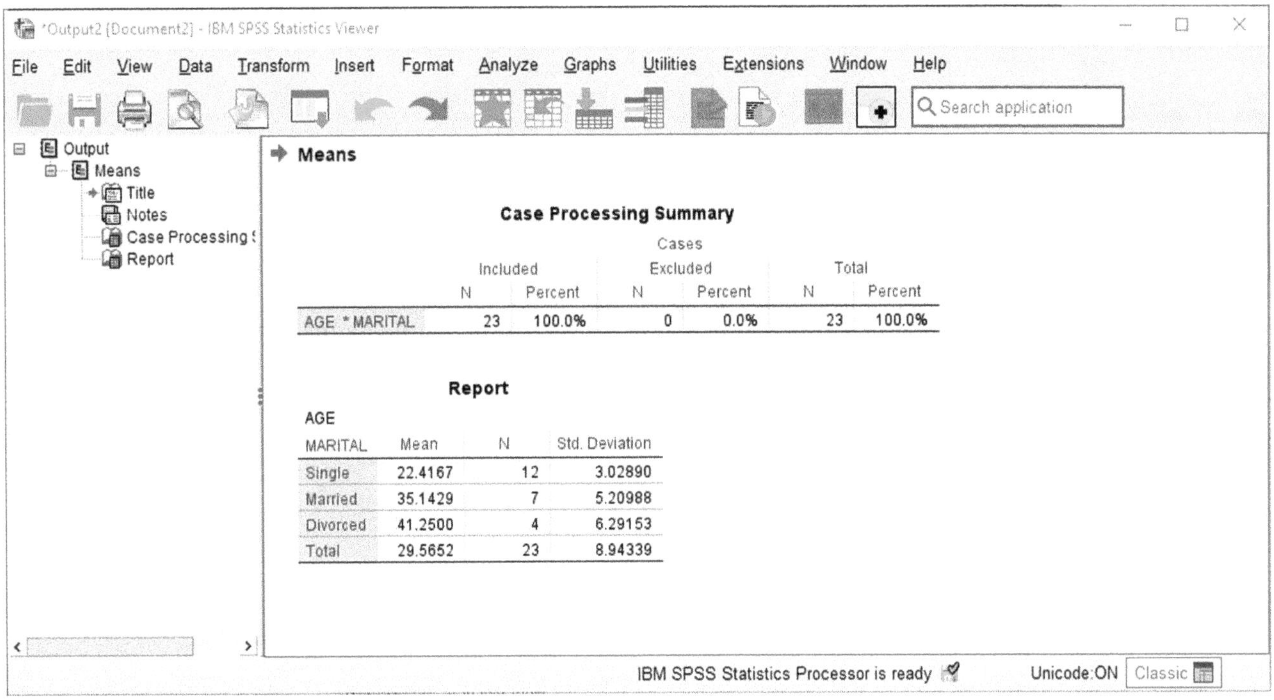

Case Processing Summary

	Cases					
	Included		Excluded		Total	
	N	Percent	N	Percent	N	Percent
AGE * MARITAL	23	100.0%	0	0.0%	23	100.0%

Report

AGE

MARITAL	Mean	N	Std. Deviation
Single	22.4167	12	3.02890
Married	35.1429	7	5.20988
Divorced	41.2500	4	6.29153
Total	29.5652	23	8.94339

Section 3.5

Use Practice Data Set 2 in Appendix B. Determine the *z*-score that corresponds to each employee's salary. Determine the **mean** *z*-scores for salaries of male employees (ANSWER: .130) and female employees (ANSWER: –.130). Determine the **mean** *z*-score for salaries of the total sample (ANSWER: 0).

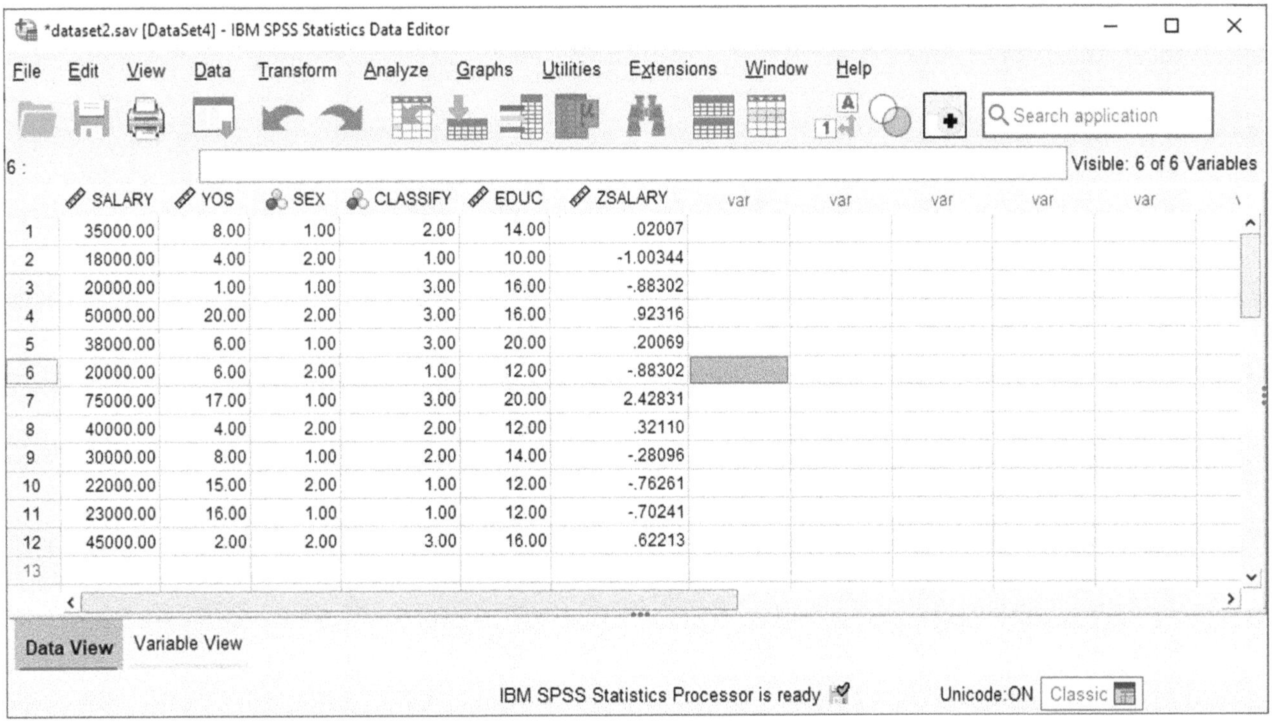

	SALARY	YOS	SEX	CLASSIFY	EDUC	ZSALARY	var	var	var	var	var
1	35000.00	8.00	1.00	2.00	14.00	.02007					
2	18000.00	4.00	2.00	1.00	10.00	-1.00344					
3	20000.00	1.00	1.00	3.00	16.00	-.88302					
4	50000.00	20.00	2.00	3.00	16.00	.92316					
5	38000.00	6.00	1.00	3.00	20.00	.20069					
6	20000.00	6.00	2.00	1.00	12.00	-.88302					
7	75000.00	17.00	1.00	3.00	20.00	2.42831					
8	40000.00	4.00	2.00	2.00	12.00	.32110					
9	30000.00	8.00	1.00	2.00	14.00	-.28096					
10	22000.00	15.00	2.00	1.00	12.00	-.76261					
11	23000.00	16.00	1.00	1.00	12.00	-.70241					
12	45000.00	2.00	2.00	3.00	16.00	.62213					
13											

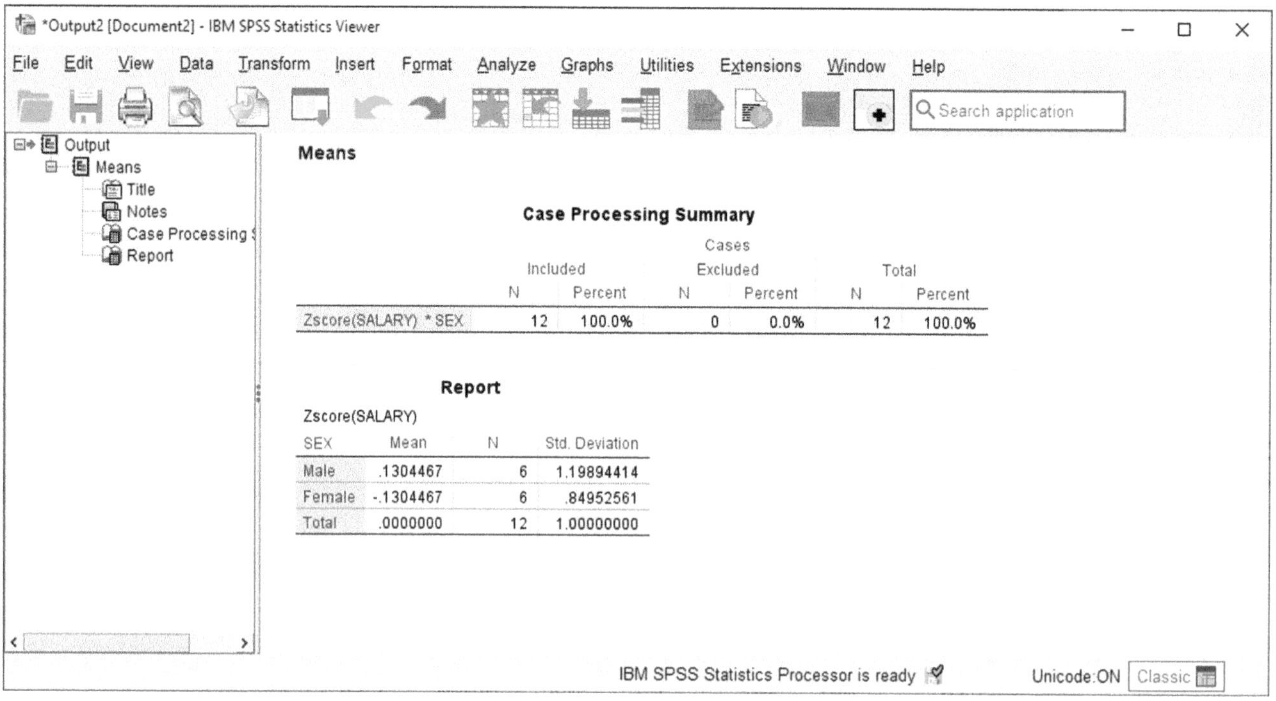

Case Processing Summary

	Cases					
	Included		Excluded		Total	
	N	Percent	N	Percent	N	Percent
Zscore(SALARY) * SEX	12	100.0%	0	0.0%	12	100.0%

Report

Zscore(SALARY)

SEX	Mean	N	Std. Deviation
Male	.1304467	6	1.19894414
Female	-.1304467	6	.84952561
Total	.0000000	12	1.00000000

Section 4.2

Use Practice Data Set 1 in Appendix B. After you have entered the data, first construct a histogram that represents the mathematics skills scores and displays a normal curve and then construct a bar chart that represents the frequencies for the variable AGE.

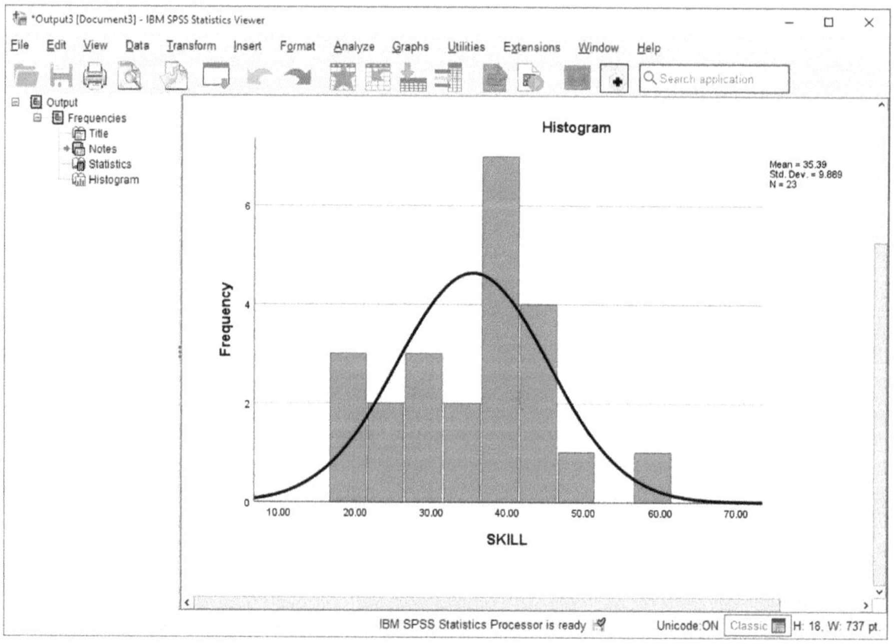

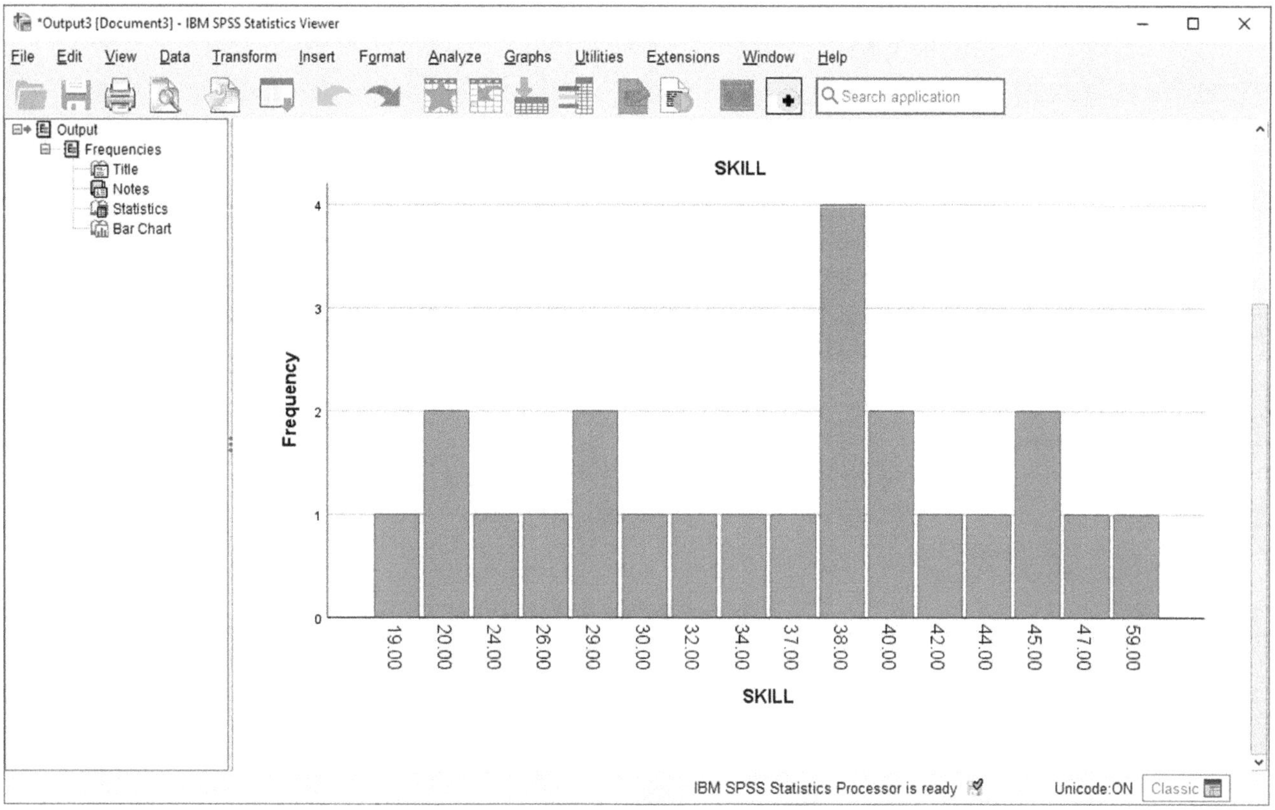

Section 4.4

Use Practice Data Set 2 in Appendix B. Construct a scatterplot to examine the relationship between SALARY and EDUCATION.

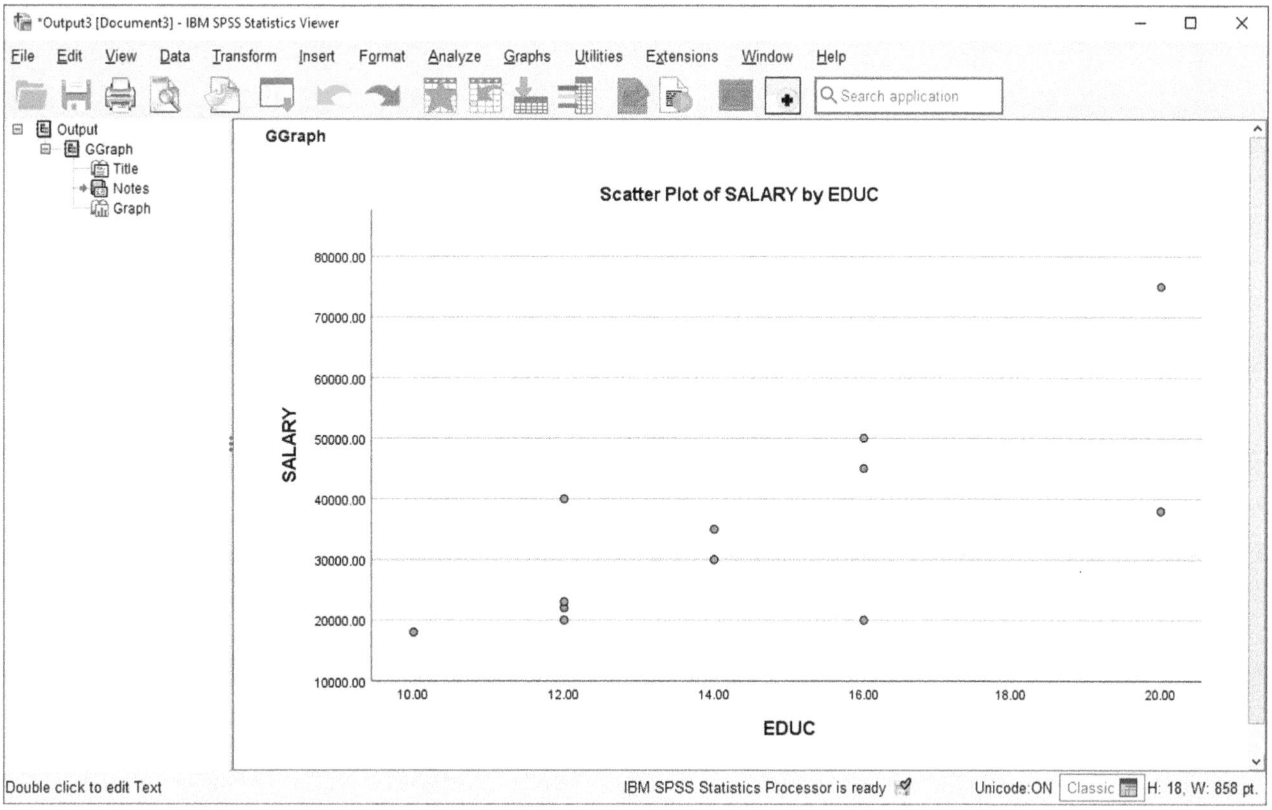

Note: It would also be acceptable to assign salary to the *X*-axis and education to the *Y*-axis.

Section 4.5

Use Practice Data Set 1 in Appendix B. Construct a clustered bar graph examining the relationship between MATHEMATICS SKILLS scores (as the **dependent variable**) and MARITAL STATUS and SEX (as **independent variables**). Make sure you classify both SEX and MARITAL STATUS as nominal variables.

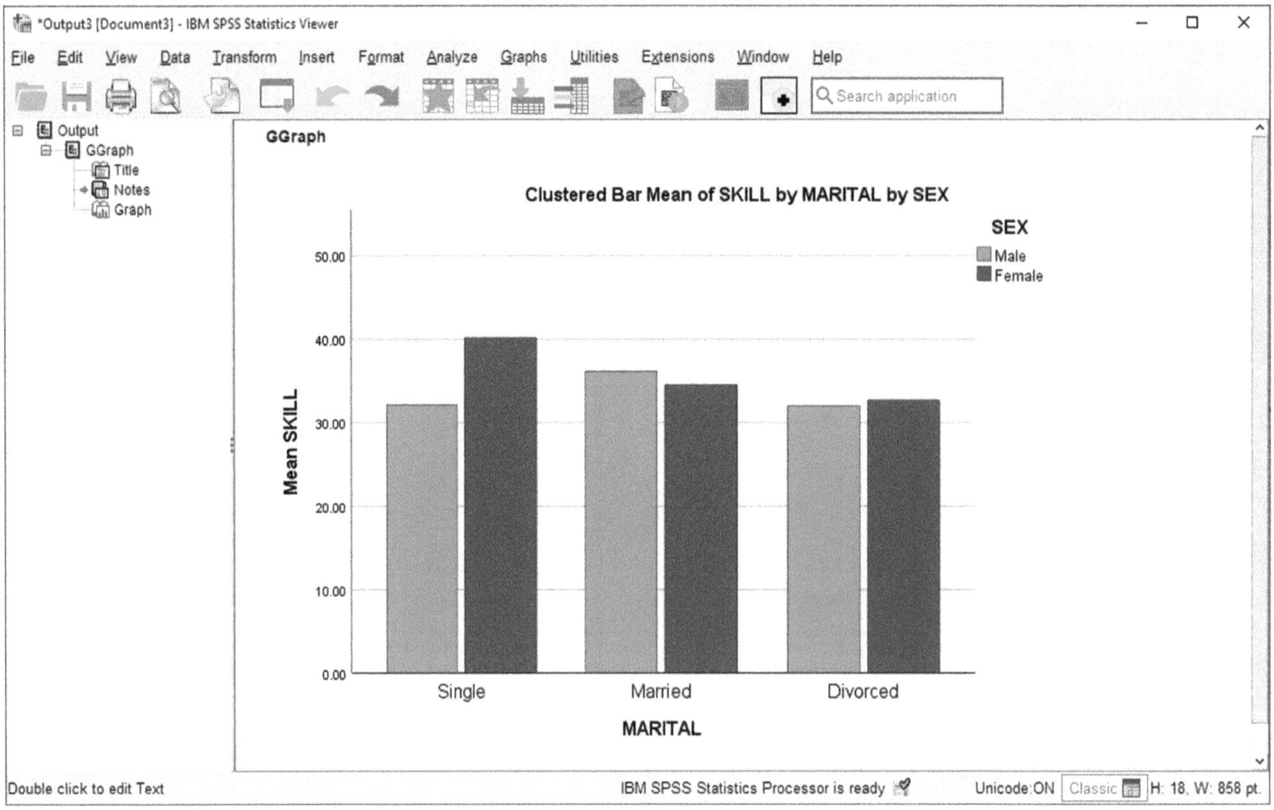

Section 5.1

Use Practice Data Set 2 in Appendix B. Determine the value of the Pearson correlation coefficient for the relationship between SALARY and YEARS OF EDUCATION and phrase your results.

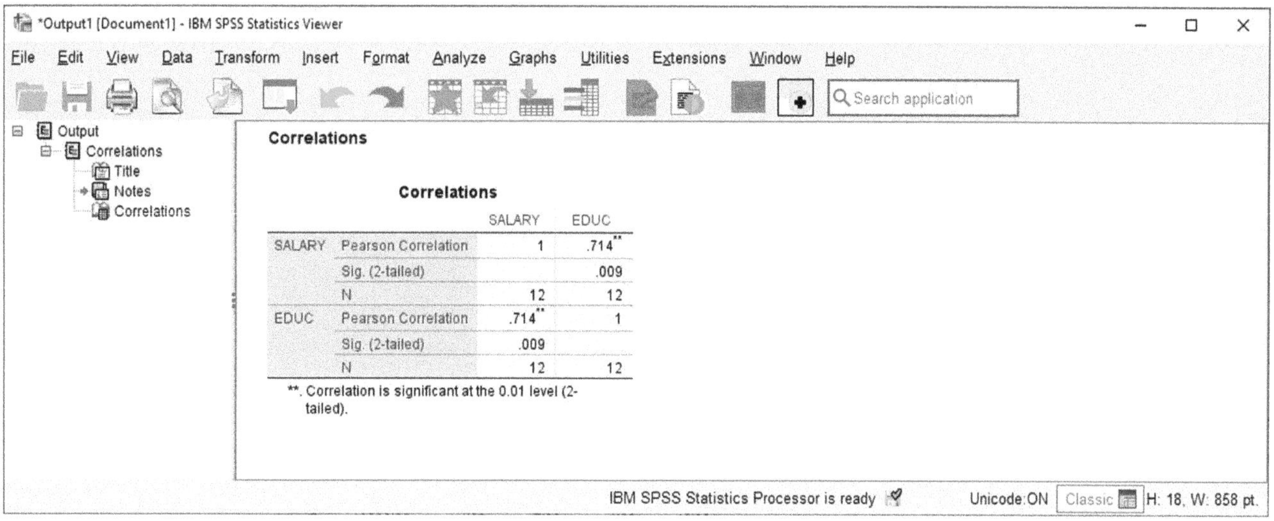

Correlations

Correlations

		SALARY	EDUC
SALARY	Pearson Correlation	1	.714**
	Sig. (2-tailed)		.009
	N	12	12
EDUC	Pearson Correlation	.714**	1
	Sig. (2-tailed)	.009	
	N	12	12

**. Correlation is significant at the 0.01 level (2-tailed).

A Pearson correlation coefficient was calculated for the relationship between subjects' educations and salaries. A strong positive correlation was found ($r(10) = .714$, $p = .009$), indicating a significant linear relationship between the two variables. Better-educated individuals tended to be paid more.

Section 5.2

Use Practice Data Set 2 in Appendix B. Determine the strength of the relationship between salary and job classification by calculating the Spearman *rho* correlation.

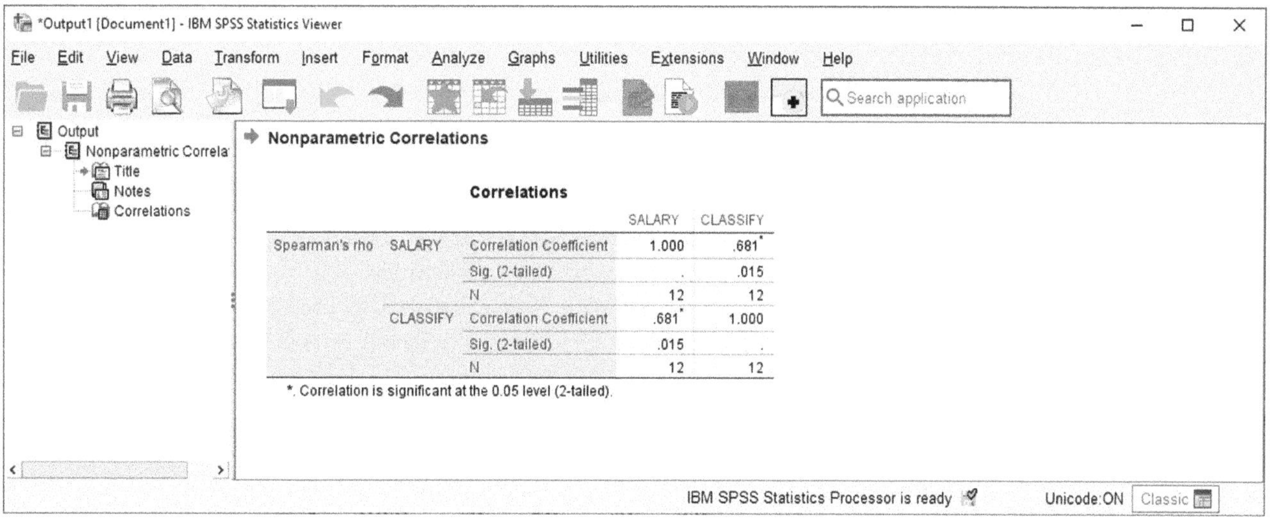

Nonparametric Correlations

Correlations

			SALARY	CLASSIFY
Spearman's rho	SALARY	Correlation Coefficient	1.000	.681*
		Sig. (2-tailed)	.	.015
		N	12	12
	CLASSIFY	Correlation Coefficient	.681*	1.000
		Sig. (2-tailed)	.015	.
		N	12	12

*. Correlation is significant at the 0.05 level (2-tailed).

A Spearman *rho* correlation coefficient was calculated for the relationship between the subjects' salaries and their job classifications. A strong positive correlation was found (*rho* (10) = .681, $p = .015$), indicating a significant relationship between the two variables. Subjects with higher job classifications tended to be paid more.

Section 5.3

Use Practice Data Set 2 in Appendix B. If we want to predict salary from years of education, what salary would you predict for someone with 12 years of education? What salary would you predict for someone with a college education (16 years)?

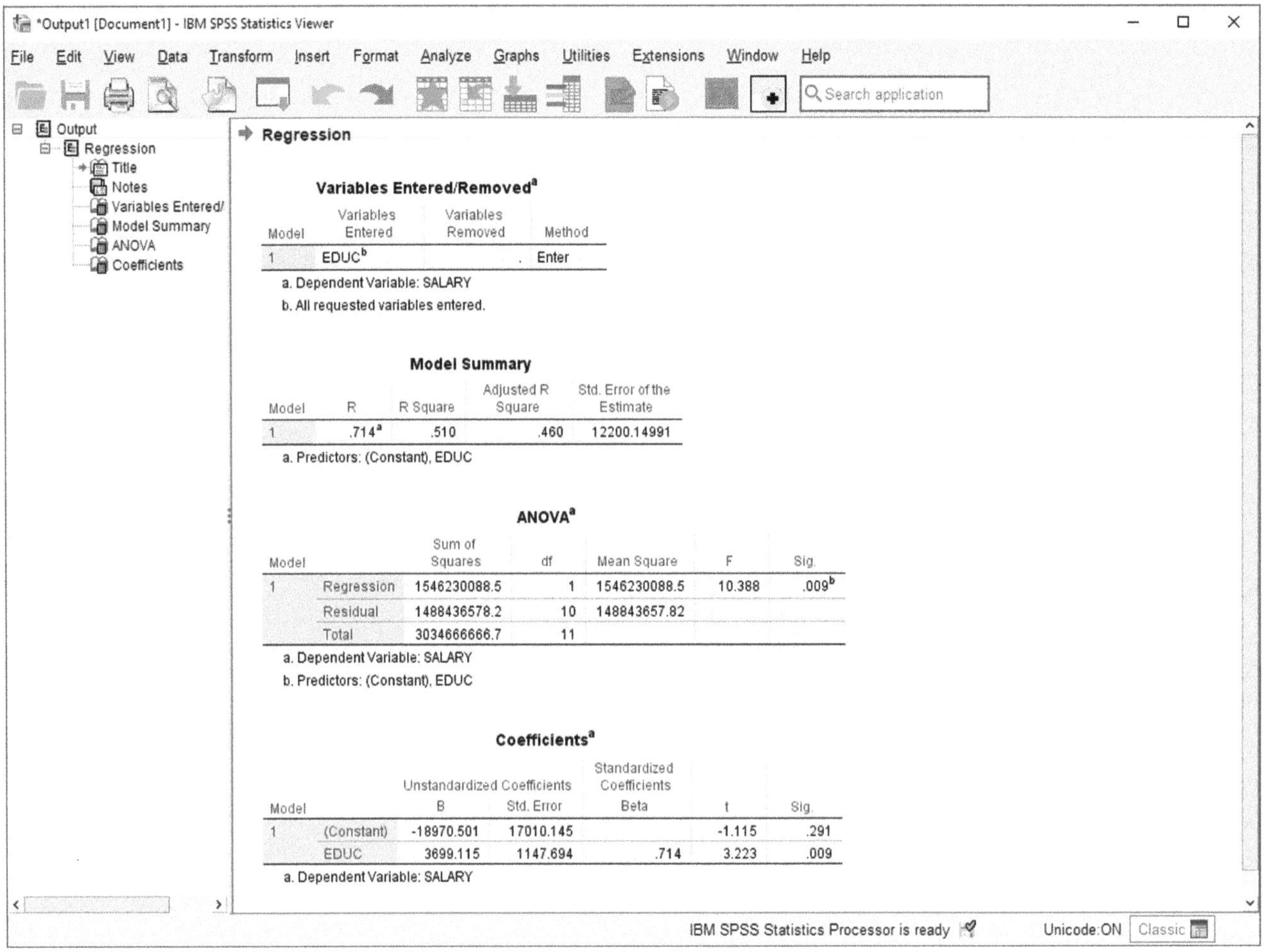

Variables Entered/Removed[a]

Model	Variables Entered	Variables Removed	Method
1	EDUC[b]	.	Enter

a. Dependent Variable: SALARY
b. All requested variables entered.

Model Summary

Model	R	R Square	Adjusted R Square	Std. Error of the Estimate
1	.714[a]	.510	.460	12200.14991

a. Predictors: (Constant), EDUC

ANOVA[a]

Model		Sum of Squares	df	Mean Square	F	Sig.
1	Regression	1546230088.5	1	1546230088.5	10.388	.009[b]
	Residual	1488436578.2	10	148843657.82		
	Total	3034666666.7	11			

a. Dependent Variable: SALARY
b. Predictors: (Constant), EDUC

Coefficients[a]

Model		Unstandardized Coefficients B	Std. Error	Standardized Coefficients Beta	t	Sig.
1	(Constant)	-18970.501	17010.145		-1.115	.291
	EDUC	3699.115	1147.694	.714	3.223	.009

a. Dependent Variable: SALARY

$\text{Salary}' = -18970.5 + 12(3699.115) = \$25{,}418.88$

$\text{Salary}' = -18970.5 + 16(3699.115) = \$40{,}215.34$

A simple linear regression was calculated to predict a participant's salary on the basis of education level. A significant regression equation was found ($F(1,10) = 10.388$, $p = .009$), with an R^2 of .510. Participant's predicted salary is equal to −18970.5 + 3699.115 (years of education). Participant's average salary increased by $3,699 for every year of education.

Section 5.4

Use Practice Data Set 2 in Appendix B. Determine the prediction equation for predicting salary based on education, years of service, and sex. Which variables are significant predictors? If you believe that men were paid more than women, what would you conclude after conducting this analysis?

$\text{Salary}' = -46257.7 + 4175.856(\text{Education}) + 8856.781(\text{Sex}) + 795.054(\text{Years of Service})$

Only education is a significant predictor.

When controlling for education and years of service, men do not get paid significantly more than women do.

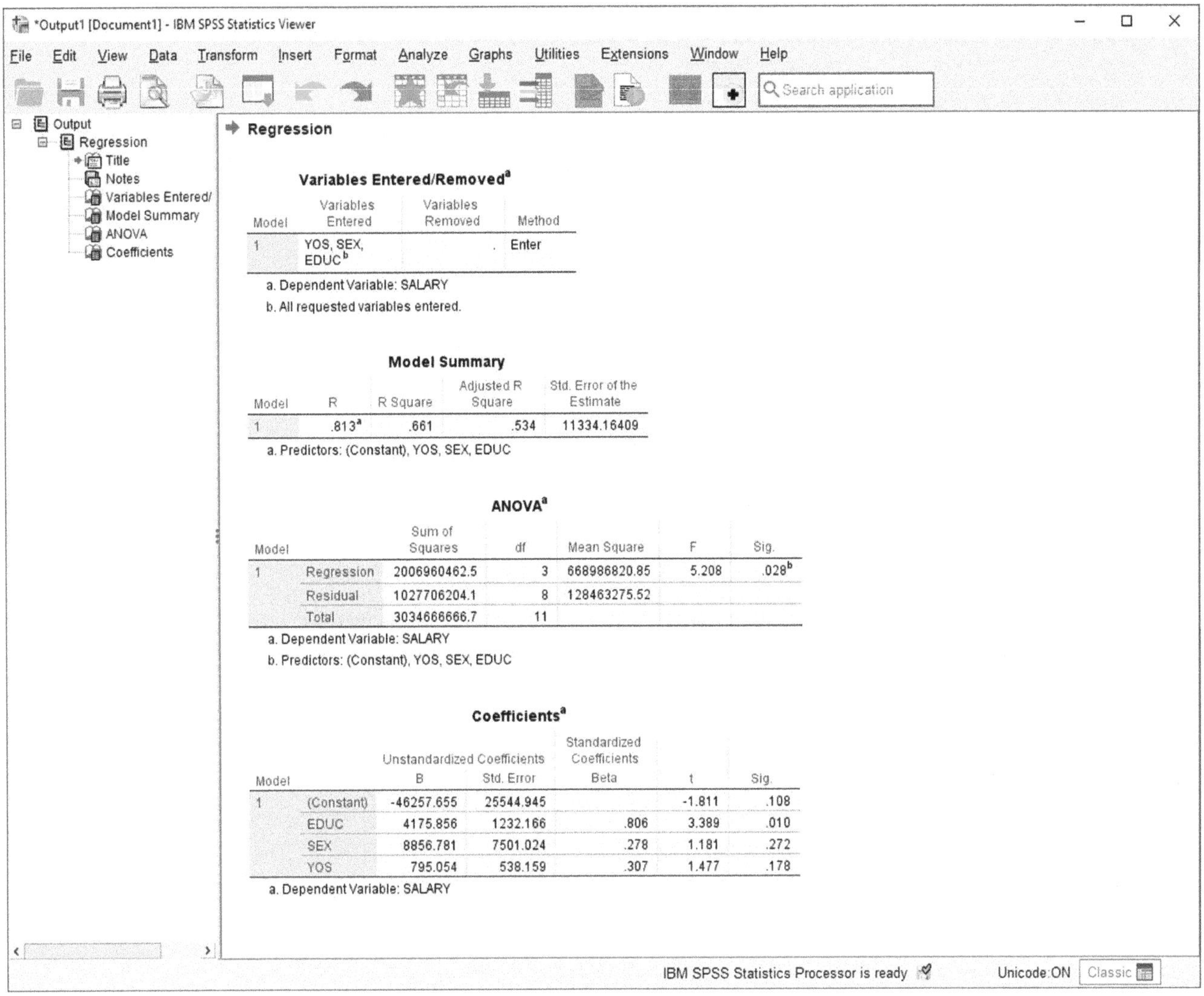

Regression

Variables Entered/Removed[a]

Model	Variables Entered	Variables Removed	Method
1	YOS, SEX, EDUC[b]	.	Enter

a. Dependent Variable: SALARY
b. All requested variables entered.

Model Summary

Model	R	R Square	Adjusted R Square	Std. Error of the Estimate
1	.813[a]	.661	.534	11334.16409

a. Predictors: (Constant), YOS, SEX, EDUC

ANOVA[a]

Model		Sum of Squares	df	Mean Square	F	Sig.
1	Regression	2006960462.5	3	668986820.85	5.208	.028[b]
	Residual	1027706204.1	8	128463275.52		
	Total	3034666666.7	11			

a. Dependent Variable: SALARY
b. Predictors: (Constant), YOS, SEX, EDUC

Coefficients[a]

Model		Unstandardized Coefficients B	Std. Error	Standardized Coefficients Beta	t	Sig.
1	(Constant)	-46257.655	25544.945		-1.811	.108
	EDUC	4175.856	1232.166	.806	3.389	.010
	SEX	8856.781	7501.024	.278	1.181	.272
	YOS	795.054	538.159	.307	1.477	.178

a. Dependent Variable: SALARY

Note: At this point, you would repeat the analysis using only the significant predictors to obtain your final answer.

Section 6.2

The **mean** salary in the United States is a hypothetical average of $25,000. Determine if the average salary of the participants in Practice Data Set 2 in Appendix B is significantly greater than this value. Note that this is a one-tailed hypothesis.

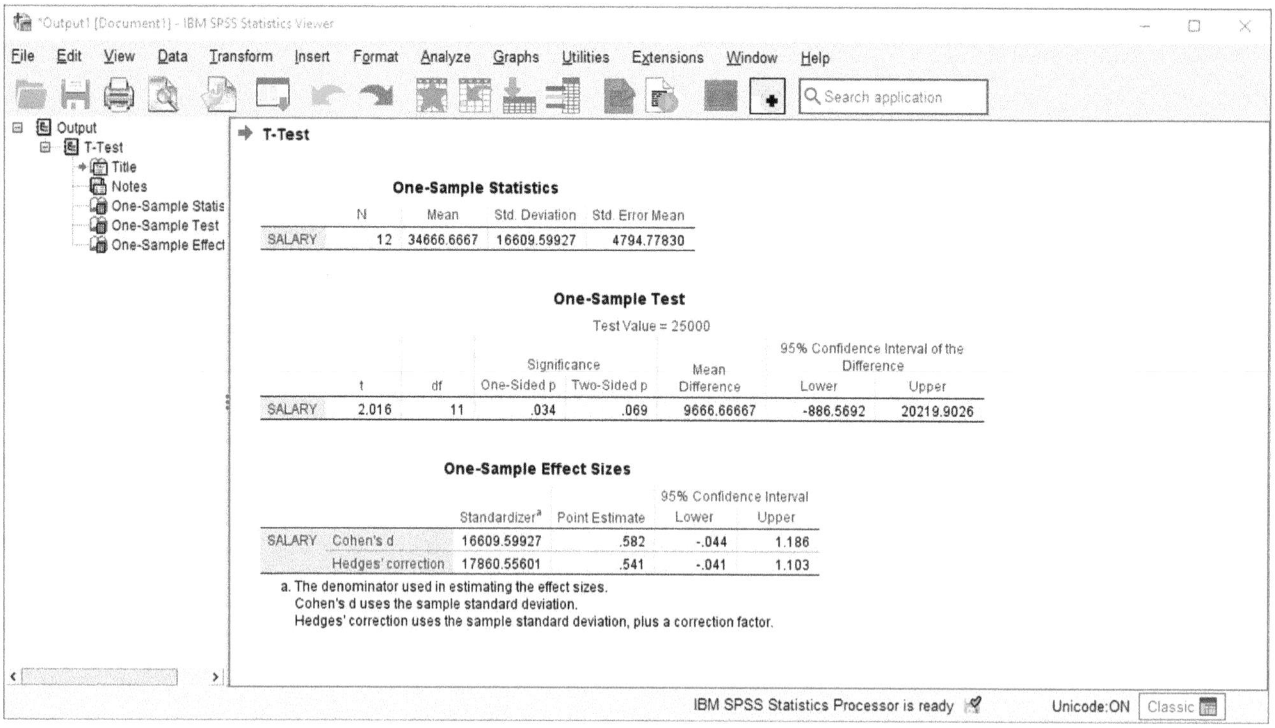

A single-sample t test was calculated to determine if the average salary of our sample is significantly higher than $25,000. A significant difference was found ($t(11) = 2.016$, $p = .034$). The sample had a significantly higher average salary (M = $34,666.67, sd = $16,609.60) than the population. This was a moderate effect size (d=.582).

Section 6.3

Use Practice Data Set 1 in Appendix B to solve this problem. We believe that young individuals have lower mathematics skills than older individuals. We would test this hypothesis by comparing participants 25 or younger (the "young" group) with participants 26 or older (the "old" group). Hint: You may need to create a new variable that represents each age group. See Chapter 2 for help.

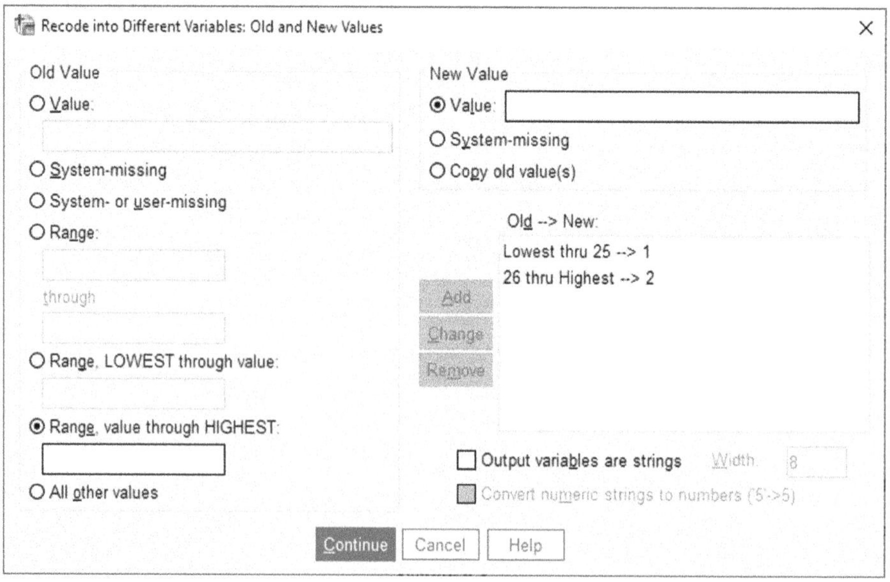

There is no significant difference. Note that we use .463 as the actual significance level because this is a one-tailed test.

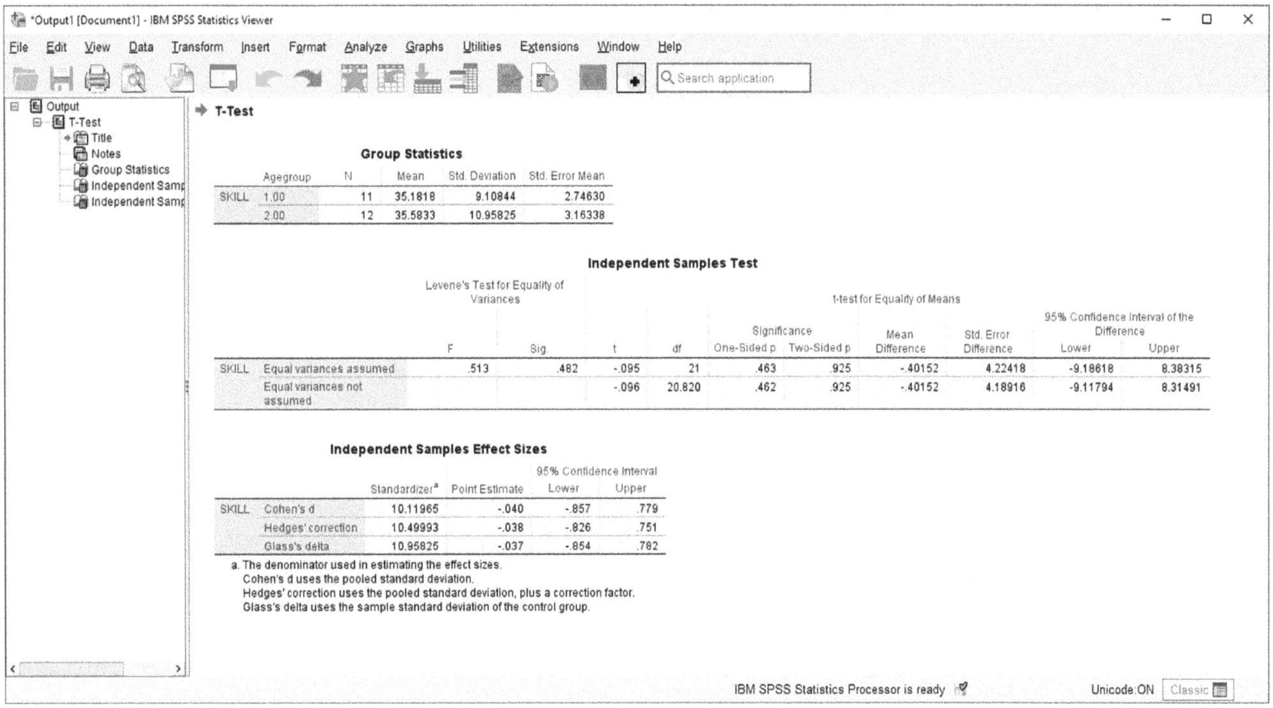

Group Statistics

	Agegroup	N	Mean	Std. Deviation	Std. Error Mean
SKILL	1.00	11	35.1818	9.10844	2.74630
	2.00	12	35.5833	10.95825	3.16338

Independent Samples Test

		Levene's Test for Equality of Variances		t-test for Equality of Means							
						Significance				95% Confidence Interval of the Difference	
		F	Sig.	t	df	One-Sided p	Two-Sided p	Mean Difference	Std. Error Difference	Lower	Upper
SKILL	Equal variances assumed	.513	.482	-.095	21	.463	.925	-.40152	4.22418	-9.18618	8.38315
	Equal variances not assumed			-.096	20.820	.462	.925	-.40152	4.18916	-9.11794	8.31491

Independent Samples Effect Sizes

		Standardizer[a]	Point Estimate	95% Confidence Interval	
				Lower	Upper
SKILL	Cohen's d	10.11965	-.040	-.857	.779
	Hedges' correction	10.49993	-.038	-.826	.751
	Glass's delta	10.95825	-.037	-.854	.782

a. The denominator used in estimating the effect sizes.
Cohen's d uses the pooled standard deviation.
Hedges' correction uses the pooled standard deviation, plus a correction factor.
Glass's delta uses the sample standard deviation of the control group.

An independent-samples t test was calculated to determine whether the math skills of subjects 25 or younger were lower than the skills of those 26 or older. No significant difference was found ($t(21) = -.095, p = .463$). The math skills of the younger group ($M = 35.18$, $sd = 9.11$) were not significantly lower than the math skills of the older group ($M = 35.58$, $sd = 10.96$). This was a small effect size (d=.04).

Section 6.4

Use the same GRADES.sav data file and compute a paired-samples t test to determine if scores increased from midterm to final.

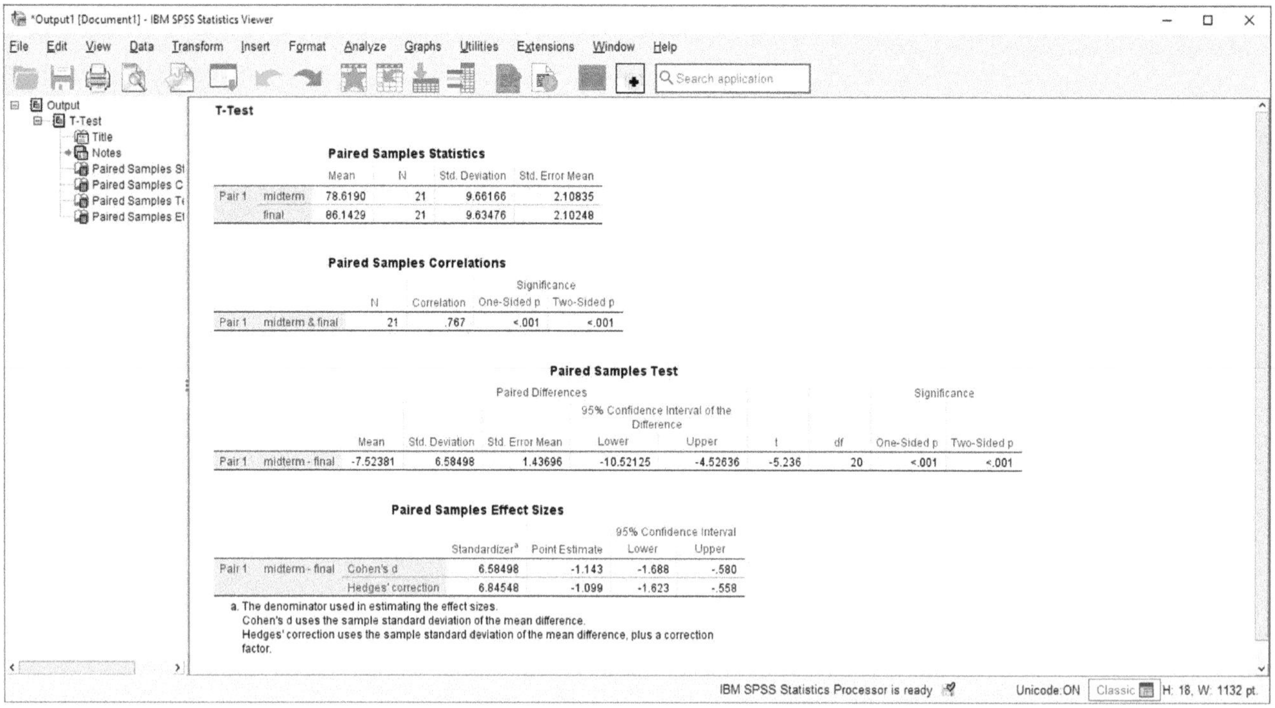

T-Test

Paired Samples Statistics

		Mean	N	Std. Deviation	Std. Error Mean
Pair 1	midterm	78.6190	21	9.66166	2.10835
	final	86.1429	21	9.63476	2.10248

Paired Samples Correlations

		N	Correlation	Significance One-Sided p	Significance Two-Sided p
Pair 1	midterm & final	21	.767	<.001	<.001

Paired Samples Test

		Paired Differences Mean	Std. Deviation	Std. Error Mean	95% Confidence Interval of the Difference Lower	Upper	t	df	Significance One-Sided p	Two-Sided p
Pair 1	midterm - final	-7.52381	6.58498	1.43696	-10.52125	-4.52636	-5.236	20	<.001	<.001

Paired Samples Effect Sizes

			Standardizer[a]	Point Estimate	95% Confidence Interval Lower	Upper
Pair 1	midterm - final	Cohen's d	6.58498	-1.143	-1.688	-.580
		Hedges' correction	6.84548	-1.099	-1.623	-.558

a. The denominator used in estimating the effect sizes.
Cohen's d uses the sample standard deviation of the mean difference.
Hedges' correction uses the sample standard deviation of the mean difference, plus a correction factor.

A paired-samples *t* test was calculated to determine if scores increased from midterm to final. A significant increase was found ($t(20) = -5.236, p < .001$). The final scores were significantly higher ($M = 86.14$, $sd = 9.63$) than the midterm scores ($M = 78.62$, $sd = 9.66$). This was a large effect size (d=1.143).

Section 7.2

Use Practice Data Set 1 in Appendix B. Determine if the average math scores of single, married, and divorced participants are significantly different. Write a statement of results.

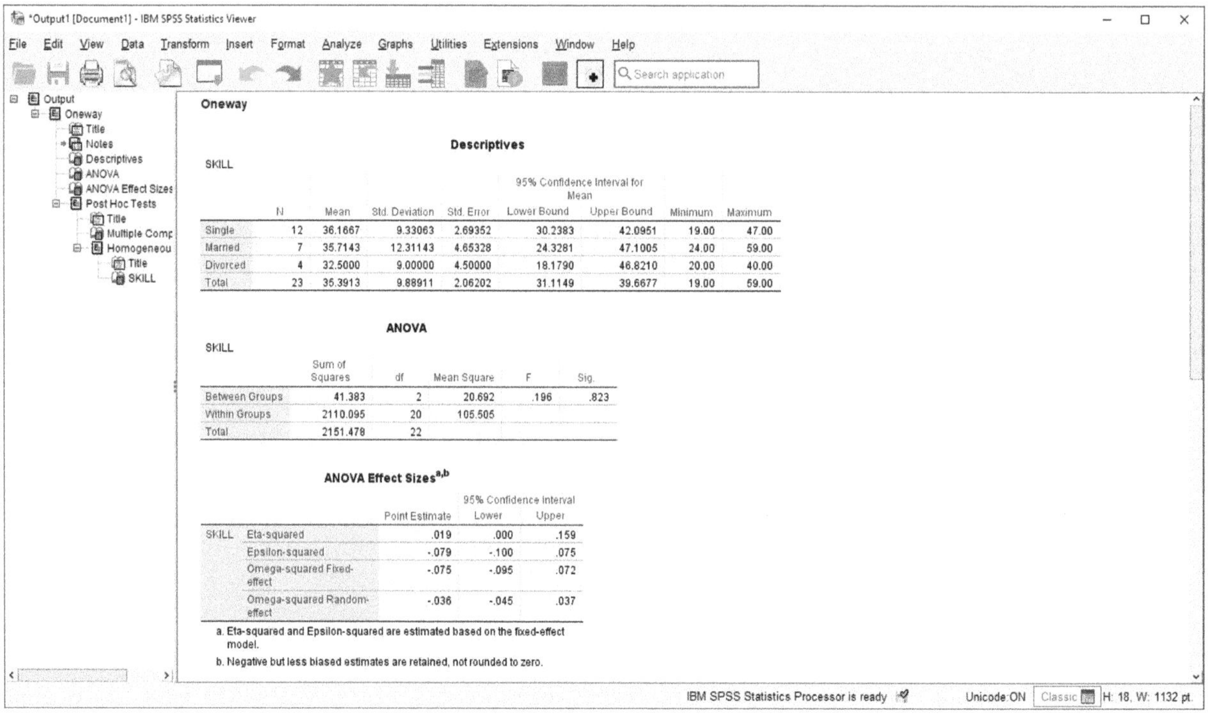

Oneway

Descriptives

SKILL

	N	Mean	Std. Deviation	Std. Error	95% Confidence Interval for Mean Lower Bound	Upper Bound	Minimum	Maximum
Single	12	36.1667	9.33063	2.69352	30.2383	42.0951	19.00	47.00
Married	7	35.7143	12.31143	4.65328	24.3281	47.1005	24.00	59.00
Divorced	4	32.5000	9.00000	4.50000	18.1790	46.8210	20.00	40.00
Total	23	35.3913	9.88911	2.06202	31.1149	39.6677	19.00	59.00

ANOVA

SKILL

	Sum of Squares	df	Mean Square	F	Sig.
Between Groups	41.383	2	20.692	.196	.823
Within Groups	2110.095	20	105.505		
Total	2151.478	22			

ANOVA Effect Sizes[a,b]

		Point Estimate	95% Confidence Interval Lower	Upper
SKILL	Eta-squared	.019	.000	.159
	Epsilon-squared	-.079	-.100	.075
	Omega-squared Fixed-effect	-.075	-.095	.072
	Omega-squared Random-effect	-.036	-.045	.037

a. Eta-squared and Epsilon-squared are estimated based on the fixed-effect model.
b. Negative but less biased estimates are retained, not rounded to zero.

A one-way ANOVA was calculated comparing the math scores of subjects who differed in marital status. No significant differences were found ($F(2,20) = .196$, $p = .823$). Single subjects ($M = 36.17$, $sd = 9.33$) were not significantly different from married subjects ($M = 35.71$, $sd = 12.31$) or divorced subjects ($M = 32.50$, $sd = 9.00$). This was a moderate effect size (η^2=.019).

Section 7.3

Use Practice Data Set 2 in Appendix B. Determine if salaries are influenced by sex, job classification, or an **interaction** between sex and job classification. Write a statement of results.

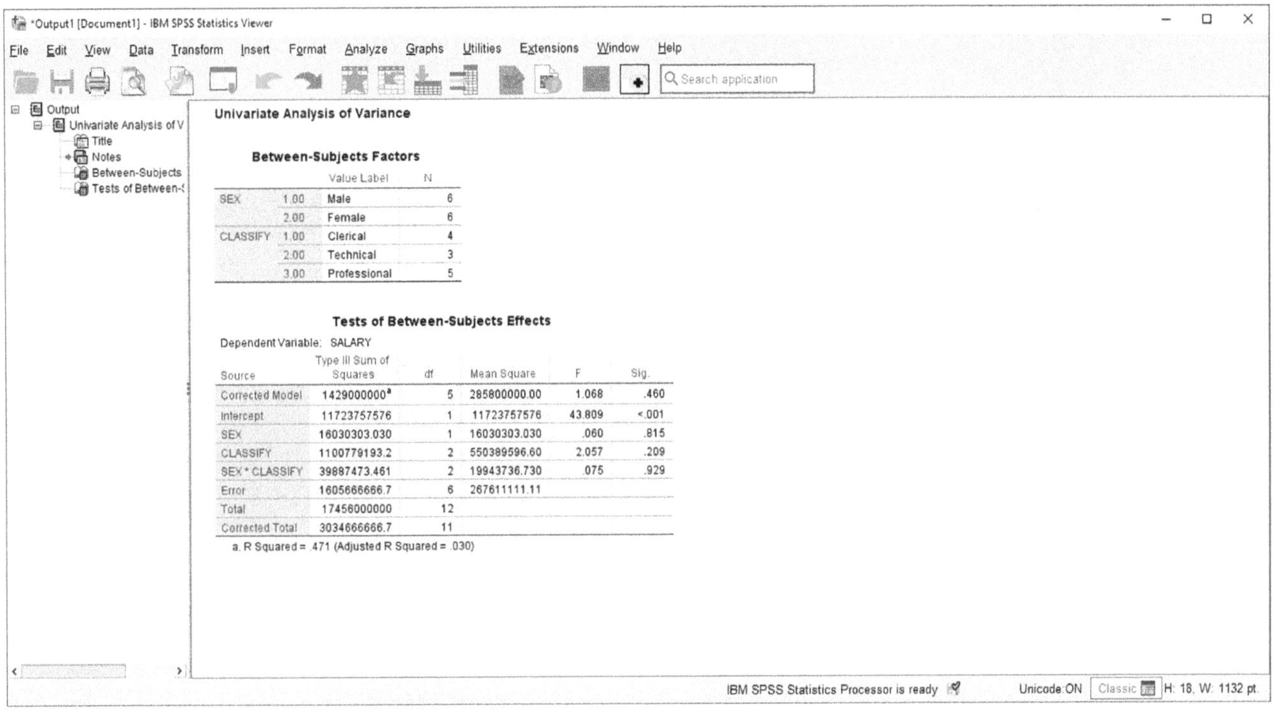

Univariate Analysis of Variance

Between-Subjects Factors

		Value Label	N
SEX	1.00	Male	6
	2.00	Female	6
CLASSIFY	1.00	Clerical	4
	2.00	Technical	3
	3.00	Professional	5

Tests of Between-Subjects Effects

Dependent Variable: SALARY

Source	Type III Sum of Squares	df	Mean Square	F	Sig.
Corrected Model	1429000000[a]	5	285800000.00	1.068	.460
Intercept	11723757576	1	11723757576	43.809	<.001
SEX	16030303.030	1	16030303.030	.060	.815
CLASSIFY	1100779193.2	2	550389596.60	2.057	.209
SEX * CLASSIFY	39887473.461	2	19943736.730	.075	.929
Error	1605666666.7	6	267611111.11		
Total	17456000000	12			
Corrected Total	3034666666.7	11			

a. R Squared = .471 (Adjusted R Squared = .030)

A 2 (sex) × 3 (job classification) between-subjects ANOVA was calculated comparing the salaries of men and women of different job classifications. No significant differences were found. The main effect for sex was not significant ($F(1,6) = .060$, $p = .815$). The main effect for job classification was not significant ($F(2,6) = 2.057$, $p = .209$). The sex × job classification interaction was not significant ($F(2,6) = .075$, $p = .929$).

Section 7.4

Use Practice Data Set 3 in Appendix B. Determine if the anxiety level of participants changed over time (regardless of which treatment they received) using a one-way repeated-measures ANOVA and protected dependent t tests. Write a statement of results.

*Output1 [Document1] - IBM SPSS Statistics Viewer

File Edit View Data Transform Insert Format Analyze Graphs Utilities Extensions Window Help

Output
- General Linear Model
  - Title
  - Notes
  - Within-Subjects Fa
  - Multivariate Tests
  - Mauchly's Test of S
  - Tests of Within-Sul
  - Tests of Within-Sul
  - Tests of Between-S
- T-Test
  - Title
  - Notes
  - Paired Samples St
  - Paired Samples C
  - Paired Samples Te
  - Paired Samples Ef

Tests of Within-Subjects Effects

Measure: MEASURE_1

Source		Type III Sum of Squares	df	Mean Square	F	Sig.
time	Sphericity Assumed	64.178	2	32.089	16.694	<.001
	Greenhouse-Geisser	64.178	1.859	34.531	16.694	<.001
	Huynh-Feldt	64.178	2.000	32.089	16.694	<.001
	Lower-bound	64.178	1.000	64.178	16.694	.001
Error(time)	Sphericity Assumed	53.822	28	1.922		
	Greenhouse-Geisser	53.822	26.020	2.069		
	Huynh-Feldt	53.822	28.000	1.922		
	Lower-bound	53.822	14.000	3.844		

Tests of Within-Subjects Contrasts

Measure: MEASURE_1

Source	time	Type III Sum of Squares	df	Mean Square	F	Sig.
time	Linear	58.800	1	58.800	31.420	<.001
	Quadratic	5.378	1	5.378	2.726	.121
Error(time)	Linear	26.200	14	1.871		
	Quadratic	27.622	14	1.973		

Tests of Between-Subjects Effects

Measure: MEASURE_1

Transformed Variable: Average

Source	Type III Sum of Squares	df	Mean Square	F	Sig.
Intercept	1933.889	1	1933.889	273.173	<.001
Error	99.111	14	7.079		

T-Test

Paired Samples Statistics

		Mean	N	Std. Deviation	Std. Error Mean
Pair 1	ANXPRE	8.2000	15	1.37321	.35456
	ANX1HR	6.0667	15	2.01660	.52068
Pair 2	ANXPRE	8.2000	15	1.37321	.35456
	ANX4HR	5.4000	15	2.22967	.57570
Pair 3	ANX1HR	6.0667	15	2.01660	.52068
	ANX4HR	5.4000	15	2.22967	.57570

Paired Samples Correlations

				Significance	
		N	Correlation	One-Sided p	Two-Sided p
Pair 1	ANXPRE & ANX1HR	15	.537	.020	.039
Pair 2	ANXPRE & ANX4HR	15	.509	.026	.053
Pair 3	ANX1HR & ANX4HR	15	.470	.038	.077

Paired Samples Test

		Paired Differences							Significance	
					95% Confidence Interval of the Difference					
		Mean	Std. Deviation	Std. Error Mean	Lower	Upper	t	df	One-Sided p	Two-Sided p
Pair 1	ANXPRE - ANX1HR	2.13333	1.72654	.44579	1.17721	3.08946	4.785	14	<.001	<.001
Pair 2	ANXPRE - ANX4HR	2.80000	1.93465	.49952	1.72863	3.87137	5.605	14	<.001	<.001
Pair 3	ANX1HR - ANX4HR	.66667	2.19306	.56625	-.54781	1.88114	1.177	14	.129	.259

Paired Samples Effect Sizes

					95% Confidence Interval	
			Standardizer[a]	Point Estimate	Lower	Upper
Pair 1	ANXPRE - ANX1HR	Cohen's d	1.72654	1.236	.544	1.903
		Hedges' correction	1.82647	1.168	.514	1.799
Pair 2	ANXPRE - ANX4HR	Cohen's d	1.93465	1.447	.702	2.168
		Hedges' correction	2.04662	1.368	.663	2.050
Pair 3	ANX1HR - ANX4HR	Cohen's d	2.19306	.304	-.219	.817
		Hedges' correction	2.31999	.287	-.207	.772

a. The denominator used in estimating the effect sizes.
Cohen's d uses the sample standard deviation of the mean difference.
Hedges' correction uses the sample standard deviation of the mean difference, plus a correction factor.

IBM SPSS Statistics Processor is ready | Unicode:ON | Classic | H: 18, W: 1132 pt.

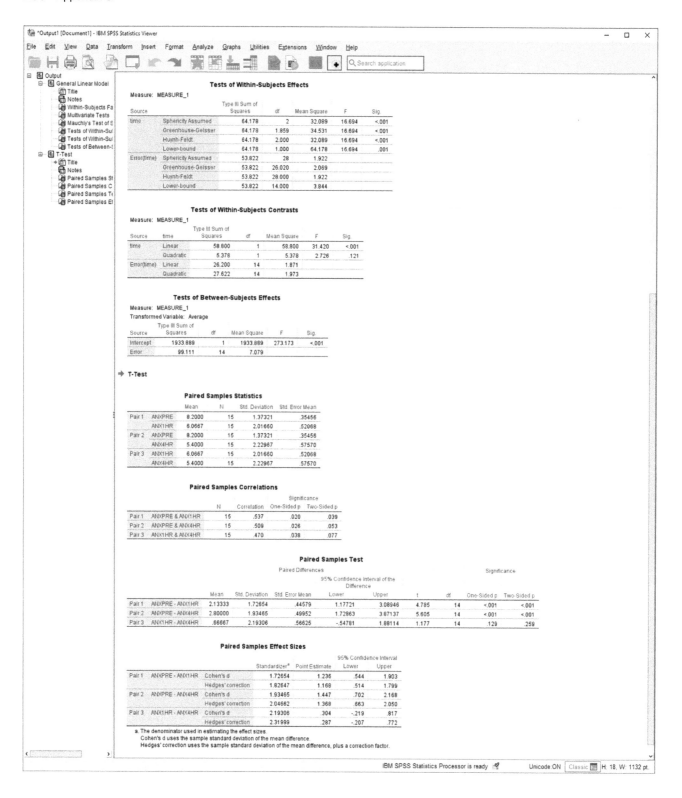

A one-way repeated-measures ANOVA was calculated to determine whether anxiety changed over time. A significant effect was found ($F(2,28) = 16.694$, $p < .001$). Anxiety before treatment was higher ($M = 8.20$, $sd = 1.37$) than at one hour after ($M = 6.07$, $sd = 2.02$) and at four hours after treatment ($M = 5.40$, $sd = 2.23$).

Section 7.5

Use Practice Data Set 3 in Appendix B. Determine if anxiety **levels** changed over time for each of the treatment (CONDITION) types. How did time change anxiety **levels** for each treatment? Write a statement of results.

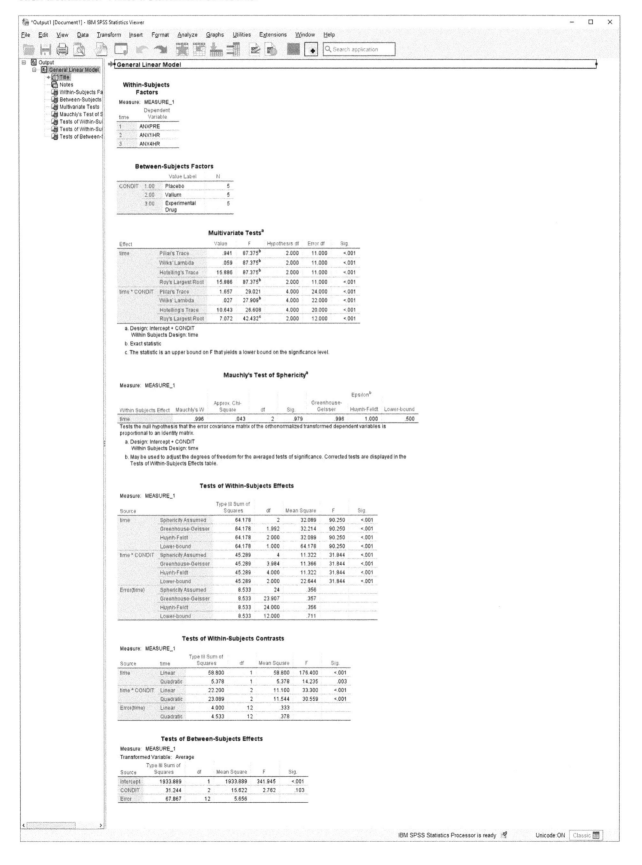

*Output1 [Document1] - IBM SPSS Statistics Viewer

File Edit View Data Transform Insert Format Analyze Graphs Utilities Extensions Window Help

Output
- General Linear Model
  - Title
  - Notes
  - Within-Subjects Fa
  - Between-Subjects
  - Multivariate Tests
  - Mauchly's Test of S
  - Tests of Within-Su
  - Tests of Within-Su
  - Tests of Between-

General Linear Model

Within-Subjects Factors

Measure: MEASURE_1

time	Dependent Variable
1	ANXPRE
2	ANX1HR
3	ANX4HR

Between-Subjects Factors

		Value Label	N
CONDIT	1.00	Placebo	5
	2.00	Valium	5
	3.00	Experimental Drug	5

Multivariate Tests[a]

Effect		Value	F	Hypothesis df	Error df	Sig.
time	Pillai's Trace	.941	87.375[b]	2.000	11.000	<.001
	Wilks' Lambda	.059	87.375[b]	2.000	11.000	<.001
	Hotelling's Trace	15.886	87.375[b]	2.000	11.000	<.001
	Roy's Largest Root	15.886	87.375[b]	2.000	11.000	<.001
time * CONDIT	Pillai's Trace	1.657	29.021	4.000	24.000	<.001
	Wilks' Lambda	.027	27.909[b]	4.000	22.000	<.001
	Hotelling's Trace	10.643	26.608	4.000	20.000	<.001
	Roy's Largest Root	7.072	42.432[c]	2.000	12.000	<.001

a. Design: Intercept + CONDIT
Within Subjects Design: time

b. Exact statistic

c. The statistic is an upper bound on F that yields a lower bound on the significance level.

Mauchly's Test of Sphericity[a]

Measure: MEASURE_1

					Epsilon[b]		
Within Subjects Effect	Mauchly's W	Approx. Chi-Square	df	Sig.	Greenhouse-Geisser	Huynh-Feldt	Lower-bound
time	.996	.043	2	.979	.996	1.000	.500

Tests the null hypothesis that the error covariance matrix of the orthonormalized transformed dependent variables is proportional to an identity matrix.

a. Design: Intercept + CONDIT
Within Subjects Design: time

b. May be used to adjust the degrees of freedom for the averaged tests of significance. Corrected tests are displayed in the Tests of Within-Subjects Effects table.

Tests of Within-Subjects Effects

Measure: MEASURE_1

Source		Type III Sum of Squares	df	Mean Square	F	Sig.
time	Sphericity Assumed	64.178	2	32.089	90.250	<.001
	Greenhouse-Geisser	64.178	1.992	32.214	90.250	<.001
	Huynh-Feldt	64.178	2.000	32.089	90.250	<.001
	Lower-bound	64.178	1.000	64.178	90.250	<.001
time * CONDIT	Sphericity Assumed	45.289	4	11.322	31.844	<.001
	Greenhouse-Geisser	45.289	3.984	11.366	31.844	<.001
	Huynh-Feldt	45.289	4.000	11.322	31.844	<.001
	Lower-bound	45.289	2.000	22.644	31.844	<.001
Error(time)	Sphericity Assumed	8.533	24	.356		
	Greenhouse-Geisser	8.533	23.907	.357		
	Huynh-Feldt	8.533	24.000	.356		
	Lower-bound	8.533	12.000	.711		

Tests of Within-Subjects Contrasts

Measure: MEASURE_1

Source	time	Type III Sum of Squares	df	Mean Square	F	Sig.
time	Linear	58.800	1	58.800	176.400	<.001
	Quadratic	5.378	1	5.378	14.235	.003
time * CONDIT	Linear	22.200	2	11.100	33.300	<.001
	Quadratic	23.089	2	11.544	30.559	<.001
Error(time)	Linear	4.000	12	.333		
	Quadratic	4.533	12	.378		

Tests of Between-Subjects Effects

Measure: MEASURE_1
Transformed Variable: Average

Source	Type III Sum of Squares	df	Mean Square	F	Sig.
Intercept	1933.889	1	1933.889	341.945	<.001
CONDIT	31.244	2	15.622	2.762	.103
Error	67.867	12	5.656		

IBM SPSS Statistics Processor is ready | Unicode:ON | Classic

A 3 (TIME) × 3 (CONDITION) mixed-design ANOVA was calculated to determine if either time or type of treatment influenced anxiety levels. The main effect for type of treatment was not significant ($F(2,12) = 2.762$, $p = .103$). The main effect for time was significant ($F(2,24) = 90.25$, $p < .001$). The TIME × CONDITION interaction was also significant ($F(4,24) = 31.844$, $p < .001$).

Section 7.6

Use Practice Data Set 2 in Appendix B. Determine if salaries are different for males and females. Repeat the analysis, statistically controlling for years of service. Write a statement of results for each. Compare and contrast your two answers.

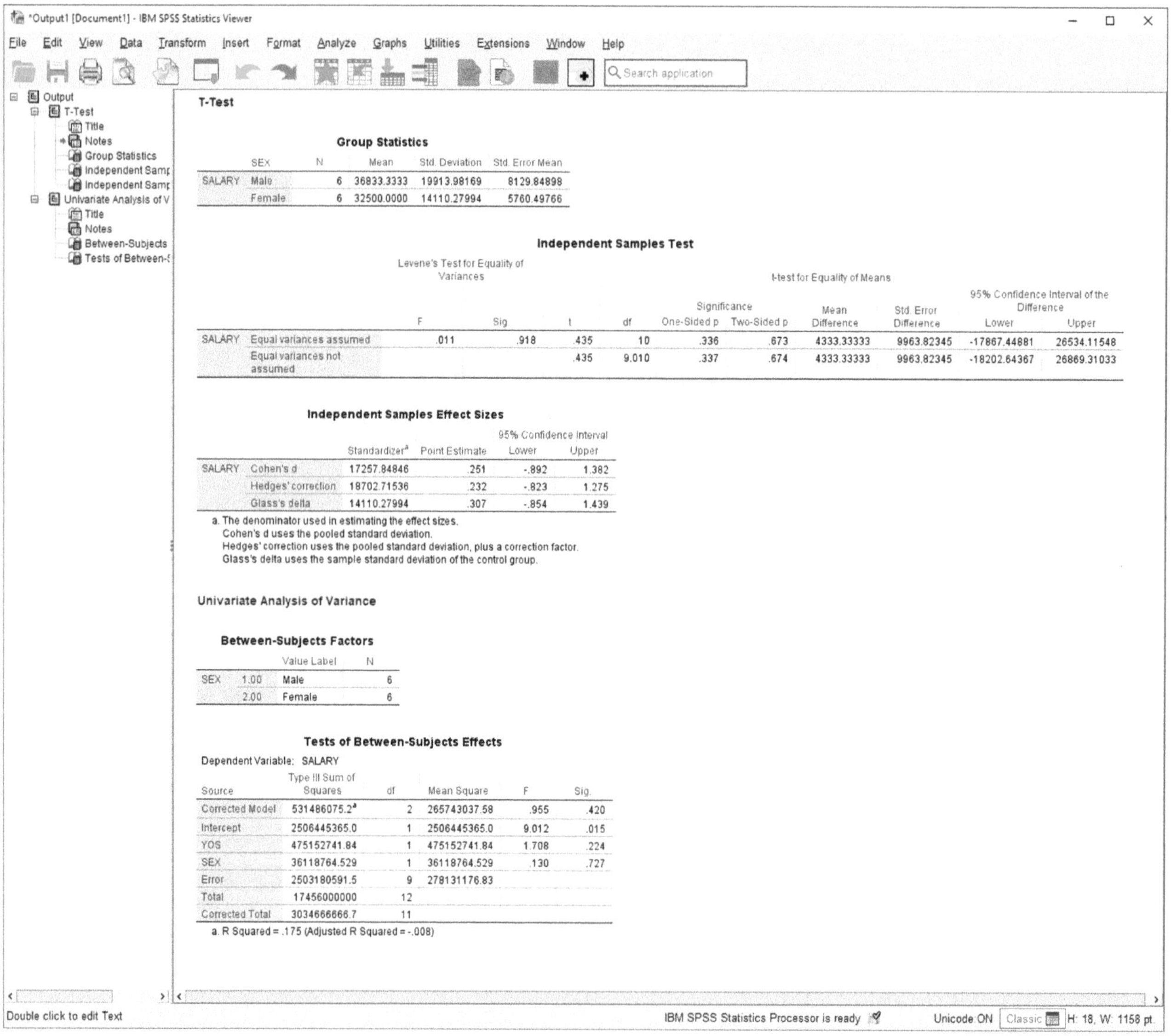

T-Test

Group Statistics

	SEX	N	Mean	Std. Deviation	Std. Error Mean
SALARY	Male	6	36833.3333	19913.98169	8129.84898
	Female	6	32500.0000	14110.27994	5760.49766

Independent Samples Test

		Levene's Test for Equality of Variances		t-test for Equality of Means							
						Significance		Mean	Std. Error	95% Confidence Interval of the Difference	
		F	Sig	t	df	One-Sided p	Two-Sided p	Difference	Difference	Lower	Upper
SALARY	Equal variances assumed	.011	.918	.435	10	.336	.673	4333.33333	9963.82345	-17867.44881	26534.11548
	Equal variances not assumed			.435	9.010	.337	.674	4333.33333	9963.82345	-18202.64367	26869.31033

Independent Samples Effect Sizes

		Standardizer[a]	Point Estimate	95% Confidence Interval Lower	Upper
SALARY	Cohen's d	17257.84846	.251	-.892	1.382
	Hedges' correction	18702.71536	.232	-.823	1.275
	Glass's delta	14110.27994	.307	-.854	1.439

a. The denominator used in estimating the effect sizes.
Cohen's d uses the pooled standard deviation.
Hedges' correction uses the pooled standard deviation, plus a correction factor.
Glass's delta uses the sample standard deviation of the control group.

Univariate Analysis of Variance

Between-Subjects Factors

		Value Label	N
SEX	1.00	Male	6
	2.00	Female	6

Tests of Between-Subjects Effects

Dependent Variable: SALARY

Source	Type III Sum of Squares	df	Mean Square	F	Sig.
Corrected Model	531486075.2[a]	2	265743037.58	.955	.420
Intercept	2506445365.0	1	2506445365.0	9.012	.015
YOS	475152741.84	1	475152741.84	1.708	.224
SEX	36118764.529	1	36118764.529	.130	.727
Error	2503180591.5	9	278131176.83		
Total	17456000000	12			
Corrected Total	3034666666.7	11			

a. R Squared = .175 (Adjusted R Squared = -.008)

Without Controlling for Years of Service

An independent samples *t*-test calculated to determine if the salaries of men and women differ. No significant difference was found ($t(10) = .435$, $p = .673$). The mean salaries of men are not significantly different from the mean salaries of women.

ANCOVA results

A one-way ANCOVA was calculated to determine if the salaries of men and women are different when controlling for years of service. Years of service was not significantly related to salary ($F(1,9) = 1.708$, $p = .224$). Salaries did not differ between men and women even when controlling for years of service ($F(1,9) = .130$, $p = .727$).

Section 8.1

Use Practice Data Set 2 in Appendix B. In the population from which the sample was drawn, 20% of employees are clerical, 50% are technical, and 30% are professional. Determine whether or not the sample drawn conforms to these values. Hint: You will need to *customize expected probabilities* and enter the category values (1, 2, 3) and relative percentages (20, 50, 30). Also, be sure you have entered the variable CLASSIFY as nominal.

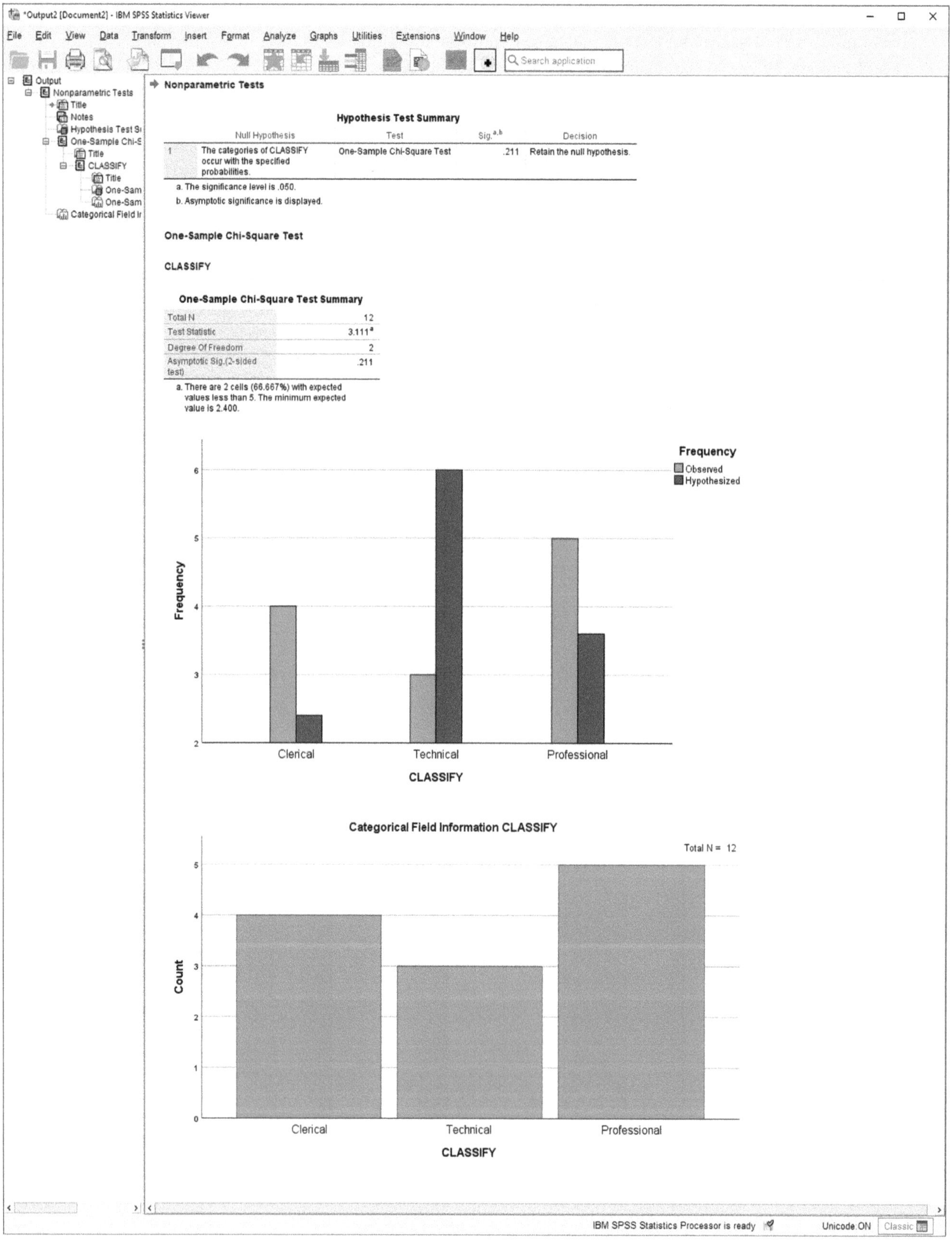

A chi-square goodness of fit test was calculated to determine if the sample fit with the hypothesized distribution of 20% clerical, 50% technical, and 30% professional. No significant deviation from the hypothesized values was found ($\chi^2(2) = 3.11$, $p > .05$). The observed distribution (4/12 clerical, 3/12 technical, 5/12 professional) did not vary significantly from the hypothesized values.

Section 8.2

A researcher wants to know if individuals are more likely to help in an emergency when they are indoors or when they are outdoors. Of 28 participants who were outdoors, 19 helped and 9 did not. Of 23 participants who were indoors, 8 helped and 15 did not. Enter these data and find out if helping behavior is affected by the environment. The key to this problem is in the data entry. (Hint: How many participants were there, and how many pieces of information do we know about each? Having SPSS give you cell percentages will help you with interpretation of this problem.)

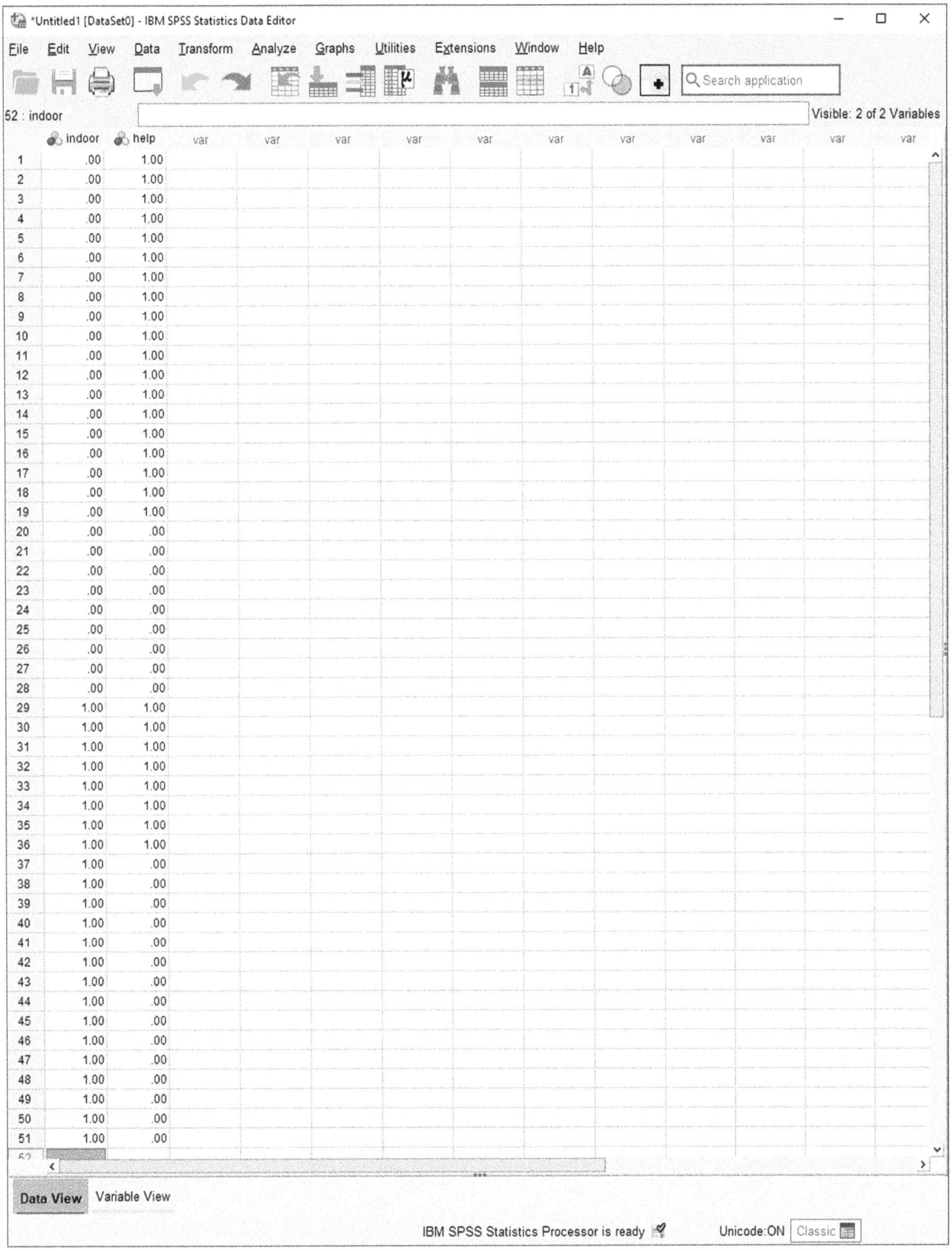

	indoor	help
1	.00	1.00
2	.00	1.00
3	.00	1.00
4	.00	1.00
5	.00	1.00
6	.00	1.00
7	.00	1.00
8	.00	1.00
9	.00	1.00
10	.00	1.00
11	.00	1.00
12	.00	1.00
13	.00	1.00
14	.00	1.00
15	.00	1.00
16	.00	1.00
17	.00	1.00
18	.00	1.00
19	.00	1.00
20	.00	.00
21	.00	.00
22	.00	.00
23	.00	.00
24	.00	.00
25	.00	.00
26	.00	.00
27	.00	.00
28	.00	.00
29	1.00	1.00
30	1.00	1.00
31	1.00	1.00
32	1.00	1.00
33	1.00	1.00
34	1.00	1.00
35	1.00	1.00
36	1.00	1.00
37	1.00	.00
38	1.00	.00
39	1.00	.00
40	1.00	.00
41	1.00	.00
42	1.00	.00
43	1.00	.00
44	1.00	.00
45	1.00	.00
46	1.00	.00
47	1.00	.00
48	1.00	.00
49	1.00	.00
50	1.00	.00
51	1.00	.00

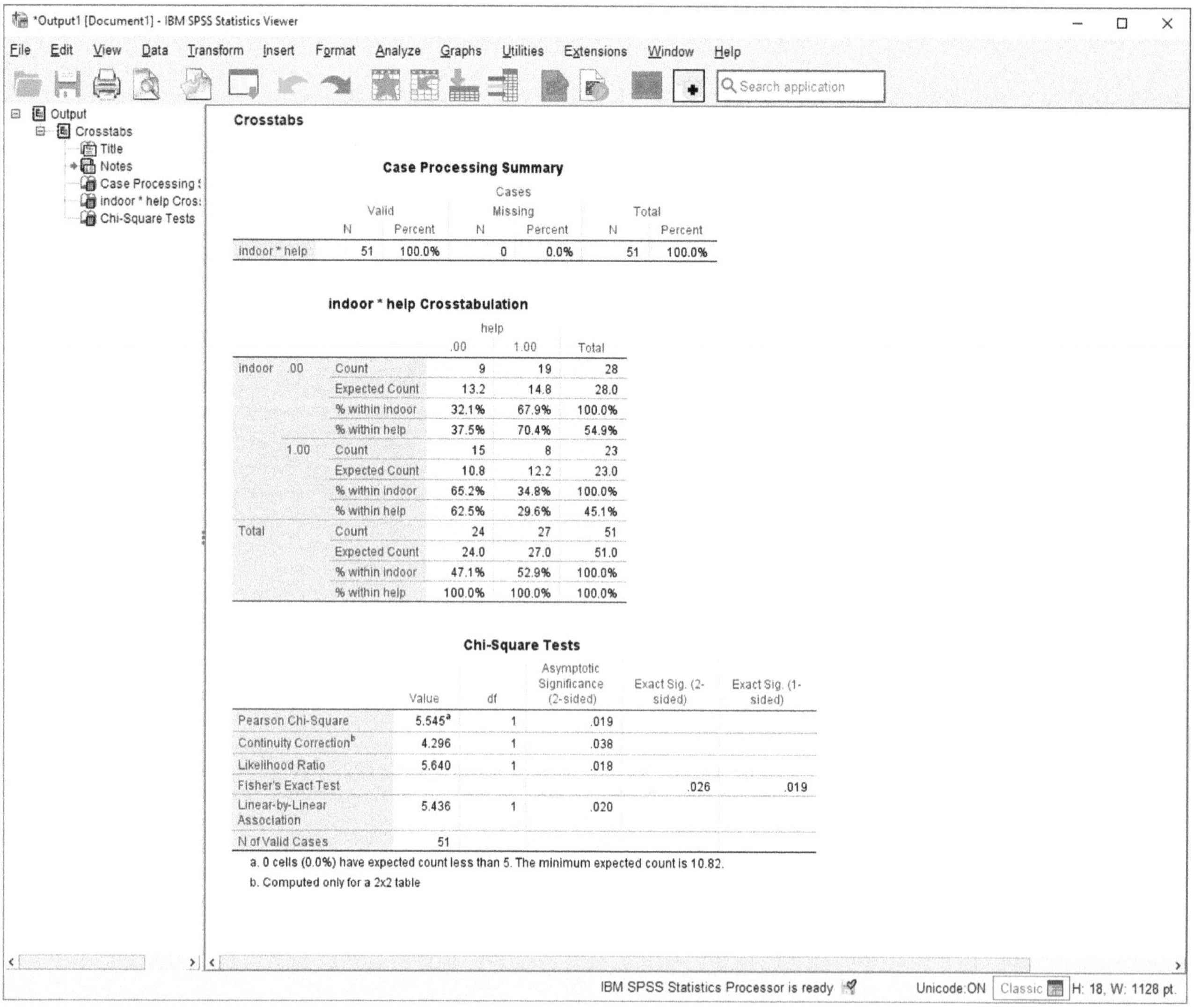

Crosstabs

Case Processing Summary

	Cases					
	Valid		Missing		Total	
	N	Percent	N	Percent	N	Percent
indoor * help	51	100.0%	0	0.0%	51	100.0%

indoor * help Crosstabulation

			help		
			.00	1.00	Total
indoor	.00	Count	9	19	28
		Expected Count	13.2	14.8	28.0
		% within indoor	32.1%	67.9%	100.0%
		% within help	37.5%	70.4%	54.9%
	1.00	Count	15	8	23
		Expected Count	10.8	12.2	23.0
		% within indoor	65.2%	34.8%	100.0%
		% within help	62.5%	29.6%	45.1%
Total		Count	24	27	51
		Expected Count	24.0	27.0	51.0
		% within indoor	47.1%	52.9%	100.0%
		% within help	100.0%	100.0%	100.0%

Chi-Square Tests

	Value	df	Asymptotic Significance (2-sided)	Exact Sig. (2-sided)	Exact Sig. (1-sided)
Pearson Chi-Square	5.545[a]	1	.019		
Continuity Correction[b]	4.296	1	.038		
Likelihood Ratio	5.640	1	.018		
Fisher's Exact Test				.026	.019
Linear-by-Linear Association	5.436	1	.020		
N of Valid Cases	51				

a. 0 cells (0.0%) have expected count less than 5. The minimum expected count is 10.82.

b. Computed only for a 2x2 table

A chi-square test of independence was calculated comparing the frequency of helping behavior for subjects inside and outside. A significant interaction was found ($\chi^2(1) = 5.545$, $p < .05$). Subjects were more likely to help outside (67.9%) than they were inside (34.8%).

Section 8.3

Use Practice Data Set 1 in Appendix B. Determine if younger participants (<26) have significantly lower mathematics scores than older participants. (Note: You will need to create a new ordinal variable representing each age group. See Section 2.2 if you need a refresher on how to do that.)

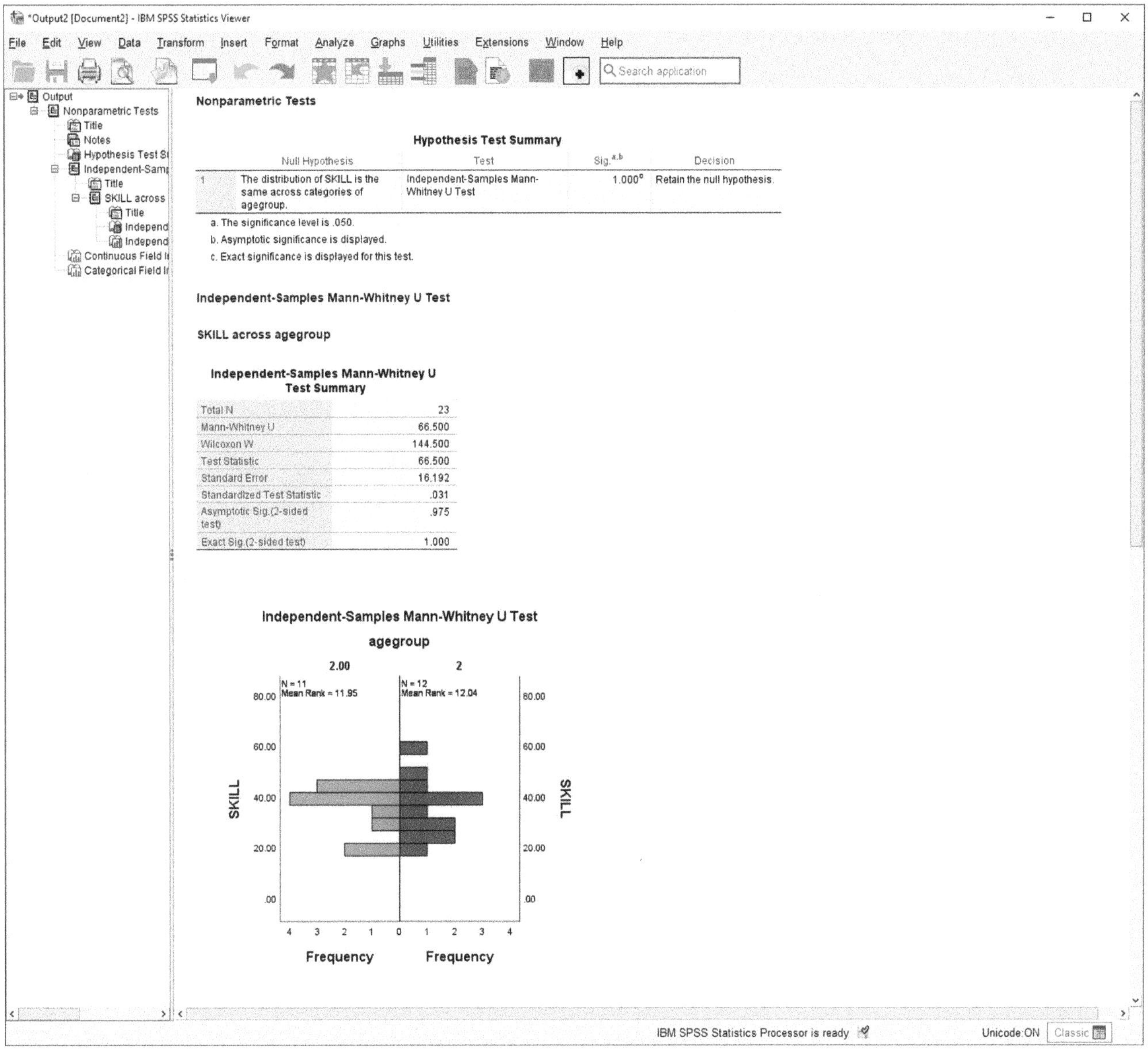

Hypothesis Test Summary

	Null Hypothesis	Test	Sig.[a,b]	Decision
1	The distribution of SKILL is the same across categories of agegroup.	Independent-Samples Mann-Whitney U Test	1.000[c]	Retain the null hypothesis.

a. The significance level is .050.
b. Asymptotic significance is displayed.
c. Exact significance is displayed for this test.

Independent-Samples Mann-Whitney U Test

SKILL across agegroup

Independent-Samples Mann-Whitney U Test Summary

Total N	23
Mann-Whitney U	66.500
Wilcoxon W	144.500
Test Statistic	66.500
Standard Error	16.192
Standardized Test Statistic	.031
Asymptotic Sig.(2-sided test)	.975
Exact Sig.(2-sided test)	1.000

A Mann-Whitney *U* test was calculated examining the math skills scores of older and younger subjects. No significant difference was found ($U = 65.5$, $p > .05$). Younger subjects had a mean rank of 11.95 while older subjects had a mean rank of 12.04.

Section 8.4

Use the RACE.sav data file to determine whether or not the outcome of short-distance races is different from that of medium-distance races. Phrase your results.

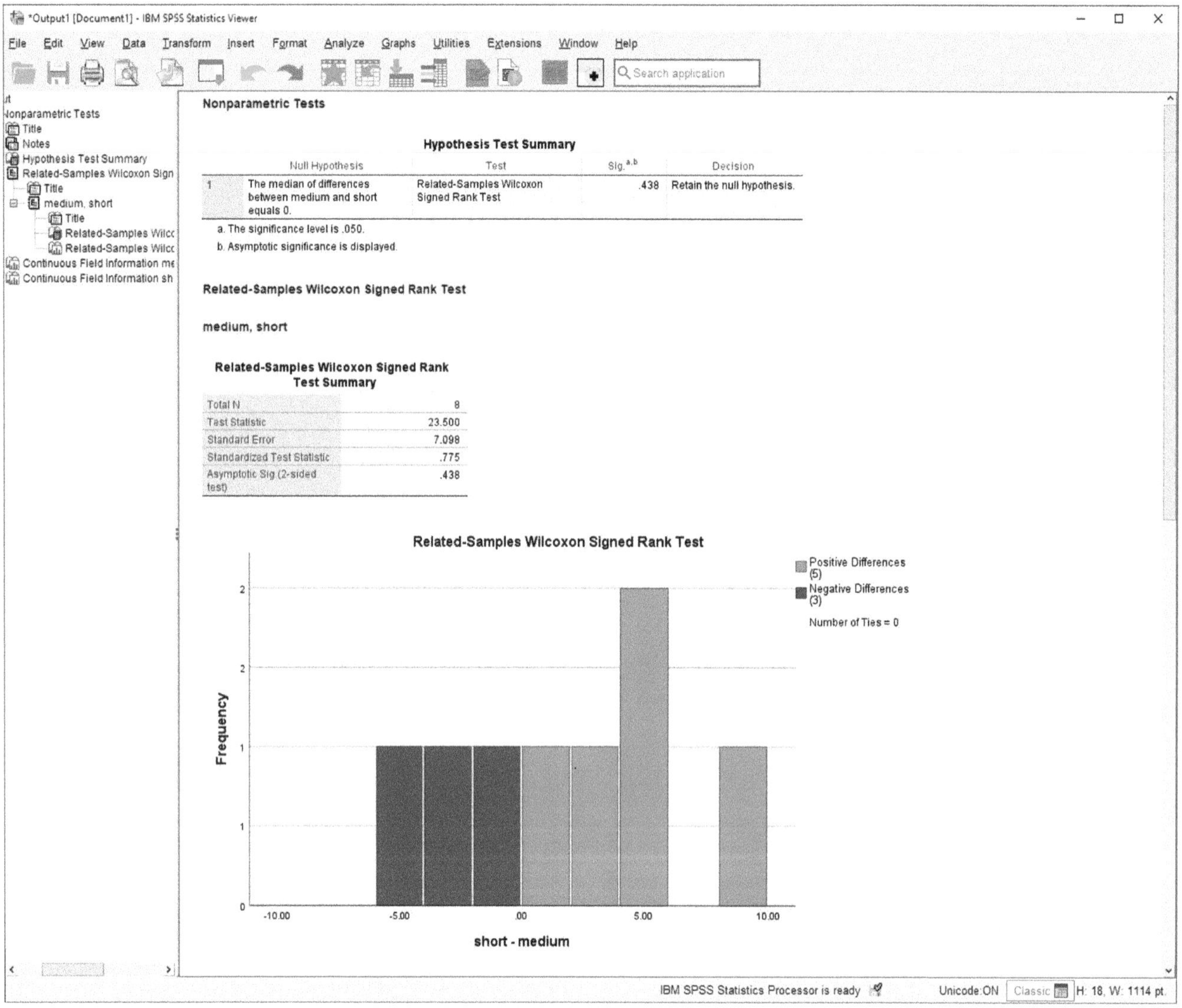

Hypothesis Test Summary

	Null Hypothesis	Test	Sig.[a,b]	Decision
1	The median of differences between medium and short equals 0.	Related-Samples Wilcoxon Signed Rank Test	.438	Retain the null hypothesis.

a. The significance level is .050.

b. Asymptotic significance is displayed.

Related-Samples Wilcoxon Signed Rank Test Summary

Total N	8
Test Statistic	23.500
Standard Error	7.098
Standardized Test Statistic	.775
Asymptotic Sig.(2-sided test)	.438

A Wilcoxon test examined the results of medium-distance and short-distance races. No significant difference was found in the results ($Z = -0.775$, $p > .05$). Medium-distance results were not significantly different from short-distance results.

Section 8.6

Use the data in Practice Data Set 3 in Appendix B. If anxiety is measured on an **ordinal scale**, determine if anxiety levels changed over time. Phrase your results.

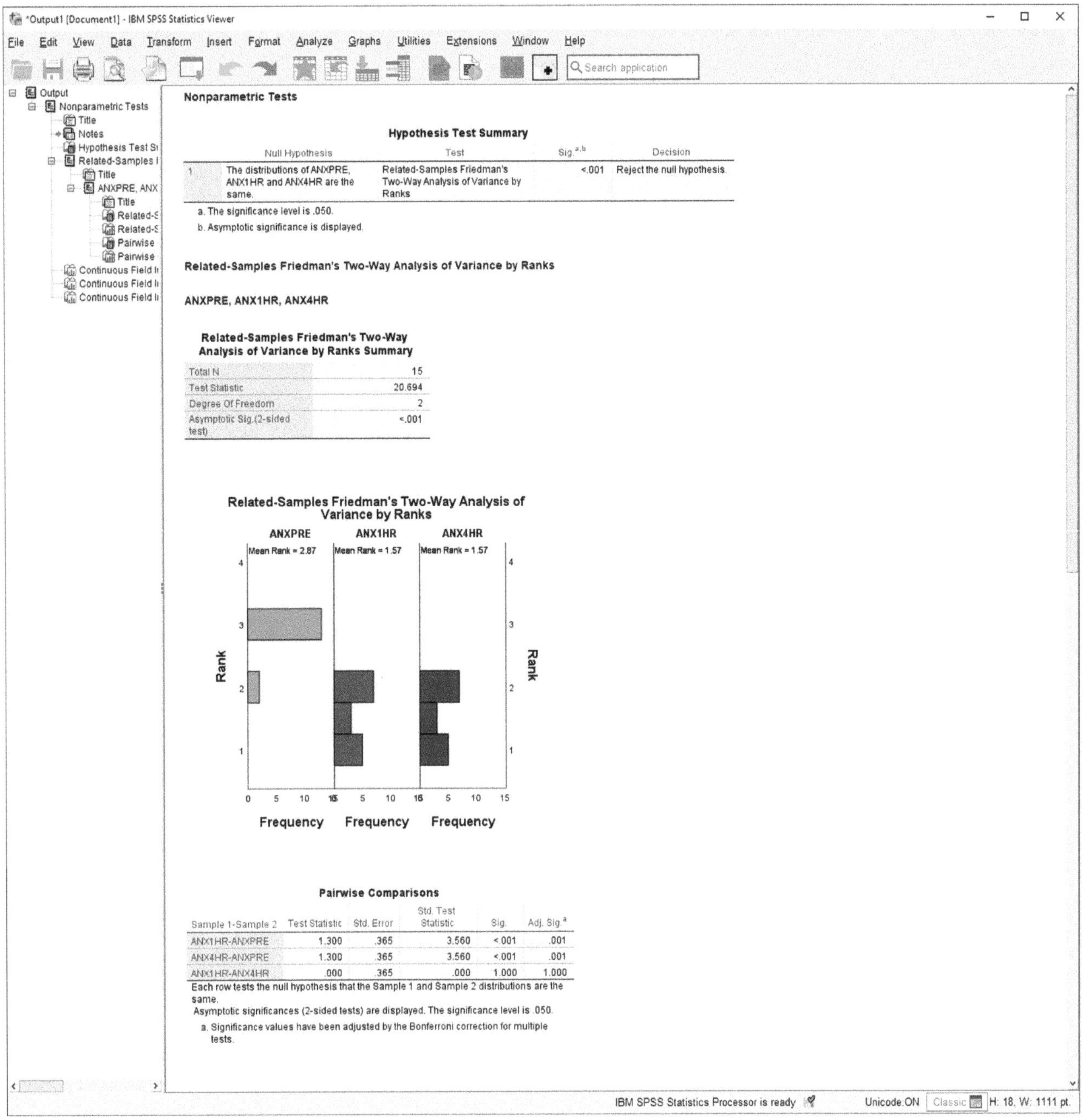

Hypothesis Test Summary

	Null Hypothesis	Test	Sig.[a,b]	Decision
1	The distributions of ANXPRE, ANX1HR and ANX4HR are the same.	Related-Samples Friedman's Two-Way Analysis of Variance by Ranks	<.001	Reject the null hypothesis.

a. The significance level is .050.
b. Asymptotic significance is displayed.

Related-Samples Friedman's Two-Way Analysis of Variance by Ranks Summary

Total N	15
Test Statistic	20.694
Degree Of Freedom	2
Asymptotic Sig.(2-sided test)	<.001

Pairwise Comparisons

Sample 1-Sample 2	Test Statistic	Std. Error	Std. Test Statistic	Sig.	Adj. Sig.[a]
ANX1HR-ANXPRE	1.300	.365	3.560	<.001	.001
ANX4HR-ANXPRE	1.300	.365	3.560	<.001	.001
ANX1HR-ANX4HR	.000	.365	.000	1.000	1.000

Each row tests the null hypothesis that the Sample 1 and Sample 2 distributions are the same.
Asymptotic significances (2-sided tests) are displayed. The significance level is .050.
a. Significance values have been adjusted by the Bonferroni correction for multiple tests.

A Friedman test was conducted comparing the average anxiety ranking of subjects before, one hour after, and four hours after treatment. A significant difference was found ($\chi^2(2) = 20.694, p < .05$). Student anxiety was significantly reduced from the pretest value at both one hour and four hours.

Notes

Preface to the Twelfth Edition

1 Holcomb, Z. (1997). *Real data: A statistics workbook based on empirical data.* Los Angeles, CA: Pyrczak Publishing.

Chapter 1 Getting Started

1 Items that appear in the Glossary in Appendix E are presented in **bold**. *Italics* are used to indicate menu items.
2 Depending on your version of SPSS, it may be displayed as 2.0E + 009.

Chapter 9 Test Construction

1 Landis, J.R., & Koch, G.G. (1977). The measurement of observer agreement for categorical data. *Biometrics, 33,* 159–174.

Appendix A Effect Size

1 Kirk, R. E. (1996). Practical significance: A concept whose time has come. *Educational & Psychological Measurement, 56,* 746–759.
2 Cohen, J. (1992). A power primer. *Psychological Bulletin, 112,* 155–159.
3 Cohen, J. (1988). *Statistical power analysis for the behavioral sciences* (2nd ed.). Hillsdale, NJ: Lawrence Erlbaum.

Index

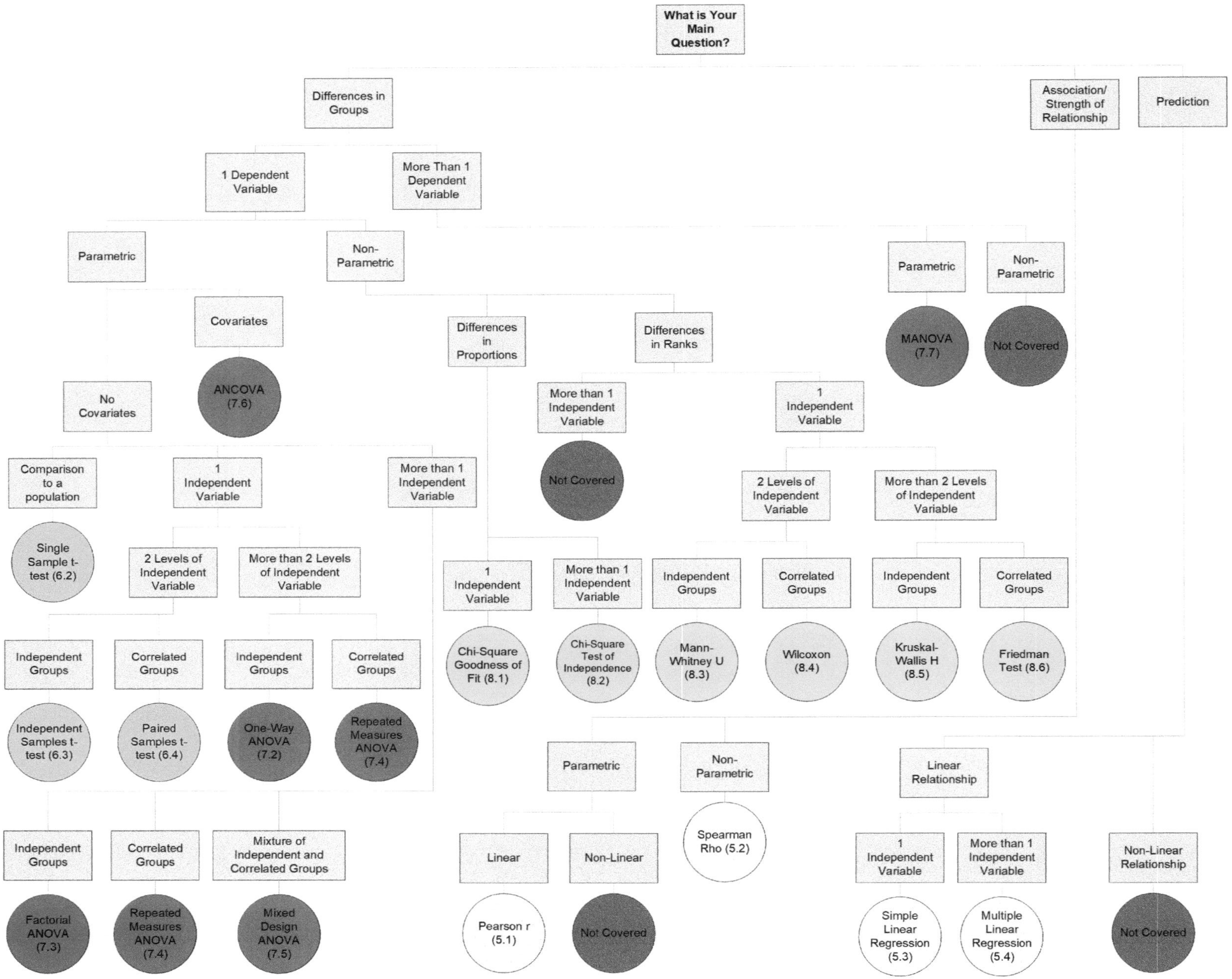
What is Your Main Question?
Differences in Groups
Association/ Strength of Relationship
Prediction
1 Dependent Variable
More Than 1 Dependent Variable
Parametric
Non-Parametric
Covariates
ANCOVA (7.6)
No Covariates
Comparison to a population
Single Sample t-test (6.2)
1 Independent Variable
More than 1 Independent Variable
2 Levels of Independent Variable
More than 2 Levels of Independent Variable
Independent Groups
Correlated Groups
Independent Groups
Correlated Groups
Independent Samples t-test (6.3)
Paired Samples t-test (6.4)
One-Way ANOVA (7.2)
Repeated Measures ANOVA (7.4)
Independent Groups
Correlated Groups
Mixture of Independent and Correlated Groups
Factorial ANOVA (7.3)
Repeated Measures ANOVA (7.4)
Mixed Design ANOVA (7.5)
Differences in Proportions
Differences in Ranks
More than 1 Independent Variable
Not Covered
1 Independent Variable
2 Levels of Independent Variable
More than 2 Levels of Independent Variable
1 Independent Variable
More than 1 Independent Variable
Independent Groups
Correlated Groups
Independent Groups
Correlated Groups
Chi-Square Goodness of Fit (8.1)
Chi-Square Test of Independence (8.2)
Mann-Whitney U (8.3)
Wilcoxon (8.4)
Kruskal-Wallis H (8.5)
Friedman Test (8.6)
Parametric
Non-Parametric
MANOVA (7.7)
Not Covered
Parametric
Non-Parametric
Linear
Non-Linear
Pearson r (5.1)
Not Covered
Spearman Rho (5.2)
Linear Relationship
1 Independent Variable
More than 1 Independent Variable
Simple Linear Regression (5.3)
Multiple Linear Regression (5.4)
Non-Linear Relationship
Not Covered

Milton Keynes UK
Ingram Content Group UK Ltd.
UKHW020837141024
449569UK00021B/784

9 781032 582351